Manufacturing
Excellence

Manufacturing Excellence

The competitive edge

Edited by
T. Pfeifer, W. Eversheim, W. König and M. Weck

Fraunhofer Institute of Production Engineering,
Aachen, Germany

CHAPMAN & HALL
London · Glasgow · New York · Tokyo · Melbourne · Madras

Published by Chapman & Hall, 2-6 Boundary Row, London SE1 8HN, UK

Chapman & Hall, 2-6 Boundary Row, London SE1 8HN, UK

Blackie Academic & Professional, Wester Cleddens Road, Bishopbriggs, Glasgow G64 2NZ, UK

Chapman & Hall Inc., One Penn Plaza, 41st Floor, New York, NY10119, USA

Chapman & Hall Japan, Thomson Publishing Japan, Hirakawacho Nemoto Building, 6F, 1-7-11 Hirakawa-cho, Chiyoda-ku, Tokyo 102, Japan

Chapman & Hall Australia, Thomas Nelson Australia, 102 Dodds Street, South Melbourne, Victoria 3205, Australia

Chapman & Hall India, R. Seshadri, 32 Second Main Road, CIT East, Madras 600 035, India

First English language edition 1994

© 1994 T. Pfeifer, W. Eversheim, W. König and M. Weck

Original German language - Wettbewerbsfaktor Produktionstechnik - 1993 VDI Verlag GmbH.

Printed in Great Britain at the University Press, Cambridge

ISBN 0 412 56780 6

∞ Printed on permanent acid free text paper, manufactured in accordance with ANSI/NISO Z39.48-1992 and ANSI/NISO Z39.48-1984 (Permanence of Paper).

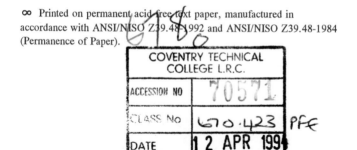

Contents

Preface   ix

Part One: Business Strategies   1

1   Competitive strategies for a global market   3
1.1 Changing demands imposed on production engineering   3
1.2 Process orientation - a comprehensive approach   11
1.3 Partners for reorganization   25
1.4. Recapitulation   33

2   Quality management as a corporate strategy   35
2.1. Introduction   35
2.2 The current sphere of influence of quality management   41
2.3 Quality management as a corporate strategy   42
2.4 Examples of the application of quality management strategies   54
2.5 Recapitulation and prospects   57

**3   Strategies for the introduction of a quality
      management system**   59
3.1 Introduction   59
3.2 Prerequisites for the introduction of a QM system   59
3.3 System of defining objectives by consensus   63
3.4 Case studies   66
3.5 Implementation of Quality Management systems - costs
      and benefits (example: Messrs Edelmann)   75
3.6 Recapitulation   77

Part Two: Product Development

**4   Supplier involvement in product development
      - risks and potentials-**   81
4.1 Introduction   81
4.2 Objectives for cooperative product development
      partnerships   84
4.3 Characterization of the product development partnership   88
4.4 Guidelines for product development partnerships   103
4.5 Summary   108

5   **The design workstation of the future - from initial design to NC programming**   109
    5.1 Introduction   109
    5.2 State of the art in the field of computer-aided product planning   111
    5.3 Trends in computer-aided product development   114
    5.4 System architecture of the design workstation of the future   117
    5.5 System modules for the design workstation   125
    5.6 Summary and outlook   140
    5.7 Glossary   143
6   **Innovative software technologies for quality management**   145
    6.1 Requirements relating to the application of data processing in quality management   145
    6.2 The deployment of innovative software technologies for quality management   156
    6.3 The deployment of innovative software technologies - a competitive advantage   168
7   **Rapid prototyping: the way ahead**   173
    7.1 Time as a competitive factor   173
    7.2 Prototype requirements in product development   177
    7.3 Methods of prototype production   180
    7.4 Prerequisites and procedures for the implementation of rapid prototyping   195
    7.5 Conclusion   198

Part Three: Production   201
8   **Resource-orientated design of production** systems   203
    8.1 Introduction and definition of the problem area   203
    8.2 Primary objectives   206
    8.3 Elements of resource-oriented production design   209
    8.4. Summary and outlook   226
9   **The key of efficient process and manufacturing sequences - technological development and innovation**   229
    9.1 Introduction   229
    9.2 Basic requirements for cost-minimized production   230
    9.3 Available resources   238
    9.4 Summary   265
10   **Superior tools as a basis for tomorrow's manufacturing**   267
    10.1 The future scope of requirements for

production engineering 267

10.2 The improvement of conventional tools via new
    processes 271

10.3 New machining processes for new materials 280

10.4 New technologies for the manufacture of innovative
    products 284

10.5 Summary 288

11   **Integrated Quality inspections** 291

11.1 Introduction 291

11.2 Production metrology today 292

11.3 Controlling processes by monitoring states 298

11.4 Sensors for process surveillance 301

11.5 Post-process measuring equipment monitors
    capable processes 305

11.6 Monitoring of quality control equipment 307

11.7 Summary and outlook 312

Part Four: Production Plants 315

12   **Fast - precise - clean: the machine tool in the
    context of conflicting interests of economy and ecology** 317

12.1 Introduction 317

12.2 Development trends in the machine tool
    manufacturing sector 320

12.3 Increasing the economic efficiency of a machine tool 323

12.4 Improving the reliability of a machine tool 331

12.5 Increasing the precision of a machine tool 334

12.6 Increasing the machining speed of a machine tool 343

12.7 Improving the environmental compatibility of
    a machine tool 349

12.8 Summary 353

13   **The open controller - key element of high-performance
    production facilities** 355

13.1 The background situation 355

13.2 The scope of functions and deficiencies of modern
    numerical controllers 359

13.3 Essential functions of open controllers 362

13.4 The implementation of open control systems 373

13.5 Summary and outlook 382

14   **Robotics in production systems** 385

14.1 Introduction 385

14.2 The course of development of robotics and current fields
 of application 387
14.3 Definition of problems for small batch sizes and suitable
 approaches for solving such problems 409
14.4 Summary and prospects 447

Part Five: Environment 449
15 Cooling lubricants - the ecological challenge facing production engineering 451
15.1 Introduction 451
15.2 Environmental problems occurring within the
 work process 454
15.3 Problems related to cooling lubricants 457
15.4 Approaches to reducing environmental pollution 473
15.5 Recapitulation and prospects 493
16 Strategic cost/benefit evaluation of product development, production and waste disposal 495
16.1 Introduction 495
16.2 Systems of objectives in changing times 497
16.3 Strategies for evaluation: tools for target-oriented
 decision support 503
16.4 Case studies 517
16.5 Conclusions and prospects 520

Appendix A: Members of the working groups 523
Appendix B: References 531
Index 565

Preface

Political events are seldom without repercussions for economic development. While the global market, the nascent Single European Market and restructuring in Eastern Europe harbour certain economic risks, these phenomena also present new opportunities and challenges. Intense competition at home and abroad, in conjunction with the recession currently stalking markets world-wide, are exerting increasing pressure on companies to protect their ability to compete. Production technology is emerging as the decisive factor in this increasingly cut-throat competition. Forward-looking companies are invited to rise to the challenge and to invest in innovative production technologies.

The results of the Aachen Machine Tool Colloquium 1993 have been condensed and are presented in this volume. Solutions to problems companies currently facing are developed on the basis of strategic considerations. Prospects for production engineering of tomorrow are highlighted against the backdrop of product development, production, machine tools and environmental protection.

Each of the 16 contributions is the result of close cooperation between representatives from industry and research, thus in a unique way combining practical experience with the latest research results. Our many thanks are due to all those who have contributed to this book.

Tilo Pfeifer
Walter Eversheim
Wilfried König
Manfred Weck

Aachen, June 1993

Part One

BUSINESS STRATEGIES: COMPETITIVE PROCESS AND QUALITY MANAGEMENT

1 Competitive strategies for a global market
2 Quality management as a corporate strategy
3 Strategies for the introduction of a quality management system

1

COMPETITIVE STRATEGIES FOR A GLOBAL MARKET

1.1 Changing demands imposed on production engineering

The factors influencing corporate action globally are changing more and more rapidly. Any analysis and development of competitive strategies as a response to changing boundary conditions must be preceded by a study of these changes. Conclusions must then be drawn as to the likely effects on the company. The significance of the current changes is best viewed and evaluated against the background of historical development and the associated changing requirements. This must be a continuous process regardless of the current economic situation, since 'accelerating change is the only thing which is stable' [1].

The changing environment in which companies are finding themselves can be characterized as follows (Fig. 1.1):

Changing social values and structures exert a decisive influence on corporate structures. This influence, however, is less visible in periods of economic downturn. The trend increasingly is towards self-fulfilment and away from values of duty and acceptance. The structure of education and training is evolving with changing age and ethnic structures among the population [2]. Companies are seeing these developments reflected in the labour market. The indirect results, for example legislative changes, exert

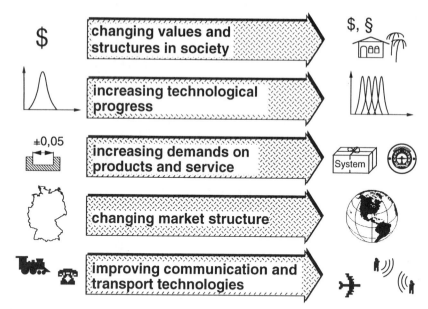

Fig. 1.1 Changing conditions for business

direct influence on corporate activities.

Product life cycles are becoming progressively shorter. Technical innovations are required to be incorporated into new products within even shorter periods of time. Speed is rapidly advancing to become the decisive factor in success. This applies to product and production alike.

Changing markets are characterized by increasingly high requirements which have to be met by products and services. The most pressing demands are for quality as the principal means of achieving customer satisfaction, short delivery periods, higher levels of environmental friendliness and integrated solutions instead of `insular' solutions.

Additionally, changing market structures are influencing corporate thinking. The globalization of markets can offer advantages in terms of sales and production but also involves dangers such as the loss of markets and areas of production due to sweeping political change. Changes like this and global shifts in the distribution of purchasing power between North America, Europe, Asia and the area around the Pacific combined with the unpredictability of the eastern European market demand farsighted corporate action. At the same time, regions with differing product demands, cultures and purchasing power continue to demand solutions

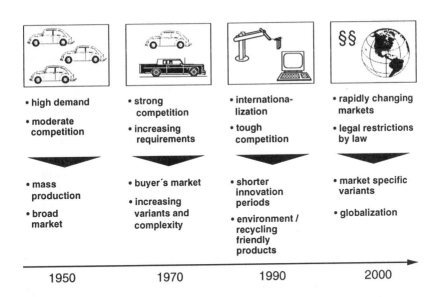

| • high demand | • strong competition | • internationa-lization | • rapidly changing markets |
| • moderate competition | • increasing requirements | • tough competition | • legal restrictions by law |

| • mass production | • buyer´s market | • shorter innovation periods | • market specific variants |
| • broad market | • increasing variants and complexity | • environment / recycling friendly products | • globalization |

| 1950 | 1970 | 1990 | 2000 |

Fig. 1.2 Historical review - product market

adapted to meet their needs [3].

Steady improvement in world-wide transport and communications technology is forcing companies to investigate and implement the resulting potential offered by, for example, improved information processing, in order to increase their competitiveness. Innovation is the strongest driving force in the market and profits are increasingly being built on the ability to reduce time to market to a minimum. Only the systematic and rapid use of new technologies can ensure that a company remains competitive in the long term.

In view of the variety and complexity of changing boundary conditions, the question presents itself as to what methods can best be employed to sustain the competitiveness of a company. In the wake of the sometimes uninspiring results achieved by computer integrated manufacturing CIM [4] and the popular discussion of `lean' strategies, there is intensive argument as to the suitability of all available methods, especially in the present atmosphere of economic gloom. In order to facilitate evaluation of the current situation and to demonstrate possible solutions, a rough outline follows of the historical development in the field of production engineering, concentrating on markets, quality, ecology, labour market, information technology and organization.

Immediately after the second world war, demand for consumer and capital goods was high; competition, much of which was regional or domestic, was limited (Fig. 1.2). The principle of mass production was regarded in many places as the one most suitable for the new seller's market. Rising customer demands combined with intensified and by now international competition resulted in a transition from a seller's to a buyer's market. The products demanded were increasingly varied and complex. The fierce competition and globalization of the past decade combined with increasing environmental awareness have led to even higher demands.

Shorter innovation cycles and recycling-oriented product design represent additional requirements. In the future, changes in both demands and market structures will accelerate even further. This development will intensify the globalization of companies' activities and will make it necessary to produce market-oriented variations of any given product.

The term quality was synonymous with function in the period immediately after the war. For this reason, periodical checks during the production process, classification and a final check on the parts and products was sufficient. Increasingly high demands were imposed, above all, by the armed forces; in order to meet these demands, statistical improvements and more accurate evaluation of defects were introduced [11]. Rising test, inspection and re-working costs prompted companies to incorporate test and inspection steps into the production stage. This development is reflected in the change in terminology from quality control to quality assurance. Aggressive competition from high quality goods made in Japan, the spiralling cost of quality and largely uniform standards of quality led to the introduction of a certification aimed at systematic, reproducible quality assurance [12]. Additionally, the involvement of quality in all areas of companies has recently been lent impetus as an integral part of holistic quality management. Growing awareness of quality will lead to a re-definition of quality which will no longer be limited to a product but will come to represent total customer satisfaction [11].

A lack of environmental awareness accompanied by accelerated economic growth resulted in an increase in pollution. It was not until a number of environmental disasters had taken place that environmental awareness began to take hold among the population as a whole. This new awareness is reflected in recent environmental legislation. Tighter legislation and market demand for environment friendly products will increase further the significance of ecology as a factor in competition [13, 14].

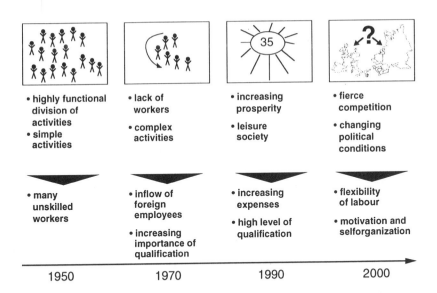

Fig. 1.3 Historical review - labour market

From the point of view of the labour market, the perpetuation of Taylor's theories made it possible for unskilled or semi-skilled workers to be employed in the manufacture of complex products in mass production (Fig. 1.3). Due to the economic development however, there was soon a lack of unskilled workers in the Federal Republic of Germany. This problem was overcome by employing immigrant labour. At the same time, however, the level of qualifications required rose as a result of the growing complexity of the products and technologies [5]. Greater prosperity and with it, more opportunity to satisfy the wishes of the individual led, in the past decade, to a change in values and indeed the term 'the leisure society' was coined in this period. Labour costs rose prompting companies to explore means of raising the level of automation. The development and operation of increasingly complex technology put further pressure on the qualifications of the work force. Toughening competition will focus attention on the question as to where best to set up a manufacturing plant [6]. The global distribution of manufacturing plants in response to this problem is already under way. The labour market in Germany must therefore offer a highly qualified work force at all levels in order to remain an attractive proposition for high-tech production.

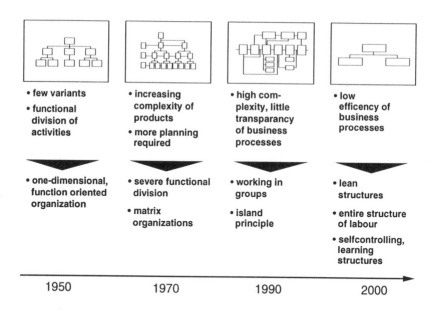

Fig. 1.4 Historical review - organization

The last few decades can be termed the computer-controlled machine tool era. The invasion of the computer into every area of corporate life heralded the advent of the NC machine and growing significance of indirect areas of production. Computer-related costs were initially relatively high, resulting in the need for central organization. The increasing efficiency of computers and substantial price cuts now permit rapid, decentralized on-site data processing. Systems now communicate with one another and transfer data via networks. However the trend towards decentralized data storage has brought with it the problem of data inconsistency. The concept of logically centralized but physically distributed data management will, therefore, become more important in the future. The actual location at which data is stored or processed will lose virtually all significance as world-wide communication via network (GAN = Global Area Networks) becomes the norm [10].

The organization of companies was, in the past, characterized to a great extent by Taylor's principles [7]. The division of the working process into the simplest movements so that the unskilled work force available could manufacture complex parts using mass production techniques certainly made sense at the time (Fig. 1.4). The products shown were well suited to this principle which depended on one-dimensional, functionally oriented organization. The increasing complexity of the products and the emerging indirect areas of production resulted in further division of functions [7]. The more complex nature of the tasks led to multidimensional project and matrix organization forms. These organizational structures became too complex and the sequences too cumbersome just as international competition became fiercer. In response, leaner organizational structures and team work were introduced. The future organizational structure will be process-oriented and will feature teams of workers, each of whom will have many skills and greater responsibility. The organization will be required to excel in its degree of flexibility and adaptability [8, 9].

These developments show that changing conditions contribute to changing demands being imposed on production engineering. The competitiveness of a company is directly dependent on the speed of the reaction to that change. In the current climate of dynamic movement, the aim of every company should be to help to lead the change by pursuing a policy of action rather than reaction and thus to gain a competitive advantage. Size is giving way to speed as the decisive competitive factor [15].

In connection with characteristics such as speed and flexibility, the question presents itself as to whether Germany is currently providing conditions conducive to a positive development. The mass of strict laws and regulations is one of Germany's main disadvantages besides high labour costs as a manufacturing site.

The option of balancing out this disadvantage by increasing productivity is rapidly running up against obstacles [16]. Higher prices abroad due to quirks in the exchange rate, high taxation and limited flexibility with regard to legislation on permissible working hours are combining to make life difficult for German companies. In contrast, the advantages previously enjoyed by companies based in Germany such as an excellent infrastructure, an education system which is held in respect throughout the world and the associated high level of qualifications are diminishing steadily. Parallel to the declining foreign investment in Germany in comparison with investments being made by German companies abroad, advantages are dwindling steadily [6].

Changing conditions affect corporate aims (Fig. 1.5). Cost limitation which has always been a priority for Germany as a manufacturing site, is likely to be joined by efforts to be one step ahead of the competition and to deploy total quality management techniques. Human resources as integrating factors and technology consumers will increasingly be the focus

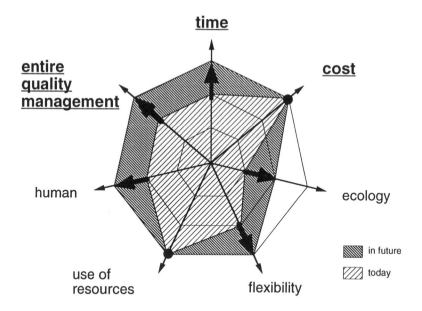

Fig. 1.5 Change of corporate objectives

of corporate attention as the d emands of the future can only be met by companies with qualified and highly motivated staff on all levels (17).

While the use to full capacity of operating resources was previously regarded as desirable, this target must now be broadened. The aim must now be to exploit all available resources to the full. The term `increased flexibility´ will no longer be restricted only to products but will also extend to include the time factor. The ability to adapt nimbly to changing market conditions will be crucial to the success of a company. The significance of ecological arguments will grow. Government legislation and regulations are instrumental in influencing companies in this respect. The environmental friendliness of products will assume even greater importance as a marketing argument [13].

To sum up, it can be said that the demands imposed on companies by the market will continue to rise (Fig. 1.6). Attention is likely to focus on enhanced quality and environmental friendliness, integrated solutions, outstanding service and shorter delivery periods. Stiffening competition is shaped to a great extent by political developments. The creation of the Single European Market for example was accompanied by an increase in the number of trade restrictions put up by other countries while, at the

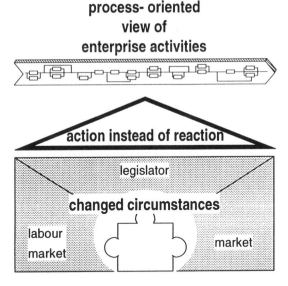

Fig. 1.6 Process orientation as a solution for the future

same time, entire markets such as Eastern Europe for example dried up. Such changes lead to changes in production engineering which should always be viewed against the background of the historical development.The conclusion to be drawn from all this can only be `act instead of reacting'. A process-oriented view of the company offers the opportunity to meet future challenges head-on and to effect any changes necessary to sustain the competitiveness of the company.

1.2 Process orientation - a comprehensive approach

The objective of maintaining or better still of exceeding a company's present level of competitiveness can only be achieved if a comprehensive analysis of the company is undertaken. In periods of economic downturn, weak points become more noticeable. These have to be eliminated if the company is to be in a position to satisfy more exacting requirements in terms of lead time, price, quality and flexibility. This, inturn, requires a change away from a functional point of view towards proess orientation. What process orientation really means, why added value can be regarded as a core process and the approach to defining competitive corporate processes will be explained in the following chapters.

1.2.1 Added value as the core process

Companies still frequently neglect to adapt their organization to both company size and to the increasingly complex production tasks [18, 19].The structure of production plants, based on Taylor's theories, have lead to an encrustment of the then valid principle of division of work inmanufacturing processes even in the growing administrative areas.

Initial efforts to improve competitiveness centred on achieving flexibility in production and rationalization. This, it was hoped, would be achieved by the deployment of computers in almost all areas of the company [21]. Due to the complexity and fascination of the new manufacturing concepts (e.g. Computer Integrated Manufacturing) and the use of computers, organizational approaches and their implementation were largely ignored, especially in the planning and administrative areas [22]. It is notable that although concepts, methods, tools and philosophies aimed at achieving and sustaining competitive advantage abound, they rarely live up to expectations (Fig. 1.7).

Strategies such as CIM and TQM (Total Quality Management) cannot simply be regarded in isolation; it is not enough to introduce only one of them into a company. One-sided measures, targeted at departmental requirements are inadequate as they are frequently detrimental to the greater good of the company. There is also the danger that buzz words such as `CIM' and more particularly `Lean Production' will raise expectations unrealistically high while no-one really knows what they stand for.

This is an example of the approach previously adopted by many companies (Fig. 1.8). Initially, an attempt was made to raise the efficiency of departments by introducing division-specific measures. The increased efficiency of computer systems, which had led to the development and deployment of CAD systems, was a major contributory factor in this move. Although problems affecting specific divisions were partially resolved, a closer inspection revealed that new problems were occurring in other departments. It is still true to say that the deployment of CAD without suitable product structure or systematic repeat part search results in a higher number of drawings and part numbers. In this case, it is more rational for the designer to re-draw a simple drawing (e.g. a shaft) than to go through a complicated search routine to find a shaft which is at least

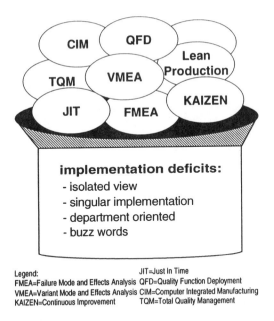

Fig. 1.7 Existing methods and philosophies

similar. This, of course increases the work level necessary in the process planning department in order to draw up new process plans, NC programs etc. Clearly, what is the optimum procedure for one department may well conflict with what is most beneficial for the company as a whole.

Further increases in the efficiency of computer systems permitted system integration in the course of time. Problems related to data storage (e.g. error-prone double entry, redundant data) were defused. Departments were amalgamated, some old organizational units were phased out and new ones formed in their place. Departmental boundaries were shifted but not eliminated. However the optimization of special fields alone is not sufficient. Only the   multidivisional management of information can achieve the level of coordination necessary and thus contribute effectively to the problems previously described [23 - 28]. This requires a departure from the traditionally function and department oriented views of corporate processes and a movement towards a process oriented view. The overall optimum is achieved only when the culture has been transformed to include process orientation and all management tools are deployed systematically to optimize the overall quality of the company. This re-orientation

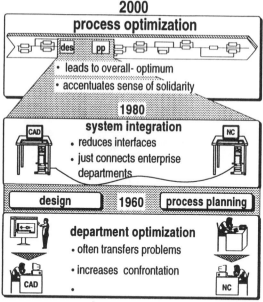

Fig. 1.8 Process-oriented view of enterprise activities - an example

intensifies the sense of solidarity within the company as each employee becomes aware of his/her role as a cog in the wheel.

The terms `function' and `process' can be defined according to `Müller' as follows [29]: A function is, as the result of a job analysis, a structural and organizational collection of one or more steps [30]. In terms of a position in the company, an area of work or a computer system this can be a single activity such as drawing up a parts list, a department such as the design department or division-related computer support such as the function dimensioning of a CAD system. System support is also related to individual activities within a position-related task regardless of who requires these functions when and in what connection with the work to be carried out. Position or department related scope of work and the actual tasks this work involves, are the priorities when a functional view is adopted [31]. The aim is the simplification of the process, i.e. how input is transformed into output and what is added during the process [29].

A sequence planning related group of elementary tasks forms a process [32]. In contrast to a function, attention does not focus in this case on the relation between input and output. The actual existence of processes, the time they require and their complex interlinkage is specified. Processes are

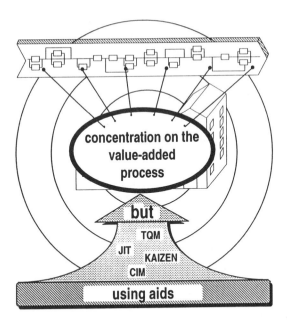

Fig. 1.9 Value added process as a core process

linked to form process chains connecting the jobs which accumulate to form processes via information flows. The formation of process chains represents the effort to integrate the functions currently existing in companies technically and organizationally [29]. This is also essential if the demand for a reduction in delivery periods is to be met [33].

Process orientation requires the concentration of the methods, tools, concepts and philosophies currently available on the optimization of the overall value added process (Fig. 1.9). The added value has, up to now, been defined according to VDMA [34] as operating income minus prepayment. In accordance with the view demanded above however, it must be remembered that the value added process begins in the Sales Department with the finalization of the order and ends in the Despatch Department when the customer's wishes have been satisfied. The aim of process orientation and the concentration on the value added process is to assess rationalization measures on their global merits and to permit their overall effects to be evaluated in terms of lead time, cost and quality. The value added process is defined here as shown in Fig. 1.10.

The production factors deployed in a company, frequently also termed

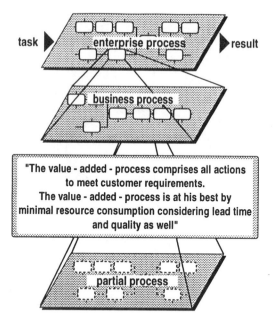

Fig. 1.10 Definition of the value added process

resources, include personnel, operating equipment, buildings, capital and computer systems. These are used to execute various tasks and determine the level of costs incurred, e.g. by order processing. In principle, processes in a company can be considered in differing degrees of detail. It is therefore possible to divide them up in various levels in order to facilitate the consideration of all the interrelationships. It must, however, always be borne in mind that a process step (e.g. a manufacturing operation) is firmly integrated within the primary area of order processing. Optimization of the process step therefore also affects the process itself.

Various examples illustrate the view of a process outlined above (Fig. 1.11). When a drawing is completed on the basis of unclear instructions and the part is subsequently manufactured perfectly in accordance with the drawing, there has not necessarily been a process of added value. If it emerges that the drawing was incorrect and the part manufactured accordingly therefore fails to meet the customer's requirements, the only conclusion to be drawn is that resources have been consumed. From this angle, it can be said that activities undertaken in the Sales Department, for example to clarify exactly what the customer wants, are also part of the

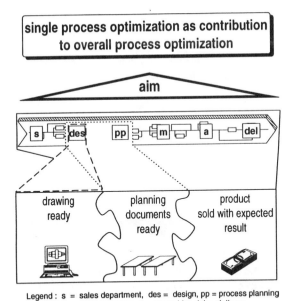

Legend : s = sales department, des = design, pp = process planning
m = manufacturing, a = assembly, del = delivery

Fig. 1.11 The process of increasing value in various departments

added value process. This process is not complete until the product has been sold with the desired result. The target of any company must therefore be to optimize individual processes in such a way as to achieve an overall optimum.

When process orientation is considered from this angle, weak points typical of current added value processes in manufacturing companies can be identified (Fig. 1.12). Depending on the size of the company, weak organizational structures can render processes powerless. A number of departments are frequently involved in order processing. Decisions can only be made on the higher echelons. By the time these decisions are implemented, the measures prescribed are sometimes out of date or no longer suitable. Large, unwieldy companies are particularly prone to this problem which is why many are currently engaged in a down-sizing process, eliminating entire layers of management. This is not an end in itself, it must, rather be part of the effort to tackle the problems mentioned above. This is one area in which small can be said to be beautiful. Smaller companies must seize the opportunity to exploit their own advantages in terms of shorter lead times and greater flexibility.

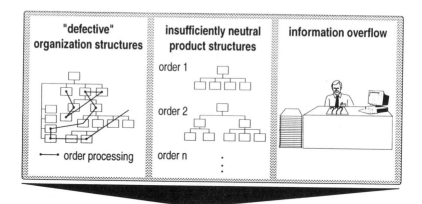

Fig. 1.12 Typical weak points of value-added processes nowadays

Another problem facing companies is data storage and product structure. Although this has long been a subject of research and great importance is attached to it in practice, a number of recent surveys show that a considerable amount remains to be done. Only a handful of companies have actually established a neutral product structure for single part and small batch production of complex products with a large number of variants. Ideally, such a structure provides a basis for an improved or computer-aided order configuration and for the optimization of many planning processes. Instead, an analysis must be carried out for each order to establish which parts have to be customized and what changes this will involve for the planning, manufacturing and assembly departments. The design department is still trying at this point to take account of the demands of various areas of the company (e.g. Sales, Assembly, Despatch) regarding the organization of the product structure since the PPS systems are unable to create an unlimited number of department-specific structures. Consequently, a compromise structure emerges which causes considerable delays in order processing and subsequent delivery to the customer. One need only imagine the problem facing fitters on-site especially those

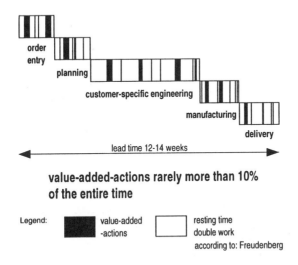

Fig. 1.13 The added value process with its value gaining and avoidable actions

working with customized machinery, when there is no indication as to which parts and structural components have been sent in which crate.

An additional weak point of current value added processes is the insufficient flow of information for the completion of various tasks. This applies to virtually all manufacturing companies. In spite of widespread use, the computer has failed to live up to its promise, in the pipe dreams of the 80's when it was believed that the paper-free office was just around the corner. In fact, just the opposite is true. Japanese companies have benefited from the fact that due to the difficulties presented by their complicated lettering, the use of computers was relatively late in spreading. The Japanese have therefore been able to learn from mistakes made in Europe and the USA. Surveys show that both in automotive companies, for example in the course of product development and in companies which manufacture complex products in a lot size of one and small batch production

- a large amount of information is made available but is not required
- important information is passed on too late but also

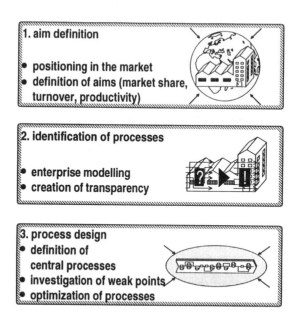

Fig. 1.14 Procedure for investigation of competition influencing corporate processes

 – due to the supply of redundant data, contradictory information is
available throughout the order processing stage.

The main weak points of current value added process described here
are reflected in order processing by the high proportion of idle time which
can be as high as 90 % of the entire lead time, by the high proportion of the
lead time in the indirect areas and by unnecessarily high resource
consumption. In this context, excessive complexity is often cited as the
problem which most needs to be remedied [4].

The case study illustrated in Fig. 1.13 underlines the statements
above. When idle time and the time spent on double work in relation to the
total lead time are deducted, the proportion of time actually contributing to
the added value process is 10 %. This shows too the enormous potential
lying dormant within a company for the application of lean production
techniques if the consumption of resources can be minimized within the
added value process.

The interim result can be expressed as follows:

- Due to changing boundary conditions, methods and tools previously in common use are no longer adequate,
- trend towards process orientation affords the opportunity to increase competitiveness
- attention must focus on the added value process
- consumption of resources must be minimized while taking full account of lead time and quality

1.2.2 Approach to the definition of competitive corporate processes

A three-phase approach is required in order to define and organize competitive corporate processes (Fig. 1.14). If measures aimed at rationalization are to be successful, it is absolutely essential that the management of the company issue a clear definition of objectives. It is important to ensure that the targets are realistic. The market position of the company must also be considered. Market chances beckon as a result of the growth of large economic units in Europe, America and Asia with increasing globalization, however the risk involved is not inconsiderable.

The second phase entails a detailed analysis of operations in the company aimed at recording how the company really works in order to define the operations at the heart of that company. Almost all companies are strongly influenced by Taylor's theories, based on his belief that processes and organization should be structured in detail. Investigations conducted by the WZL show that even in smaller companies, departmental managers are not familiar with all operations since communications structures have developed in addition to fixed organization structures. Appropriate steps which extend beyond setting out yet another list of guide-lines concerning reorganization can be defined only when the processes which actually take place are known. This applies equally to product development in the automotive industry and to order processing in a company specializing in customized machinery.

The third phase builds on the new clarity of definition of both the central processes in the company and the weak points with regard to the targets specified. The conditions required to optimize processes can be classified as human, technical and organizational parameters.

There are a number of approaches to company analysis and modelling, each accompanied by different methods and with a variety of objectives [36, 37]. As there was no practice-oriented method specially for process analysis, a new method was developed at the WZL and has since been

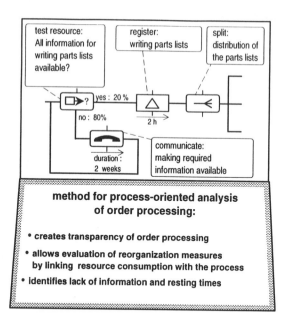

Fig. 1.15 Process analysis of order processing

deployed with success in various companies which had identified problems with double work and idle time. The field of application extends from the analysis of product development processes in the automobile industry to the analysis of order processing in companies specializing in customized machinery (Fig. 1.15).

In the case study under discussion, the actual process of drawing up a parts list was preceded by a resource test in order to establish whether or not all the information required was available. This was found not to be the case in 80 % of all orders. Consequently, a time consuming process of collecting the relevant information (symbolized by the communication element of 2 weeks on average) had to be undertaken.

This example shows that the method contributes to the process of identifying areas in which insufficient information is available. Although the staff in charge have a considerable amount of information at their disposal, the data they require in order to carry out their work is frequently missing. Once the weak points have been pin-pointed, conclusions can be drawn as to the steps which require to be taken in order to improve the situation. Since the method developed is based on 14 different process

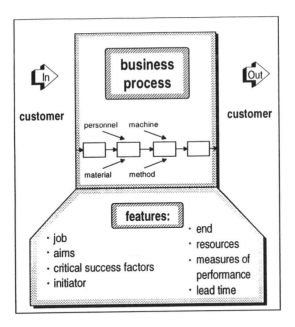

Fig. 1.16 Features of a business process

elements, it is easy to grasp, provides graphical back-up for the result of the analysis and avoids misunderstandings which otherwise arise as a result of unclear terminology in most of the standard analysis methods.

This method does not only permit the routines (processes) in the company to be presented clearly and analyzed, it also makes it possible to carry out simulation-assisted lead time calculations [29, 38]. Additionally, by identifying process-related resource consumption, new more accurate methods of cost accounting extending beyond classical actively based cost accounting can be applied.

The following case study shows the company-specific definition of a business process according to Scheidt and Bachmann (Fig. 1.16). A business process is performed between the orderer and the recipient of services. This general definition shows that a business process can also remain within one company. Departments are therefore associated internally with a customer-supplier relationship. The main characteristics of the business process are mission, aims, critical success factors, initiator, end, resources, measures of performance and lead time. Based on this general definition there are six business processes which affect the

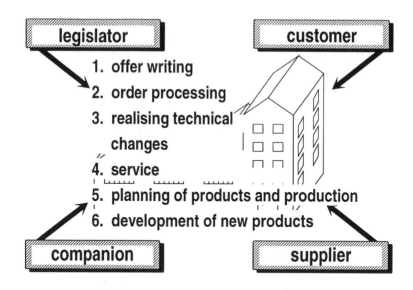

Fig. 1.17 Definition of processes which are vital to a company's ability to compete

competitiveness of a company. In addition to the significance of order processing which has previously been illustrated, service, the execution of technical modifications and the preparation of offers have been identified as important processes. It is therefore essential to ensure that these processes are optimized in terms of low resource consumption, at the same time taking equal account of lead time and quality (Fig. 1.17).

To sum up, it can be said that process orientation is a very promising approach to meeting the challenges currently facing industry. It entails investigating which processes exert the greatest influence on the competitiveness of the company with regard to the targets set by that company. The decision as to which processes could profitably be made more competitive depends on the size of the company, the product range, product structure and organization. In general terms it can be said that order processing is of central importance to companies with single part and small batch production and that product development is of pivotal importance to companies engaged in mass production. The target, in each case, is to increase the proportion of activity resulting in added value. The organization of company-specific processes in conjunction with the

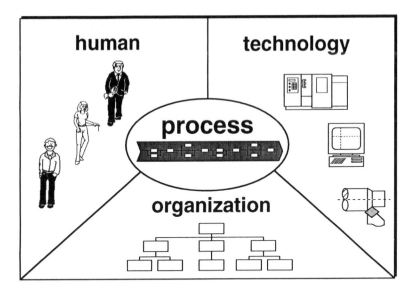

Fig. 1.18 Areas for the structuring of processes

identification of decisive parameters is the subject of the next chapter.

1.3 Parameters for reorganization

Human, technical and organizational aspects must all be taken into account when planning process-oriented reorganization (Fig. 1.18). Human resources as well as concentration on the added value process are of particular importance in view of increasing complexity and automation.

1.3.1 Human resources as the key factor

Staff motivation is a key factor (Fig. 1.19). Previous experience shows that management and the members of a project team which has assumed responsibility for reorganization are already sufficiently highly motivated since they are familiar with the overall interrelationships. In order to increase the motivation of employees on the operative level, it is therefore important to inform them in good time of any changes planned and of the reasons for change. The successful implementation of process-oriented

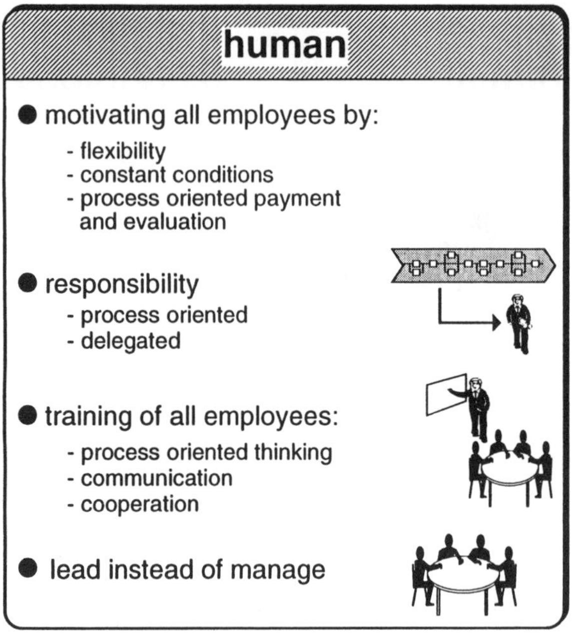

Fig. 1.19 Recommendations for structuring in the human area

measures requires a high degree of flexibility, a secure environment and dependable boundary conditions - all of which must be provided by management. Once process-oriented structures are in place, an incentive strategy and an assessment system must be developed. The guiding principle must be success and improvement in relation to the whole process, not merely the optimization of details.

This requires staff training on all levels of the hierarchy. All employees must have process-oriented attitudes `in their blood' and must act accordingly. The achievement of total quality is a long, on-going process in which communication between staff is a key element. Communication training is, therefore, of vital importance if employees are to learn to work more closely than before with their co-workers. Parallel to the introduction of incentives and assessment, areas of responsibility must be oriented towards processes. Examples of change and willingness to change must be exemplified, not simply organized at managerial level. Managerial staff are required to reflect on leadership as `doing the right thing' rather than management as `doing something right'.

These guide-lines regarding training in process-related thinking have

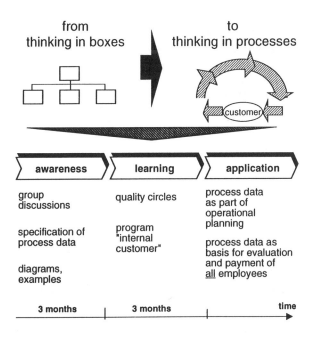

Fig.1.20 Training employees for "thinking in processes"

been implemented in three-phases in the company Freudenberg (Fig. 1.20) [40]. In the first phase, the teams in charge of putting the plan into action present their concepts to the other employees in so-called infoshops. The approaching changes are explained and illustrated using charts. The staff are informed not only about the most pressing long-term corporate aims but more particularly about what steps will be necessary to achieve these aims. During the second phase, the measures introduced in order to implement the plans are discussed and optimized in monthly team meetings and in weekly quality circles. In the mean time, learning has become a continuous process. Experience shows that active staff cooperation can frequently be engaged by involving them fully in the process from an early stage. Setting standards regarding process data for operative planning and for assessing and rewarding all employees demonstrates commitment and therefore aids the process of change away from thinking in `little boxes' towards thinking in terms of processes [41].

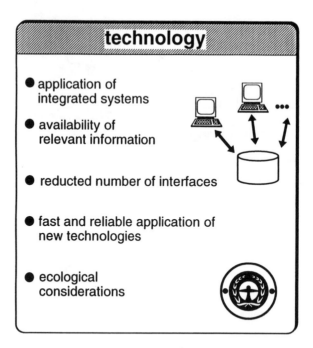

Fig. 1.21 Recommendations for structuring in the technology area

1.3.2 Technology

The technology deployed must also be adapted to the process-oriented organizational structures (Fig. 1.21). Primarily, this involves the use of integrated systems which support entire processes rather than individual functions. It is essential to ensure that the information supplied is up-to-date in order to guarantee a process which is consistent throughout.

The rate at which innovative technologies are introduced and applied is also of major importance. The reliability of the technology is just as decisive a factor in the success of the quality effort. From the point of view of production, recycling-oriented design is also important as it is at this point that recycling options and costs are defined. The manufacturing process itself must be environment-friendly. Efforts here must be directed at reducing the amount of production waste and energy and to promoting the use of FFC cooling lubricants.

Hertel's TESS (Tool Expert Software System) is an example of the successful development and introduction of a process-oriented system (Fig. 1.22). It supports the entire offer and order process for the majority of

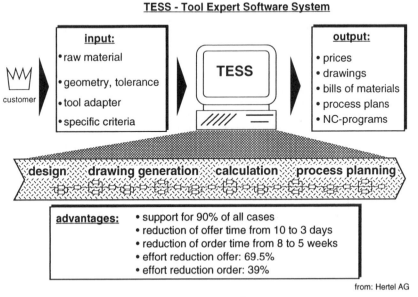

Fig. 1.22 Process oriented offer and order processing

customized cutting tools supplied by Hertel. The structure of the system is modular, permitting all the material required in order to prepare an offer including not only drawing and calculation, but also process plans and NC programs to be drawn up on the basis of sparse parametric input such as initial and finished state of the workpiece and tool holding fixtures. The system was largely planned and implemented by the employees involved which is reflected in the level of acceptance demonstrated and in a number of other advantages. Of the operations to do with drilling, 90 % now benefit from the use of the system. The time required to complete an offer has been reduced from 10 to 3 days and manufacturing time from 8 to 5 weeks. In urgent cases, a complete offer can be submitted within a few hours. The working time needed to prepare an offer has decreased by almost 70 % and order processing now takes 39 % less time than before [44].

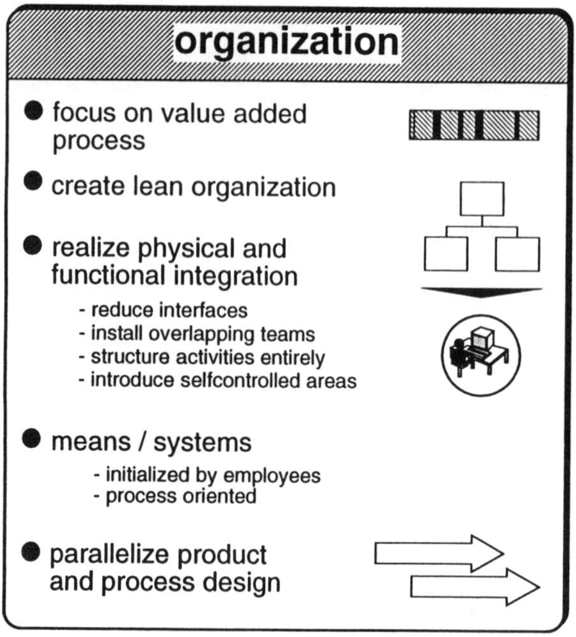

organization

- focus on value added process
- create lean organization
- realize physical and functional integration
    - reduce interfaces
    - install overlapping teams
    - structure activities entirely
    - introduce selfcontrolled areas
- means / systems
    - initialized by employees
    - process oriented
- parallelize product and process design

Fig. 1.23 Recommandations for structuring in the organization area

1.3.3 Organization

The changes also focus on the organization of a company as this is the area which must ensure that the conditions are in place for process-oriented sequences (Fig. 1.23). The re-orientation of functions towards a holistic process approach means in concrete terms that the aim is not to make savings at one individual place but to strive to achieve lean organization and with it, improved lead times. The corresponding processes must be represented in spatial and functional integration so as to decrease the number of interfaces. Additionally, interdepartmental teams should be formed, the work should be condensed into holistic activities and employee emowerment should become a reality. The teams should be organized as self-controlling loops in order to ensure both that interruptions in the flow of events can be ironed out autonomously and that reaction can be rapid and flexible [8]. In accordance with this principle, tools and systems should be planned decentrally by the employees and should focus consistently on the process. Experience shows that more use is then made of the tools and systems available. Additionally, from an

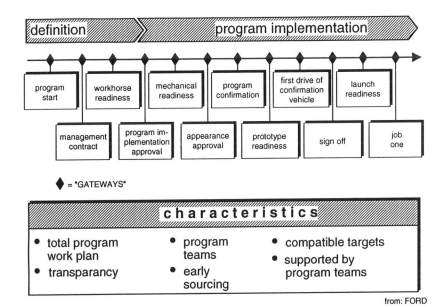

Fig. 1. 24 Structurization of the development

organizational point of view processes, product and production process design should run parallel in the sense of concurrent engineering in order to save time. The organization must concentrate strictly on added value.

The measures described were implemented by FORD of Europe for the development of cars (Fig. 1.24). The aims of the systemization of passenger car development are the improvement of product quality, a shorter development period, cost savings and greater concentration on customer requirements. A comprehensive development process divided into two main phases, definition and execution, was defined. The progress of interim aims is measured in the form of so-called `gateways'. In conjunction with clear, stable standards and aims the entire process is thus transparent to all. Further characteristics are the involvement of suppliers at an early stage and the organization of the work in teams. The development process is backed up throughout by the other processes such as that of the development of business strategies and by the technical process [42].

Order processing as practiced in the Carl Freudenberg company is an additional example of a process-oriented business area (Fig. 1.25). The initial situation was a complex and time-consuming procedure involving

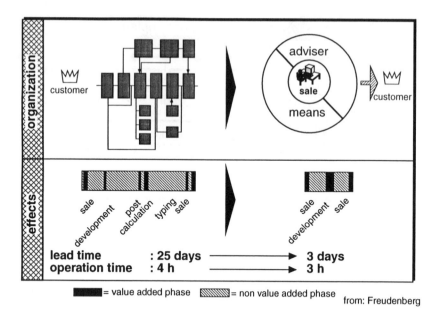

lead time   : 25 days ——————→ 3 days
operation time   : 4 h ——————————→ 3 h

■■■■ = value added phase ▨▨▨▨ = non value added phase

from: Freudenberg

Fig. 1.25 Structurization of the offer process

sales, development, accounts and typing pool and spanning several levels of the hierarchy. The integration of the activities within one offer process processmeant that time and effort spent on in-company post could be saved thus increasing the capacity to provide the customer with clear, correct information. Employees from other areas in the company are available to these teams as advisers and efficient aides helping, for example with computer-assisted cost calculation. While offers involving six-digit sums were previously treated like those to the value of a few hundred marks, employees now sort out the offers according to significance as they are received.

This has permitted the lead time for important offers and the time required to process them to be reduced considerably [42].

The company Carl Zeiss has likewise achieved notable improvements by reorganizing the manufacturing process. The manufacture of oculars was transformed from a workshop to an island principle. Various machining processes comprise an island, thus reducing the number of operations from 18 to 4 and the lead time from 27 to 3 days. Additionally, unit costs were slashed by 23 % to 65 % depending on the part. These cuts are due to reduced operating ti achieved by carrying out a complete

- efficiency of existing methods and philosophies is reduced because of changed circumstances

- process-orientation permits the overall view of lead time, quality and costs

- education and motivation of the resource "human being" are important factors for process design

- integrated view of human being, technology and organization is required in order to design competitive enterprise processes

Fig. 1.26 Summary

machining process on one fixture, the reduction of set-up times by using similar parts and manufacturing-oriented ocular design. The employees were trained both to operate the new machines and about the sequences and work within the island team in good time, thus ensuring the effectiveness of the manufacturing island. In future, island personnel will also assume responsibility for observing time schedules [43].

1.4. Recapitulation

In order to remain successful in future, companies must rise to the challenge presented by the difficult world-wide economic situation and respond to the rising demands. It has been demonstrated that lead time, quality and costs are still important parameters as regards the increased competitiveness. These, however, must increasingly take account of ecological aspects. It is also becoming clear that methods, tools and philosophies must be regarded from a historical angle and cannot simply be passed on to the next decade. Changing patterns are compelling companies to adopt a process-oriented view of themselves (Fig. 1.26). Individual measures aimed at improving economic efficiency must cease to be department-oriented and must in future be examined and assessed in terms of their contribution to the optimization of the entire value-added process. The value added process can be regarded as the central process. It embraces all activities required to satisfy the wishes of the customer. The consumption of resources, lead time and quality are all significant optimization criteria.

An analysis must be carried out separately for each company, taking account of the corporate aims in order to identify the individual processes which determine the competitiveness of that company in the global market. This requires a detailed analysis of sequences in order to achieve the degree of transparency necessary to design the processes thus identified.

The parameters human resources, technology and organization can all profit from reorganization. Attention focuses on the human element. Process orientation cannot simply be prescribed by executive staff. Instead, this way of thinking must be adopted throughout the company, at all levels. This requires motivation and training. Technology and organization must then follow suit in adopting process orientation.

Process orientation presents a means of utilizing systematically the potential for rationalization available and thus increases the ability of companies to compete. An opportunity not to be missed.

2

QUALITY MANAGEMENT AS A CORPORATE STRATEGY

2.1 Introduction

Companies must respond to change. Any company which is seriously aiming at success has to address the issues of changing social values and structures, accelerating technical progress, increasingly high demands on products and services as well as evolving market structures. Social norms of the past years, characterized by the necessity to strive for ever higher levels of wealth, security and prestige are giving way to a more holistic way of thinking. Ecological and social aspects such as conservation are assuming more importance as measures of satisfaction in society. Changing customer awareness is rapidly being reflected in new sales arguments. Quality, functionality and stability of value have recently taken over from the pure satisfaction of need which has been the main motivation for buying goods or services throughout the previous two decades. These developments are paralleled by a global market which is in a state of continuous flux as is illustrated by events in the Single European Market, the countries which formed the former Eastern Bloc and above all those in the Asiatic-Pacific area. Europe has always been particularly important to German companies. Markets in, for example the USA or Japan account for under 10 % of German exports (1991) [1] although German products are

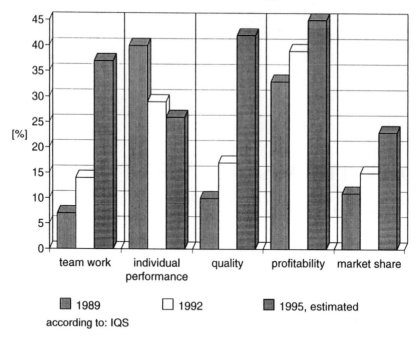

according to: IQS

Fig. 2.1 Growing importance of Quality Management

highly thought of in those countries. Quality, or rather, characteristics such as durability and performance are the principle sales arguments in the American and Japanese markets [2]. The significance of quality as a reason to buy is confirmed by the fact that in 1989, 89 % of 500 European executives questioned reported that quality was the primary reason for buying [3]. The reputation of 'Made in Germany', for decades a guarantee of excellence is, however, now on the wane. Surveys, especially those dealing with the automotive industry show that the former superior quality of German products can no longer be relied upon [4, 5].

This situation is one of the reasons for the increasing significance of Quality Management as a corporate strategy. Whereas the strategic advantages it offers the entire company were previously frequently neglected, there is a growing awareness in industry of the benefits to be reaped (Fig. 2.1). Less importance is now attached to the capabilities of the individual employee which used to be an important criterion by which to evaluate staff. Ability to work within a team features increasingly in the holistic evaluation of a company. Significantly, the ability to produce 'quality' holds the key to profitability. Consequently, it enjoys virtually the

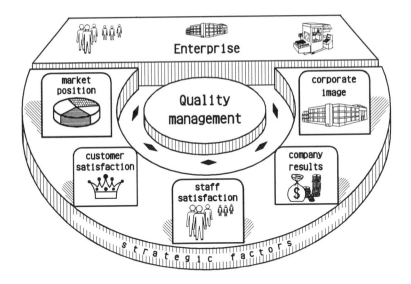

Fig. 2.2 Initial situation

same status.

The satisfaction of customers' individual desires is now, more than ever before, the key to recapturing and sustaining the competitive advantage in a market characterized by cut throat competition and customer-orientation. Customer satisfaction is obtained by offering products which meet these demands and which therefore help to achieve the corporate objectives of individual companies. These objectives are determined largely by five strategic success factors. They underpin the existence and development of the company (Fig. 2.2).

The image projected by a company and above all public opinion about the attitude of the company regarding matters of public interest such as issues involving the quality of life, environmental protection or resource depletion. In contrast, customer satisfaction reflects the opinion of external customers of the company in general and about its products and services in particular. In addition to external quantities, important criteria are those on which the assessment of a company's ability to produce in-company quality are based. Internal characteristics affect business results and employee satisfaction. The four characteristics listed in Fig. 2.2 are in accordance with the criteria on which companies in contention for the European Quality Award 1992 [6] are judged. This prize is awarded to companies

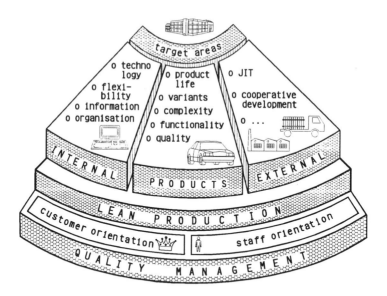

Fig. 2.3 Objectives and the development of appropriate measures

demonstrating that their approach to total quality management has contributed significantly to satisfying the expectations of customers and employees.

The scope of these characteristics must of course be extended to include market position since this is the decisive factor in the survival and continued development of the company.

As has just been outlined, quality management is shaped more than any other area within a company by both internal and external influences and objectives. As an element extending beyond the confines of any one company, quality management is caught up in the conflict between efforts aimed at ensuring success on one hand, and on the other hand changing market parameters, altered customer-supplier relationships and the resultant demands imposed on products and on the manufacturing process. Quality management, therefore, involves exploiting the tools available to the full and, above all, developing and implementing new methods of meeting customers' demands.

Production structures hitherto have been unable to cope with growing demands from the market place in general and from the customers in particular. The main features of this gradual development include the trend towards mass customization, shorter innovation cycles and shorter product life cycles. Those changed boundary conditions call for strategies for the adaptation of production structures to meet these current demands. Companies tend to react by adopting one of three, mainly technically oriented, approaches (Fig. 2.3).

The field of action which has, until now, been given priority is the optimization of internal resources. Reduced lead times and costs in conjunction with increasingly agile production are the principle objectives in this field. Attempts at rationalization are therefore concentrated on automating manufacturing and assembly operations. In this context, questions related to improved logistic flows are of major importance.

Companies are increasingly availing themselves of tools and other aids in their efforts to implement their strategic corporate aims. Computer integrated production (CIM) and the voguish term `lean production' which embraces new personnel management and general management concepts is frequently bandied about whenever the talk is of the `optimization of production structures' [7, 8, 9]. These new strategies which are currently receiving much attention are being expanded to include the `fractal' factory [10]. They are intended to help close the gap between purely technical orientation and the organizational and human aspects within the company. In spite of intense activity in these areas there have been no major success stories up until now [11].

Alongside the improvement of internal resources, there is an additional area in which efforts are being made within the company to optimize products. Shorter product development cycles, greater variety, more functionality, higher levels of complexity and better use of the potential of a product are all helping companies to meet the demands of individual customers through increased flexibility. Optimization of external efficiency and thus of external resources forms the third field of action. It is characterized above all by changing forms of cooperation between customer and supplier. The trend towards manufacture by outside specialists in conjunction with the increasing tendency to leave development to the supplier are typical of this strategy. The reduction in the number of suppliers is significant. The concept of just-in-time manufacturing has also contributed to the new relationship between customer and supplier.

So far, the success of these methods has fallen short of expectations. This is due mainly to the following facts:

- The methods listed were conceived as instruments and are generally regarded as tools which can be used to enhance the technological principles and technical boundary conditions of product development.

Companies frequently disregard the fact that such optimization and the methods involved affect not only individual technological or organizational aspects but result in changes throughout the company in terms of the deployment of technology, organization and employee involvement. Instruments which promote and accelerate evolving employee awareness and structural changes in the organization as well as developing technology have previously been virtually ignored.

- The introduction of methods contributing to change to both the technical environment and structural boundary conditions is a protracted process. Evolving employee awareness invariably results in changed corporate culture. The success of such methods is, therefore, determined largely by interrelationships between these methods and between different methods. Generally speaking, previous approaches have failed to take account of dependencies between measures and their effects. Consequently, the effects of individual measures can be contradictory or can even cancel one another out. Long term cooperation between customer and parts supplier makes for low purchase prices and ensures a steady supply over a given period but reduces flexibility in the Purchasing Department. Objectives can be said to be in competition with one another when one can be reached only at the cost of an other. High stock levels ensure that the manufacturing department never runs out of parts, but involve high levels of capital investment and are thus a strain on liquidity [12].

To recapitulate, companies are currently imposing individual measures, frequently neglecting to include and take due account of the other areas affected. The reason for this is the high degree of technological orientation. Little thought is given to the significance of human resources as necessary to the successful implementation of any measure or to a harmonization of individual measures.

This situation, coupled with experience of tried and tested approaches in past years, demonstrates that future corporate strategies will have to be effective beyond the confines of an individual company and will have to take equal account of technological and human-oriented aspects. They will also be required to coordinate individual activities.

Lean production is certainly a decisive step in the right direction. Notable core elements of this concept are, alongside parallel product development and global manufacturing, the following principles:
- concentration on the customer as the focal point of all efforts
- integration of the supplier in the production system
- transferral of responsibility for quality to the individual.

One concept which embraces each individual activity while focusing on employees and customers is that of Total Quality Management (TQM). The aim of improved quality and efficiency of processes and communication is inherent in this concept. It is, as its name suggests, a sweeping concept. To date however, it has not been adopted in many companies, nor has it been possible to apply quality assurance methods prior to manufacture with the degree of efficiency anticipated. Any analysis of the reasons for this always stumbles over the after-effects of Taylor's work principle, still trapped in fossilized departmental walls which have not yet been surmounted [13].

2.2 The current sphere of influence of quality management

The areas influenced by the concept of quality management still extend over various phases of product development. The basic model signifying the elements of quality assurance is the quality spiral. Its main elements are concerned with ensuring quality in terms of sales, development, planning, purchasing, production and service (Fig. 2.4). Manufacturing quality, previously often regarded as the only aspect of quality is now regarded as a major contributory factor [13]. Quality assurance has thus progressed from its origin as a mere collection of test and inspection results and the determination of test findings.

The quality of distribution and marketing is characterized by the relationship with the customer, the manner in which customer wishes are treated in negotiations and by consultation offered. Market studies can be useful sources of information in this context. In the development and production planning phases, customer wishes received by the Sales Department form the basis of new definitions and design of manufacturing, assembly and environment oriented products and operations. Significant factors which must be considered are deadlines and costs as well as quality criteria. Estimated risk, determined, for example, by the failure mode and effects analysis (FMEA) is used to safeguard planning results. High parts and materials quality is ensured by close cooperation with the suppliers who are assessed in terms of the quality of their goods and service.

It is characteristic of this department oriented quality management technique used for a variety of tasks, that like the effort outlined above to optimize internal and external resources, it frequently fails to live up to its promise. It targets mainly the classical criteria on which optimization is based (time, quantity and cost). The benefits of the tendency towards a

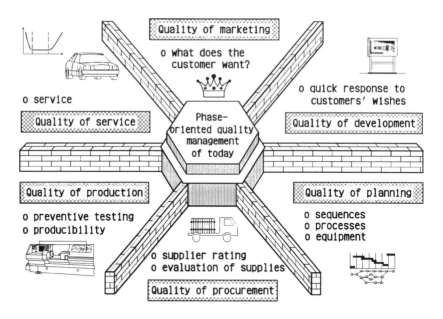

Fig. 2.4 Phase-oriented quality management of today

quality circle with departmental and hierarchy oriented optimization are limited.

2.3 Quality management as a corporate strategy

Previous approaches at increasing in-company quality concentrated on different aspects. The technologically oriented approach focused above all on the methods and tools of phase oriented quality management whereas the holistic approach homed in on the company and its processes. The second approach was dominated by the demand for social and methodical empowerment of the employees rather than expert and technically oriented ability to ensure and direct the quality of products and services. In future, quality management must unite these two approaches harmoniously.

If quality management is to be successful as a corporate strategy, the following principles must be observed:

– Quality management is a continuous process involving executive staff and work force alike. As a continuous process it extends from the definition of aims through to implementation and is thus characterized

largely by the wishes of the customer and the satisfaction of these demands by staff and suppliers.
- Quality management provides a number of extremely effective tools which have an impact on diverse elements in the company. Their deployment is the fundamental prerequisite for the success of quality management.
- Quality management can be divided into various basic elements which provide both strategies for achieving changing awareness within the company and the conditions which are essential for this change.
- Certain rules must be observed concerning the perception of executive functions if quality management is to be implemented successfully. These rules are indeed inherent elements of quality management. Provided they are applied consistently, they create the conditions required for an effective link between technologically oriented strategy elements and elements representing the changed awareness.

These principles will now be explained in detail.

2.3.1 Quality management as a continuous process

Quality management within the company is a continuous and dynamic process (Fig. 2.5). The satisfaction of customer wishes is the primary aim, whereby it must be remembered that all efforts must focus on both internal and external customers. On the principle that quality management should begin with management, this is the level at which corporate aims are defined. These are oriented towards the five strategic factors for success: customer satisfaction, employee satisfaction, business results, corporate image and market position.

The responsibility for the quality of products, services and actions must be transferred to the individual. It is up to the individual to define internal aims on the basis of the principle stated above and in accordance with the function assigned, and then to work to reach them. This process increases the motivation of the employees involved by giving them more responsibility for the results. The fact that aims defined in this phase can be achieved only by the employee vested with responsibility, means that a change in awareness is essential at managerial level to bring about change in the style of management and to ensure that responsibility is distributed downwards to the individual members of the work force.

The implications and the consequences of this approach must be clear to all involved. Management must be aware that the shift in responsibility

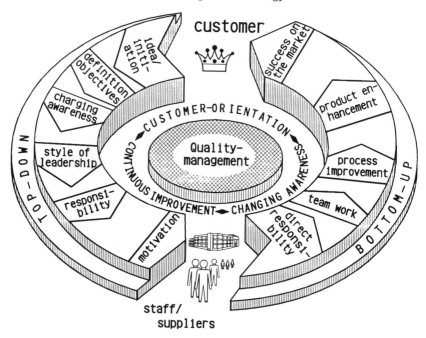

Fig. 2.5 Quality management as a continuous process

does not represent any loss of power on their level. It is, rather, the case that employee abilities previously dormant are activated thus leaving management free to explore their own capacities. Employees must be aware that they are now free to use their abilities to the full, that their responsibility extends beyond the mere execution of their duties and that they contribute directly to the security of the company and in the long term to their own security. This means not only that the individual employee is granted more responsibility but also that each employee is required to show progressively more initiative, to feel responsible for the quality of products or operations and to rectify defects instead of accepting them. The need for the employee to take the initiative is a major part of the leading functions to be adopted. Staff education and training therefore becomes an integral element of quality management.

Such approaches to quality management have, until now, focused on top-down decision making. Such a process promotes the attitudes of management to quality but neglects to involve staff and the deployment of innovative techniques of defect avoidance. The priority of these approaches is to honour the guide-line `quality management begins with management'. This approach must give way to the bottom-up approach.

This requires a radical change of opinion within the company. The implementation of both approaches to quality management means above all that the prevailing views regarding the quality of products and of the company must change. What this amounts to is a cultural revolution within the company.

This reason behind this change in perception is that a change in the style of management is required. There must be a distribution of responsibility; individuals must be encouraged both to accept direct responsibility and to work in a team, they must be prepared to motivate staff, thereby reaching success targets via continuous product and process improvement in conjunction with a high level of customer orientation in all areas.

2.3.2 Tools of quality management

Tools must be made available if quality management is to be established as a continuous process moving from a departmental and phase oriented approach to the holistic quality management outlined. These tools must be capable of increasing the individual contributions made by operations, processes or staff to the quality of the results and of the products. The objectives listed are measured at each stage of product development and are independent of the quality management tool used. On the basis of the objectives of quality management specified in phase oriented quality management there are three classes of goals (Fig. 2.6).

Each product evolves as a result of activities. It is, in effect, the result of these activities. These begin with product definition and end in maintenance or in waste disposal. Material products such as parts or planning documents and immaterial values such as service contribute to the finished product. Each activity involved, i.e. any activity with a direct effect on the material development of the product or any indirect activity such as those in the field of quality management can be called a process. The term process encompasses more than the direct manufacturing process as a sequence of manufacturing steps and operations. Assembly or service within the company are among the typical processes in a company [15]. The process consists of cooperation between the seven M-factors (huMan, Machine, Material, Method, Management, Measureability, Mankind's surroundings). It is characterized by its interfaces via which is receives input information and passes on output information to the processes

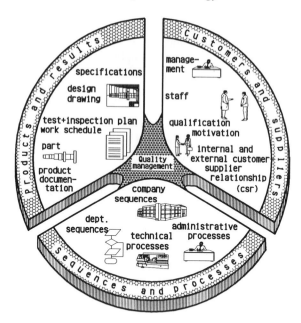

Fig. 2.6 Objectives of quality management

downstream [16]. The input interface provides a supplier with input information and the output interface gives a customer output information. Suppliers and customers are thus involved not only outside but also inside a company. All employees in a company must be aware that they have a direct internal customer since responsibility can only be assumed by employees who know exactly who they are working for [17]. When the customer - supplier principle is extended from the internal and external customer to potential customers, the success factor `corporate image' gains in significance. The company's attitude to and treatment of the environment is more important now than ever before. This aspect is inherent in the objectives of quality management.

New objectives independent of phase emerge from the process of classification itself. These are then integrated into the quality management strategy as elements not directly related to any particular problem or function. As such, they should always be considered regardless of the task in hand. Quality management now has a large number of methods and tools at its disposal with which to effect improvements in the following areas: `products and results', `customers and suppliers' and `sequences and processes' taking account of the corporate objectives. The objects of

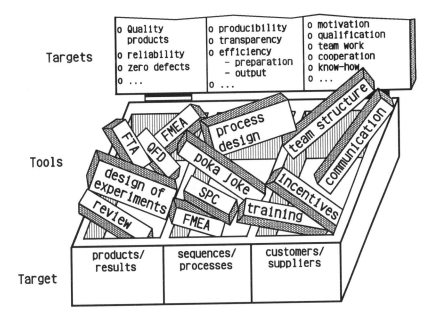

Fig. 2.7 Quality management tools

quality management can be optimized only by being divided into various details which should then be optimized with the help of the quality management tools described (Fig. 2.7).

Companies are however only partially able to apply these tools successfully. The reasons for this are diverse. The degree of success is usually very limited due to a lack of knowledge of the technical aspects of individual applications on one hand and about strategies and methods of introducing these tools on the other. This applies particularly to the application of preventive quality assurance. It, like quality management itself, is characterized by the fact that it can be effective only by involving all those who are affected. An additional reason is that application of the tools is generally directed at the enhancement of one factor. The user frequently neglects the accompanying aspects and their interaction with one another, thereby failing to exploit the potential for improvement to the full. There is frequently a lack of strategic aims and visions concerning the use of these tools within the company. It is essential in a further step to formulate elements of corporate quality management incorporating both details and general aims.

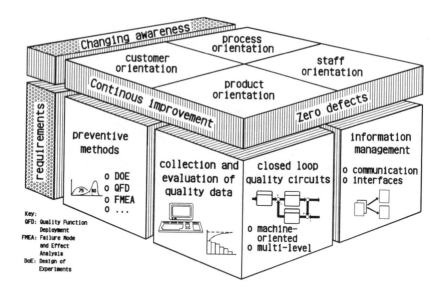

Fig. 2.8 Elements of quality management as a corporate strategy

2.3.3 Basic elements of corporate quality management

Some of the tools described above can be reduced to strategic elements representing measures which promote changed awareness in the company. These are complemented by strategic elements whose operative orientation actually pave the way for the change outlined to take place (Fig. 2.8). The most important strategic elements of quality management are customer, employee, product and process orientation. They must undergo a process of continuous improvement which extends from the deployment of preventive quality assurance methods to the application of closed loop quality circuits.

Customer orientation means a shift in emphasis away from purely divisional or departmental interest to focusing on customer satisfaction. The term customer refers here to internal and external customers and must include potential customers. This means that each employee must know and want to meet the requirements of both internal and external customers. Each employee must understand that the benefit perceived by the customer depends on his or her performance [18]. This, of course, is possible only when each employee has been informed as to what the customer's requirements are and how these can best be met. Due to the competitive environment, however, no quantity remains constant.

Employee orientation places responsibility for the quality of manufacturing and commercial processes firmly in the hands of the operative. Employees are empowered through education and training to bear this responsibility, to think about their work and to take an active part in the process. Employees on the spot know best where improvements are necessary - and possible. They can contribute to increased quality only when they have been prepared to a sufficient degree for this function. In-company education and training is a continuous process which must, of course, involve all executive staff so that they can ensure that their staff are always in touch with the latest developments [18]. If individual employees are to assume responsibility for their own actions, there must be a clear definition of rights and duties in the relationship between staff and leaders which allows for a sufficient degree of freedom and conveys recognition.

Perfect processes result in perfect products and services. The target must therefore be to pin-point faults and eliminate the causes, to shorten cycles and to improve the process continuously (Kaizen) [18, 19]. The line of thought necessary to achieve this, process orientation, stands in sharp contrast to the purely result-oriented way of thinking. The readiness to embark on a course of personal change is a prerequisite and is an integral part of processes which are to be effected by the individuals responsible for the process. It begins with self-appraisal by staff at operative level. As an element of the quality management strategy this process must be universally applicable, regardless of department and subject and must be aimed at continuous improvement of results and therefore at the long-term improvement of all corporate processes. Regular appraisal, in some cases in the form of reviews or audits, is a crucial element of continuous improvement at system level.

Continuous improvement means relentless efforts on the part of all employees to improve each individual process, i.e. to avoid making mistakes and to reduce the number of obstacles. In this context, continuous improvement applies not only to products and services which focus on product orientation, but also to the way in which machines are operated, how people work or how systems and guide-lines are implemented [19].

Preventive methods in quality assurance serve to prevent error in the areas preceding manufacturing and assembly. A system under which all sources of error are made known to all employees, within the framework of FMEA for example, aims to avoid design faults and faults within the manufacturing process. QFD helps to transform customer wishes into product characteristics in various steps. Design of experiments methodology for the determination of optimum product and process parameters is applied in the development and production planning phases.

All of the preventive methods listed can be applied in team work. They are based on knowledge of the employees in the firm. The efficient flow of the knowledge gained from experience is made possible by the introduction of a down-the-line information flow system [14]. Quality assurance methods are thus frequently effective only when they are integrated into a closed loop quality circuit. The principle behind systematic feedback into various levels of the closed loop quality circuit is that the use of historical data will prevent the same mistakes from being repeated, for example at the planning stage. The four most important levels in the closed loop quality circuit are the managerial level, at which corporate aims are defined and concepts developed, the planning level at which the original concept evolves into concrete plans for realization, the administrative level which bears responsibility for control functions and realization and the operative level at which manufacturing and test and inspection orders are carried out. Four categories of closed loops are at work in these levels, those concerned with the machine itself, those which are machine-related, those concerning one specific level and inter-level loops. The component linking them is the quality data base containing historical quality data. The zero defect concept based on the Poka-Yoke principles [20] is a special form of closed loop quality circuit. Processes based on this principle are designed in such a way as to render mistakes virtually impossible and to implement a system of error discovery allowing any errors which do occur to be nipped in the bud.

The process of collection and evaluation of quality-related data is the most important prerequisite for closed loop quality circuits and additional elements of quality management. It provides most of the information filed in the quality data base about the quality status. The operative prerequisites of a quality management strategy are connected with one another via ordered information processing. The term `ordered' applies not only to the technical side but more particularly to the smooth processing and transfer of information between the individual in-company interfaces.

Various national and international surveys confirm that the application of these strategic elements and their operative prerequisites fall far short of success [4, 5] (Fig. 2.9). International comparison (of the automotive industries) indicates that the strategic elements characterizing evolving awareness (customer, employee, process and product orientation) are insufficiently represented in Germany. Whereas the Japanese companies

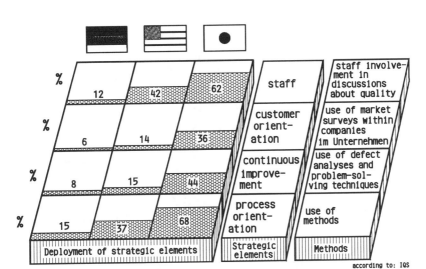

Basis: 500 companies in Canada, Japan, Germany and USA

Fig. 2.9 Current deployment of strategic Quality Management

reported that 62 % of their employees took part in regular quality meetings, the corresponding figure in German companies was only 12 %. Identification of new product and service characteristics is the almost exclusive domain of developers' and engineers' imaginations. Only a few companies back up their ideas for new developments with market analyses (6 %). Only 8 % of the German companies questioned claimed to apply the principle of continuous improvement reinforced by the deployment of error analyses and problem-solving techniques. The response was similar regarding the use of methods of promoting process-orientation. Methods such as brain-storming, statistical process control or Pareto analysis have been established in fewer than one fifth of the companies which responded.

2.3.4 Management functions in quality management

Certain rules must be in place if the corporate strategy quality management is to be put into practice. These rules are an integral part of quality management. Provided these are observed resolutely, awareness changes as required and the strategic elements of quality management are applied consistently Fig. 2.10).

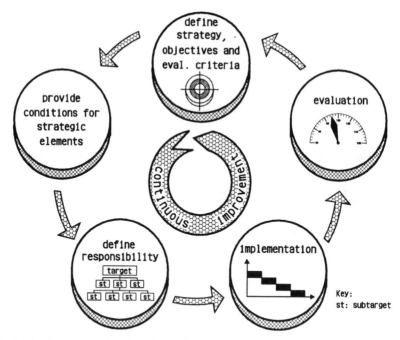

Fig. 2.10 Management functions in quality management

A clear definition of objectives in the run-up to the introduction of new quality management processes gives a definite direction to the guide-lines for steps still to be taken in the pursuit of quality. Unambiguous evaluation criteria are allocated to the targets. The notification of each employee throughout the company of these criteria, gives those responsible for applying them the opportunity to define and subsequently evaluate sub-targets of their own. A clear consensus on objectives is crucial for the adoption of responsibility at various levels of the company. It is evident that the functions of leadership outlined here must be assumed at all levels. The only difference is that each individual level addresses a different aspect of one common hierarchical system of objectives.

In order to ensure that the aims can be put into practice, certain conditions must be in place so that those responsible for the process have at their disposal both the means and the knowledge required to control the process. Six fields of action require preparatory work. The six fields of action are oriented towards the 7 M's of quality management which must undergo process-oriented specific adaptation (Fig. 2.11).

In creating the prerequisites for the deployment of quality management

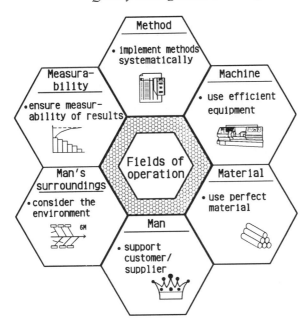

Fig. 2.11 Prerequisites for an efficient quality management

techniques, the following aspects are of particular importance:

– The introduction of new methods requires careful planning, taking account of the future application envisaged. Like the planning of products of manufacturing processes, boundary conditions governing the deployment and potential error sources and obstacles must be pin-pointed and eliminated.

– As illustrated by the definition of managerial functions, it is necessary to ensure that results are measurable. This is the only means of assessing the quality of products and services. The quality of process results can only be measured when consensus has been reached regarding the definition of standards required of processes [21].

– The deployment of efficient resources means that the resources must meet the requirements of the manufacturing of commercial process in hand. The deployment of resources which do not have the capacity to achieve the process results planned is just as detrimental to quality as resources designed to satisfy much higher demands.

– Unambiguous definition of areas of responsibility reinforces staff motivation. The existence of clear instructions and guide-lines enables employees to define sub-targets within the framework of their scope of

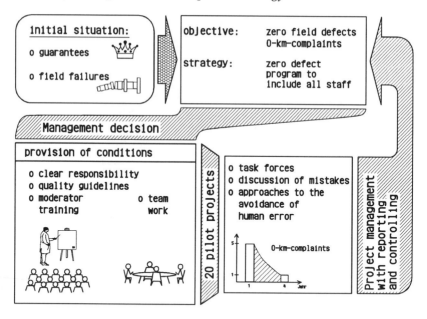

Fig. 2.12 Application of quality management strategies exemplified by the implementation of a zero defect program

responsibility and authority and to pursue these targets. In the last two steps, implementation and evaluation, the results are collected and evaluated. Employees are supervised constantly within the closed loop quality circuits allowing any corrections or redefinition of goals necessary to be undertaken.

2.4 Examples of the application of quality management strategies

An example now follows of the adoption of the approaches to the systematic application of corporate quality management strategies presented in the preceding chapter. The example is based on the implementation of a car component supplier of a zero defect program. Attention will focus on the significance of the assumption of managerial functions in general and on the creation of the conditions necessary in particular. The example shown in Fig. 2.12 demonstrates the approach adopted to the initiation and execution of quality management processes concerned with field errors. The company was prompted to introduce the

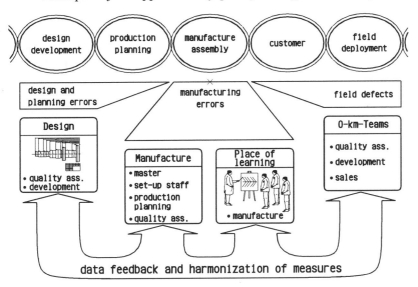

Fig. 2.13 Measures involved in a zero-defect program

program when external studies as well as internal surveys carried out as a service by the quality department revealed that the problem of field defects and complaints had reached a level requiring them to be treated as a matter of urgency.

When the situation regarding field defects and O-km complaints was brought to the attention of the management, the decision was made to increase efforts to reduce the defect rate and to ensure that the conditions necessary for such a program - in this case moderator training -   were created.

The unambiguous definition of areas of responsibility and of quality guide-lines served the dual purpose of informing all employees involved both about the objective of the program and about the individual steps involved. The stages were as follows (Fig. 2.13):

−   In problem-solving groups (0-km teams) consisting of employees from the areas of manufacturing, process planning, quality and, if necessary, development and sales, each fault discovered in the field is analyzed and made the subject of a problem-solving process. The groups have been meeting once a week on a regular basis for approximately four years.

−   Another meeting takes place, likewise once per week, to analyze manufacturing faults which have occurred in any of the manufacturing areas. The participants in this group are representatives from the manufacturing departments (supervisors - with the German `Meister' qualification, set-up staff) and specialists from the process planning and quality assurance departments.

- The involvement of the employees in the continuous improvement process was facilitated by the establishment of a so-called training shop which is under the control of the manufacturing staff. There, the group uses problem solving techniques to demonstrate exactly what quality-related problems are occurring, to find out why they are occurring and to rectify them.
- In order to intensify the involvement of the areas whose work precedes that of the department in which the fault is actually occurring, the establishment of a new work group including design staff is planned.

The aspect of improved punctuality is also using quality management tools to bring about an improvement in their ability to observe time schedules. Reports on the extent to which targets have been reached and the introduction of steps towards optimization are fundamental elements of these approaches. If the term `ability to observe time schedules' is defined as meaning `meeting delivery dates', then delivering too early is just as bad as too late. The ability to observe time schedules is regularly defined as a characteristic of quality for departments upstream of manufacturing (development, design, documentation), for contractors in and outside the company and for the manufacturing and sales departments.

One characteristic all solutions have in common is that they begin with clear definitions and consensus, they are regulated by unambiguous allocation of responsibility and their results are evaluated regularly in order to reinforce responsibility within the closed loop quality circuit and to redefine basic strategies. In a process spanning several years in which twenty pilot projects have been initiated, the number of field breakdowns has been reduced to one fifth.

The program has been successful because care was taken to ensure that the appropriate prerequisites were in place before individuals were assigned functions in the pursuit of continuous improvement. This applies particularly to moderator training. Alongside purely technical knowledge, additional capabilities usually not developed within the framework of training programs are essential. Social and methodical skills are the most important factors. These include the following:

- Communicative abilities and the ability to present results
- Knowledge about why and how conflict arises and about how to resolve conflict

- Ability to apply problem-solving techniques in team work
- Ability to apply methods of project management

The team leaders must have these abilities in order to ensure the success of the groups mentioned above which rely largely on team work. The other - untrained - team members lack technical knowledge and are frequently locked in unresolved conflict among themselves. Their expectations of the team leader are correspondingly high. This frequently results in the initial selection of very complex problems which have plagued the company for a long time, thus endangering the program start-up phase. In accordance with the principle of small steps, easily solved problems must first be dealt with in a progression towards solving the actual quality problems within the framework of the process of continuous improvement.

The problems arising in team work can be ironed out when a systematic approach to quality management is adopted. On the basis of clear objectives defined by management, areas of responsibility are assigned within the group by a trained moderator who has both the technical and social skills required. The introduction of a quality management program targets the application of QM methods to a concrete case, beginning with a pilot project. The improvement achieved is then hailed as the achievement of the group. The moderator maintains a neutral position and, ideally, becomes superfluous to requirements.

2.5 Recapitulation and prospects

Satisfied customers and satisfied, committed employees are the key to corporate success. It is, after all, the employees who determine the extent of gains made in all the important measures of corporate success - business results, corporate image and market position. Employee satisfaction is, then, of central importance to a company. Quality management provides the prerequisites and strategies with which to increase corporate internal and external efficiency. It mobilizes and activates resources already present in the company, resulting in fundamental long-term change in awareness. In addition to customer and employee orientation, the elements which exert most influence on this evolving awareness are process and product orientation and the principle of continuous improvement. Continuous improvement means relentless attention to quality, the analysis of problems and their elimination in individual processes. Improvement in the strategic elements of success becomes possible when the necessary

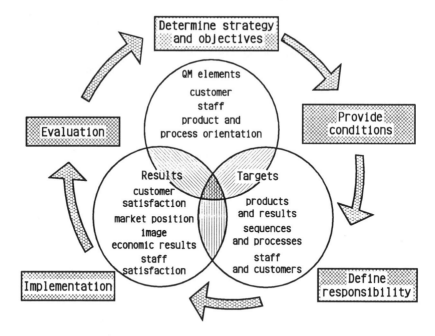

Fig. 2.14 Corporate strategy quality management.

conditions such as preventive quality assurance methods have been established.This chapter underlines the importance of quality management as a corporate strategy. It has been demonstrated what the main corporate objectives must be and which approaches are most effective in implementing this corporate strategy which influences staff, customers, products, results, sequences and processes (Fig. 2.14). In contrast to conventional approaches which target either only the promotion of methods and tools of phase-oriented quality management or only social and methodical skills, the solution presented here unites these elements to one common strategy.

Holistic quality management systems, effective throughout the company and embracing preventive methods of error avoidance are prerequisites for the implementation of other corporate strategies. Besies quality management itself, strategies for the implementation of quality management represent, in this context, an additional mainstay of corporate strategies. This will be treated in the following chapter.

3

STRATEGIES FOR THE INTRODUCTION OF A QUALITY MANAGEMENT SYSTEM

3.1 Introduction

The need to maintain the quality of a certain product or service has become a decisive argument within corporate strategy. It has thus advanced from being regarded as a tiresome obligation and is now widely accepted as a determining strategic policy factor.

Although the label `Made in Germany' was long renowned as a guarantee of quality, German companies are now finding themselves confronted with the need to react to a growing demand for quality and are having to take account of the maxims of preventive quality assurance [1 - 2]. In this chapter, fundamental strategic approaches to the introduction of a quality management system (QM system) will be presented.

3.2 Prerequisites for the introduction of a QM system

A quality management system is not, as still frequently believed in industry, an instrument used to check and safeguard the quality of a product at certain stages of its production. It is, rather, a tool which can be used to ensure that both the quality of the product as demonstrated by

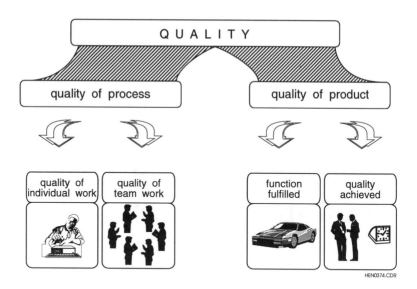

Fig. 3.1 Process and product quality

performance characteristics and the quality of production, the process quality in other words, is of consistently high quality and is tailored to meet customer demands. Process quality in this context means the quality of technical processes and, more particularly, the efficiency of service and planning processes. Quality management, therefore, has assumed a considerably more comprehensive character, and has become an interdepartmental and interdivisional function. Quality management focuses both on the continuous improvement of the quality of work performed by the individual (personal work quality) and, to an even greater degree on the improvement of the quality of cooperation of all the employees of a given company (Fig. 3.1).

All operation sequences within a company have to be viewed as a chain of overlapping processes. The term 'processes' is not confined here to technical processes such as milling but is also applied to services such as drawing up instructions for work. Efforts to bring about improvement must be directed at the enhancement of each individual stage in a process and it must always be remembered that the target is not merely the optimization of individual processes but rather the continuous improvement of the chain of individual processes.

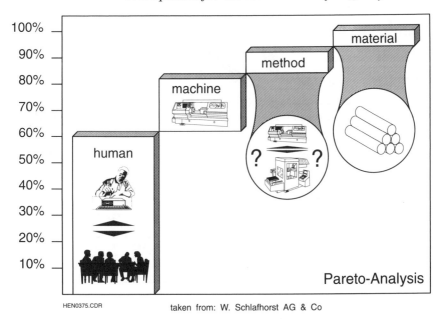

HEN0375.CDR   taken from: W. Schlafhorst AG & Co

Fig. 3.2 Reasons for nonconformity

A review of activities within a given company shows that each employee is the customer of the colleague who is the supplier of data, guide-lines, source materials etc. The `customer' in turn becomes the supplier of a product of service. Each employee, therefore, contributes directly to quality.

It is now generally agreed that a high level of quality is reached only when each employee is committed to the process of continuous quality improvement. An investigation conducted in a mechanical engineering company but applicable to other branches bears this out (Fig. 3.2). The purpose of the analysis was to determine the main causes of nonconformity. Over half of all causes were found to stem from mistakes made by employees. Other influencing factors such as machine, methods and material were of considerably less significance. These results confirm the importance of the human factor in achieving a high level of quality.

When the findings of this analysis are applied to set priorities for the introduction of a QM system, it emerges that strategies for the introduction of a QM system must focus on humans as the central element of such a system, targeting the pool of untapped potential which frequently lies dormant.

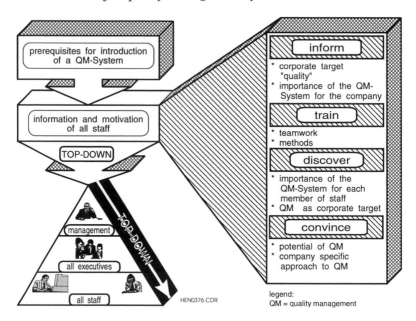

Fig. 3.3 System of defining objectives by consensus

All employees must become co-workers in the true sense of the word, independently pursuing aims reached by consensus with the corresponding superior.

In any given company, there is normally only one `position' which has direct access to each employee and which can exert direct influence on quality of work, namely management. Management, therefore occupies a key position in the introduction and maintenance of a QM system.

An analysis of possible means of implementing a Quality Management system shows that, like an `internal audit', it must be introduced and put into effect throughout the entire company. This applies just as much to planning procedures as to process sequences and to product characteristics. In an internal audit, the effectiveness of quality assuring measures implemented in different areas of the company is investigated and evaluated; additionally, suggestions are made regarding improvements which could be made to the QM system and progress is monitored. This involves exerting influence on various areas of the company - a process which would run into serious difficulties within the company or at least fail to achieve the desired success without the full support of the management. Similar rules apply to quality assuring measures.

For this reason, the initial function of an effective strategical approach must be not only to include management and executives in the process of quality improvement but to encourage them to become the driving force behind it (Fig. 3.3). Experience shows that if this fails, the efforts of, for example, a quality leader are frequently less effective than they would otherwise have been.

3.3 System of defining objectives by consensus

One major problem connected with the introduction or reorganization of a quality management system is the involvement of all employees in the process of improvement. One means of developing a sense of responsibility and of maintaining it in the long-term is to establish a system whereby all levels in the hierarchy are involved in consensus on targets (Fig. 3.4).

The initial step is the establishment of corporate aims by the company management (in the form of a quality policy) and the definition in detail of these aims as they affect individual levels. Agreement on details is reached in consensus with those involved. The result should be a set of aims which are both unambiguous and measurable (e.g. the reduction of an error rate), which are accepted by those involved and to which they are committed. The definition of targets in concrete terms inevitably involves a great deal of work. However this is an essential element for the success of each individual measure.

The main purpose of this procedure is to ensure that employees assume responsibility themselves for the process in which they are involved and do not simply expect their superiors to bear responsibility. They must be aware that they are expected not merely to perform individual functions in the sense of Taylor's division of work but to invest all their abilities and skills in improving the competitiveness of their company both in general terms and in detail.

In order to live up to this expectation, employees require considerably more room for manoeuvre than envisaged in Taylor's division of work. They must be able to participate in the decision making process and to develop their own initiative. This implies a change in awareness on the part of all staff and a change in the culture of leadership since it runs contrary to principles which have long been valid; in other words - a `cultural revolution'.

Executive staff frequently fear that their subordinates might achieve

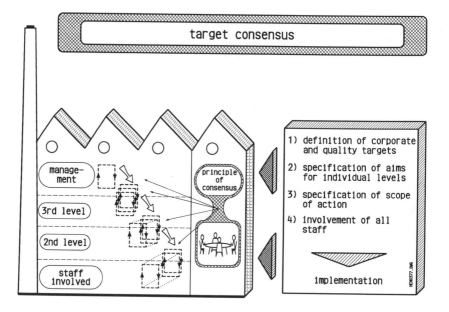

Fig. 3.4 Target consensus

goals or develop solutions which they feel they ought to have done themselves. Those in higher positions simply assume in such cases that they are technically superior. Such assumptions have no place in the participative style of leadership required.

The previously clear advantage enjoyed by those in higher positions over their staff in terms of information, must be significantly reduced, or completely dismantled in order to meet the new demands for a participative style of leadership. It is, therefore, hardly surprising that the delegation of responsibility and increased freedom often engenders fear on the part of managerial staff that they will lose their authority (more information means more power).

This problem tends to occur particularly at junior management level where managerial staff are in charge of operative staff. At this level, those responsible behave rather like a layer of clay soil, preventing the flow of revised working practice from penetrating down to the lower levels. This problem is all the more serious as the junior managerial level represents the actual multiplier for involvement of all staff. However, by initiating training courses specially tailored to suit this group, the problem can be effectively countered. The training courses should cover aspects such as:

- Presentation techniques
- Communication techniques
- Decision support tools
- Problem-solving techniques and
- Creativity promoting techniques

Experience shows that a considerable amount of intensive discussion is necessary before agreement can be reached on objectives. This applies particularly when the corporate aims, still expressed in general terms have to be applied to individual processes or structural components. The aims of the individual levels must always be oriented towards the satisfaction of customer wishes, customer in this sense meaning internal as well as external customers. The ISO 9000 certification which is currently the subject of much discussion, legislation, norms, specifications etc. provide additional guide-lines for the definition of aims and for the introduction of measures.

Measures aimed at achieving improvement can be introduced as part of projects working towards the accomplishment of aims of the individual levels.

The following steps are characteristic of the course of such projects:
- comparison of the actual situation with defined aims
- analysis of existing weak points, allocation of possible measures for improvement to individual weak points
- identification of synergies and combination of measures to form packages of measures
- prioritization of measures designed to address individual corporate aims
- implementation of measures and
- verification that aims have, indeed, been achieved and introduction of measures to ensure the continuous improvement of the QM system.

The process of continuous improvement is described succinctly by the Plan Do Check Act (PDCA) cycle, described by Deming, who was referring not only to the continuous improvement of individual measures but to the underlying attitude of each employee.

One aim within the system of structured consensus on objectives must be to bring about a culture of shared aims within the company. This new culture must strive to empower those involved, transforming them from their original position as assisting personnel into people bearers of responsibility acting on their own initiative (Fig. 3.5).

Once this has been achieved, a plethora of methodical approaches and tools such as QFD, FMEA, Poka Yoke can often be applied successfully,

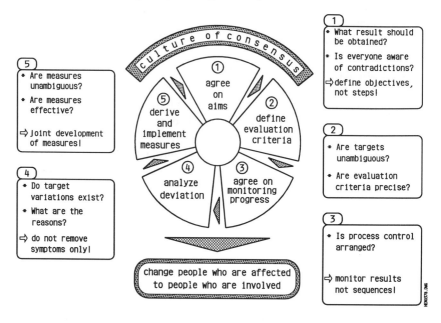

Fig. 3.5 Culture of consensus

perceived not as tiresome bureaucratic obstacles but as useful aids in the performance of one's own work.

The structure of the consensus system forms the base and the framework for the introduction of a quality management system. It gives the employees the opportunity to participate in the development of the QM system.

The loss of authority, which it was previously feared would follow in the wake of the delegation of responsibility and power and the surrender of the information advantage (knowledge is power) can be balanced out by improving the social and methodical skills of the junior management who must be encouraged to seize the opportunity thus afforded to spend more of their time on the real functions of leadership such as the development of objectives.

3.4 Case studies

The following three case studies show methods of implementing a consensus system (Fig. 3.6).

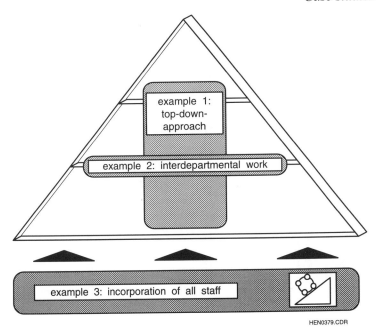

Fig. 3.6 Structure of case studies

The first example represents a consensus system extending throughout the company. It is a basis for the introduction and long-term maintenance of a QM system. A further example describes the agreement on targets by task forces. This method has proved successful in the solution of numerous problems, particularly those affecting several areas within the company. In the third example, a successful approach is presented to dealing with problems whose solution relies on the participation of all employees (e.g. tidiness and cleanliness at the work place, routine maintenance work or checking test and inspection equipment).

3.4.1 System of structured consensus

Within a company-wide strategy for quality improvement, a consensus system was introduced as a support-element (Fig. 3.7) defining concrete targets for all organizational levels from management down to the individual employee, focusing on the following factors in success:
- Competitiveness
- Quality

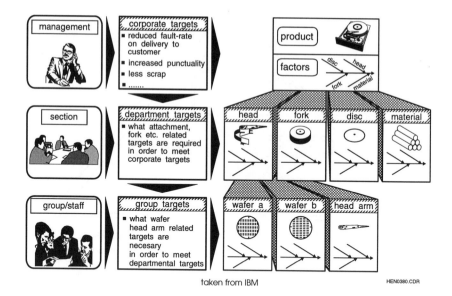

taken from IBM   HEN0380.CDR

Fig. 3.7 System of consensus, exemplified by IBM plant Mainz (taken from IBM)

- Fulfilment of terms of contract of delivery
- Production technology
- Staff and organization

This approach involved asking each organizational unit to define five aims to be given priority for one year, each of which would be likely to increase the satisfaction of external and internal customers alike.

Success achieved at each level is measured monthly in accordance with a standard method of evaluation in PPM (parts per million) (Fig. 3.8). The ISO 9001 certification of quality was awarded in 1991. Additionally, an internal evaluation in accordance with the Malcolm Baldrige Assessment criteria is performed annually.

One main problem facing quality management is that of sustaining the effectiveness of a measure and continuing to improve it once it has been introduced.

In an effort to overcome this problem, measurable targets such as defect and waste rates are kept under constant scrutiny by means of monthly reports which have to be submitted to the appropriate member of the management. Factors which impede progress towards a set target are continuously analyzed and documented. In order to inject further impetus

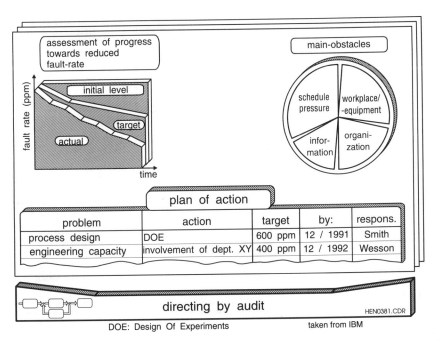

Fig. 3.8 Pursuit of aims (taken from IBM)

into the area of quality management all activities were summarized under the title `market driven quality'

3.4.2 Working with consensus in task forces

Fig. 3.9 demonstrates how common quality improvement measures can be developed and common aims defined in order to tackle problems affecting more than one area of a company.

Appropriate employees, depending on the task in hand, are called together. Each task force includes employees and a team leader from each of the areas concerned, a moderator, e.g. from the quality department and, if required, experts.

It is the function of the team leader to select individual members. It is frequently quite easy to imbue volunteers with initial enthusiasm. It is rather more difficult to motivate them to persist until the problem has been solved.

The formation of task forces grants those charged with solving the

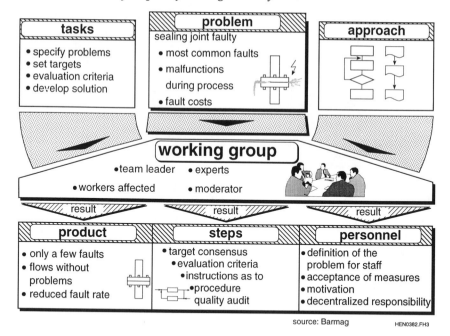

Fig. 3.9 Task forces with agreed targets (taken from Barmag AG)

problem a right to shape the decision making process. This right is conferred in the knowledge that employees are normally, by virtue of years of experience, specialists at their own work places and can therefore be expected to know them better than anyone else.

The intention is, that individual employees will assume personal responsibility for aims or measures and will commit themselves officially to working towards the solution. The groups are issued with a flow chart and forms outlining the steps to be taken and prompting the documentation of results Fig. 3.10).

It has been shown in practice that this approach promotes a broader view of the company thus encouraging more comprehensive thought and action as well as the decentralization of responsibility.

During the phase in which such measures are introduced, it is crucial that the first groups are successful. It is, therefore, advisable to select problems which are likely to be solved successfully within a relatively short period of time and without undue difficulty. It is also recommended that efforts be made to ensure that a step by step approach is adopted towards the improvement process. The attempt to solve major problems,

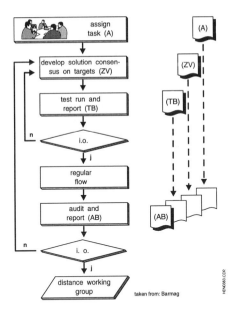

Fig. 3.10 Principles of the problem-solving process (taken from Barmag AG)

which may already have been targeted several times, is frequently too complex and insufficiently flexible and adaptable to be tackled at once.

Employees with a good track record as regards participation in successful projects within a QM program can be used selectively as 'multipliers' to promote the process of continuous improvement.

Work in successful groups is characterized by the development in team work of ways of solving the problem and by the consensus system of reaching agreement on the objectives to be pursued.

Major components of this process include trials involving the measures to be introduced in a test run, integration of the changes into company procedures and investigations into new methods of doing things in a behaviour audit.

The process of developing both solutions and the measures by which their success will subsequently be judged, often provides the group with an incentive to work out practice-oriented solutions which will cause a minimum of disruption and which will be broadly accepted (Fig. 3.11).

A prerequisite of the consensus system is a bottom up information flow in the hierarchy of the company. Management must react immediately to recommendations and suggestions from the operative level either in the

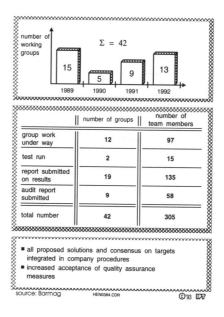

Fig. 3.11 Status and results of team work (taken from Barmag AG)

form of, for example, a rejection stating reasons or swift introduction of the corresponding measures. A systematic reporting system is just as important in this phase as public announcement of success on conclusion of the project.

3.4.3 Consensus including all employees

Quality management today is required to include all employees in the effort to ensure a high level of quality. The documentation of quality results, cleanliness in the plant or, to mention a problem which has both a technical and an organizational aspect, the monitoring of test and inspection equipment are examples of processes which require contributions from all members of staff.

Even these apparently minor problems which are all part and parcel of a customer audit often prove difficult to overcome since each employee is required to make a real contribution - permanently. In this situation, management is often caught up in the crossfire between external demands and their own limited means of meeting these demands. The limitations

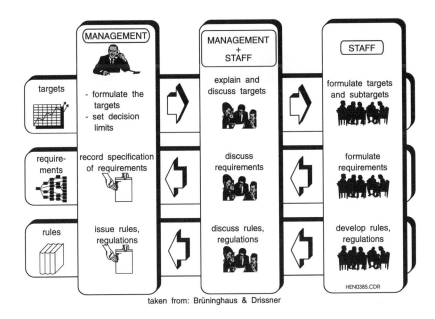

taken from: Brüninghaus & Drissner

Fig. 3.12 Introduction of a staff audit, part 1 (taken from Brüninghaus & Drissner)

stem from the fact that requirements can only be met by the employees themselves. Warnings and requests combined perhaps with regular inspection rounds can only produce short-term solutions to the problem. What is needed is a broadly based approach which gives a measure of responsibility to the employees affected.

In the case example under discussion (Fig. 3.12), great emphasis is placed on aspects such as quality data documentation, monitoring test and inspection equipment and cleanliness since these feature in the external audit.

In order to involve all employees, the demands which must be met in time for the external audit were explained and discussed in detail with them (in some cases in the presence of the external auditor). The meeting ended with the message that the staff should formulate measurable aims and sub-aims for the individual departments. Additionally, they were told to draw up rules for measuring the success of their efforts to achieve those aims. The result was a questionnaire with a clear evaluation plan showing what demands had to be met by which of the departments.

The suggestions were discussed at a meeting of management and staff and once consensus had been reached the plan of action agreed upon was

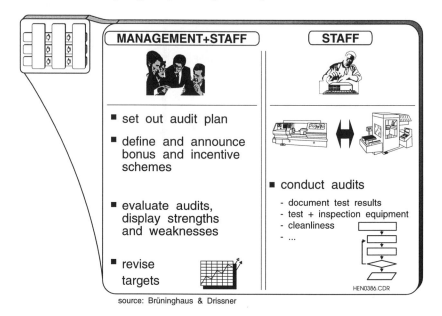

source: Brüninghaus & Drissner

Fig. 3.13 Introduction of a staff audit, part 2 (taken from Brüninghaus & Drissner)

incorporated into official company policy and is still being implemented.

It was also agreed that the success of the effort to meet demands would be measured in the form of a regular `internal audit'. Internal audits have been taking place at regular intervals, on a reciprocal basis between different departments and are monitored by management in spot checks which are carried out from time to time (Fig. 3.13).

The results are evaluated in accordance with a points system agreed with the employees. As an additional incentive, the department with the highest total number of points at the end of the year receives a bonus in the form of dinner at a good restaurant.

Of course the points system itself, which contributes to the decision as to which department has earned itself an evening out together is a potential source of discord. In the company under examination, this leads to lively discussion of the rules and numerous suggestions for improvement. The `staff audit' thus assumes a dynamic force of its own, exactly as intended. Only permanent change ensures that this tool remains effective.

The fact that the `internal audits' are performed on a reciprocal basis has proved positive. Each member of staff is thus both auditor and audited and therefore knows exactly what attempts might be made to sidestep

criterion	results of the audits	
	status July 92	status Feb. 93
quantifiable success		
■ number of measuring instruments not included in test-measurement monitoring	50	2
■ number of incorrectly marked containers	10	1
■ number of report sheets incorrectly filled in	30	4
qualitative success ■ improved order and cleanliness ■ less encumbered by detail management ■ higher staff motivation		

© IPT 1993 source: Brüninghaus & Drissner HEN0387.GEM

Fig. 3.14 Impact of staff audit (taken from Brüninghaus & Drissner)

individual demands.

It is now recognized that the interdepartmental cooperation in the development of the `internal audit' in conjunction with the consensus method of arriving at decisions regarding requirements and targets, makes it an effective quality management tool (Fig. 3.14).

3.5 Implementation of Quality Management systems - costs and benefits (example: Messrs Edelmann)

The demand made on all measures within a company is that they should, in the medium to long term, improve the competitiveness of the company. Costs incurred as a result of the introduction of QM measures are measured by the same yardstick. Fig. 3.15 shows the development of measures and the results they yielded in a successful mechanical engineering company. These measures could easily be transferred to other companies with equal effect.

There is frequently no direct correlation between individual measures such as staff training and positive effects e.g. business results. It is, rather,

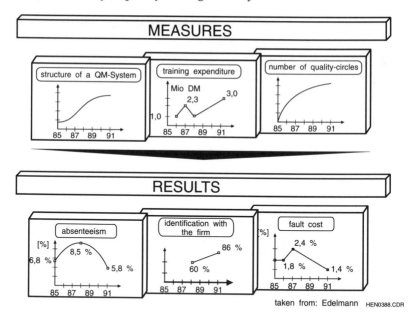

Fig. 3.15 Costs and benefits of the introduction of a quality management system
(taken from Edelmann)

the case that the combination of a number of activities has a positive effect such as a drop in absenteeism after a certain time-lag.

This is illustrated by the following example: In 1985, the company under examination set about implementing a comprehensive quality management system. This move was accompanied by a considerable increase in the number of training courses offered. Since the driving force behind this development was the desire to bring about a substantial reduction in the defect rate, extensive efforts were made to pin-point defects and their causes. The results after the first two years were sobering, to say the least. The defect rate was unchanged in the first year and then rose in the second year. A development like this would normally be regarded as the kiss of death in such a project. It was not until the following years that the defect rate dropped sharply. An analysis subsequently carried ·out showed that the apparent rise in the number of defects had been due to more thorough search and detection of faults. The initial defect rate had not reflected the true situation as mistakes made by colleagues had often gone unreported or had been covered up. A realistic picture emerged only when employees were made aware both of the consequences of undiscovered faults and of the fact that management regarded the information passed on to them by staff as providing an opportunity for the improvement of the process.

Had the project been abandoned when the first, apparently bad results emerged it would without doubt have been a disastrous decision for the medium-sized company. The annual reduction of around DM 700,000 in costs arising from defects achieved since then is encouraging.

This example shows that the success of measures implemented as part of quality management is not always immediately recognizable. There is frequently no direct correlation between any individual measure and a positive trend. It is, rather, the case that a number of interacting activities concur to promote a positive development.

The decision in favour of active quality management is a fundamental strategic change of course for a company and can therefore not be made by one person, e.g. the quality manager alone. It is first and foremost a challenge to which management is invited to respond.

3.6 Recapitulation

A holistic quality management system, effective throughout the company and emphasizing the importance of active employees, willing to bear and share responsibility is a prerequisite if companies are to continue to meet rising demands for quality. Manufacturers can sustain their market position only as long as they are successful at meeting and surpassing the expectations of their customers who demand high quality reliable products, efficiently manufactured. This involves quality assurance throughout the entire product life cycle, from customer wishes through the development phase, production and use to recycling (Fig. 3.16).

A fundamental strategic approach to the systematic introduction of a quality management system is explained within the scope of this contribution. The main steps leading up to introduction and the organizational rules to be observed in order to ensure that the development of a quality management system does not result in additional bureaucracy but in recognizable benefit for a company are described. Case studies demonstrate concrete approaches to introducing such a system and benefits which stand to be gained are outlined.

It is crucial to involve each employee in the responsibility for quality and to motivate each of them to participate actively in ensuring that a high level of quality is reached.

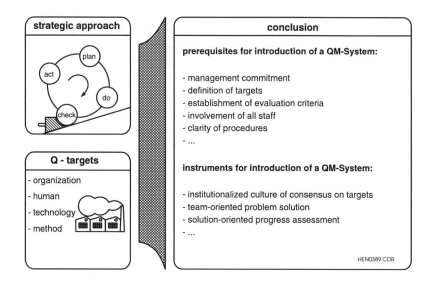

Fig. 3.16 Recapitulation and prospects

The aim must be to introduce a process of continuous improvement. This can only succeed when clear attainable aims are formulated and progress is monitored continuously. The consensus model presented here can make a major contribution in that it offers a solid framework for the continued development of the company. Individual sub-targets such as meeting the requirements of ISO 9001 or the introduction of special methods such as FMEA or QFD can thus be effectively initiated and implemented.

Quality management, directed at clearly defined corporate aims involving all employees is a tool which aids decentralized development of products and processes. The approach presented here is an effective, strategic means of promoting continuous adaptation and permanent change, since especially in market economies, the adage:
`Nothing is constant, only permanent change'
applies.

Part Two

PRODUCT DEVELOPMENT: REDUCING DEVELOPMENT CYCLES WITH RESPECT TO MARKET AND MANUFACTURING

4 Supplier involvement in product development
  - risks and potentials-
5 The design workstation of the future
  - from the initial design to NC programming -
6 Innovative software technologies for quality management
7 Rapid prototyping: the way ahead

4

SUPPLIER INVOLVEMENT IN PRODUCT DEVELOPMENT - RISKS AND POTENTIALS-

4.1 Introduction

Industrial competition is characterised by an increasing internationalization of markets, combined with strong segmentation into inhomogeneous groups of buyers. The intensified competitive situation results in a marked increase in the pressure on individual companies to produce innovations [1]. The need to satisfy customers' fastidious requirements via innovative measures has led to short technological cycles. At the same time, an increasing degree of complexity is to be observed in the products concerned. As a result of the need to gear production more closely to special customer requirements the three factors of costs, time and quality, which are of central importance to success on the market, have become primary criteria in product development.

Studies have shown that companies which are successful on the market today enjoy superiority with regard to all three factors, whereby efforts are undertaken to attain clear superiority in one of the three target areas [2]. Whereas these respective factors were accorded varying levels of priority

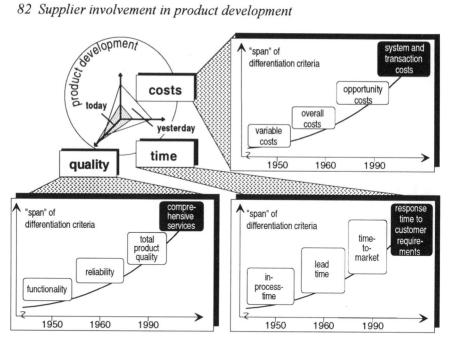

Fig. 4.1 Course of development for product differentiation criteria - the general guideline for product development

in the past, today all three must be regarded as inter-related criteria of product differentiation. Product differentiation involves distinguishing one's own products from rival products in positive terms, from the customer's point of view. The answer to the following question must therefore provide"What is it that induces the customer to prefer one's own product to rival products and subsequently to purchase one's own product?"

In this context, the factors determining success have acquired a broader scope of importance in parallel with the course of economic development (Fig. 4.1).

Cost minimization remains the classic objective of manufacturing companies. Low costs can be implemented as competitive advantages on the market in the form of low prices, as the most plausible differentiation criterion. Originally, the emphasis was on the assessment of expenditure, initially in terms of variable costs and later, as a result of increasingly severe competition, in terms of overall costs. To assist in management decision-making processes, so-called "opportunity costs" were also included in cost assessments from the middle of the 1980s, in order to quantify the marginal income lost as a result of the failure to select alternative courses of action. Most recently, the costs of system integration and internal and external transactions have acquired increased importance in the context of the practice of assessing all value-creating productive processes as a whole.

Since the 1970s, quality has become increasingly more firmly established as a standard measure of compliance with functional requirements and reliability. Today, the concept of quality is expanded to embrace so-called "Total Product Quality" [3]. This describes the customer's satisfaction with the product as a whole, i.e. including the characteristics which the consumer perceives in a subjective manner, such as product image. The near future will witness a further expansion of the concept of quality beyond the product itself. In addition to differentiation via product characteristics, a further aim is customer satisfaction via comprehensive services. Supporting services on the market, from financing through to servicing and maintenance, will thus be deployed to an increasing extent as a differentiation criterion, especially for up-market products.

Into the 1980s, time was considered a secondary success criterion with regard to costs. As long as measurable production times could be compared against clearly defined hourly rates, costs were practically time quantified in monetary terms. As a result of the increased level of innovative pressure, the development time then acquired increased importance as a differentiation criterion. "It is not the best and cheapest which will survive, but the fastest" [4]. However, the ability to set higher prices as the first supplier on the market is not the sole factor deciding success.

The shortened product life cycles mean that there is no certainty of development costs being recouped at all, when products are launched onto the market at too late a stage. In the future, the time required to respond to customers' requirements will become a decisive competitive criterion.

The presented changes in the competitive environment result in substantially increasing risks in the field of product development. Development risk is composed of technological risk, marketing risk and the closely related financing risk. Technological risk concerns the uncertainty as to whether a new product or process technology will actually function and whether it can be implemented. Financing risk results from the reduced development times in conjunction with shortened life cycles. In addition to the increasing expenditure resulting from the higher level of product complexity, shorter development times initially also necessitate a higher level of development expenditure per time unit. As a result of the shortened product life cycles, a shorter period is then available in which to recoup the development expenditure. In order to reduce the marketing risk, broad acceptance for the developed products must therefore be available on the market from the outset.

The conflicting requirements of the market in which companies operate thus result in a need to pursue two contrary strategies [5]. The successful placement of high-tech products is dependent on a rigorous system of quality assurance which involves a high level of investment in terms of time, resources, costs and know-how. Such a system requires "calm" product development processes without any pressure of time (safe to market). On the other hand, the predatory competitive situation compels companies to launch their products on the market more quickly and with a higher level of innovation. But to achieve this, exceptionally high acceleration of product development times is necessary (fast to market).

These conflicting aims impose high requirements on effective and efficient product development. On the one hand, the development costs per time unit must be increased, while on the other hand the attendant substantial risks must be minimised. Cooperation with external partners in a so-called product development partnership provides great potential here.

4.2 Objectives for cooperative product development partnerships

The capacity of a company with regard to product development, its so-called "research and development competence" is normally measured primarily according to the product- and process- know-how which is available within the company. This know-how may have been implemented in the form of existing products and processes, or it may involve new knowledge which is presently still at a theoretical stage. To date, too little attention has been paid to know-how on product development as an operational process within the company. This know-how enables effective organization of the product development process in an efficient manner, closely in line with the requirements of value creation.

In all cases, the know-how carriers are people, contributing to the success of product development efforts as individuals and in organised teams. Competence thus involves more than the purely quantitative development capacity, which can be measured, for example, in the form of engineering hours. In addition to the purely quantitative personnel capacity, the general resources contribute equally to successful product development, primarily in the form of systems (e.g. EDP-systems) and

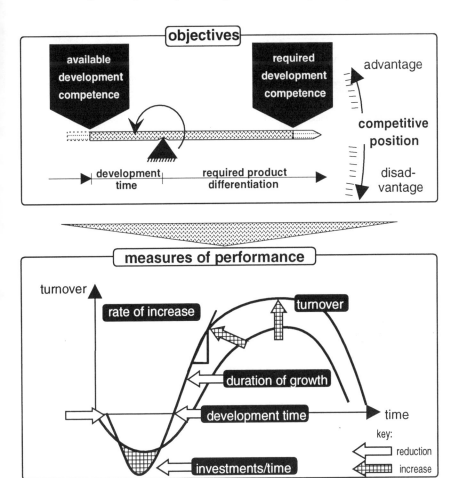

Fig. 4.2 Objectives and measures of performance for product development

facilities (e.g. test stands) to support specific development tasks, and financial resources.

Research and development competence is therefore to be seen as a central factor determining competitive product development. For every company, this competence must be kept in balance, in both qualitative and quantitative terms, with the capacity requirements resulting from the demands of the market and with the diverse risks which apply. Only when this balance is achieved a company can maintain its competitive position (Fig. 4.2). A reduction in development time must be accompanied by an

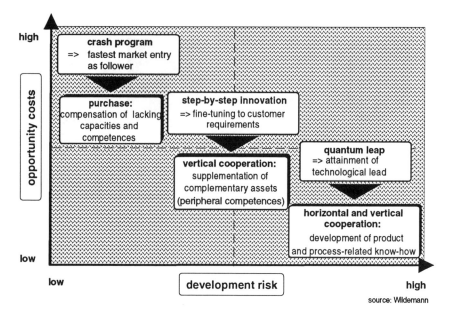

Fig. 4.3 Cooperational requirements in relation to development strategies

increase in research and development competence, otherwise the competitive position will deteriorate substantially. The increasing need for product differentiation results in the same requirement.

The objective of any cooperation in the field of product development must be to positively influence the success factors of time, costs and quality in the interests of customer satisfaction.

Time benefits: The development process is to be shortened by dividing up individual tasks and carrying out the attendant work simultaneously, to enable customer-oriented production on the basis of longer assessment of the market, or to attain sales benefits as a pioneer on the market.

Cost benefits: Cost benefits are directly attainable by involving external partners in the development work, when specialized partners possess more favourable cost structures. Such favourable cost structures result directly from reduced overheads and the utilization of synergetic effects by means of multiple cooperation. Indirect cost benefits are attained via reduced investments, i.e. capital expenditure is lowered.

Quality benefits: Via cooperation with external partners, specifically required product- or process-related knowledge can be acquired, in order to supplement a company's capabilities and thus enable it to fulfil the requirements of total product quality. The acquisition of knowledge from various sources must be coordinated in an appropriate manner in accordance with the system competence to be developed at the company concerned.

Attainment of the objectives must be manifested in appropriate success factors. These success factors concern:

- the duration of growth,
- the growth of turnover (absolute) and the
- investment rate.

Whether and to what degree cooperation with partner companies is necessary can be established on the basis of the competitive pressure which determines the company's development strategy. The need to increase a company's development competence may depend on the company's development strategy which, in turn, is determined by the factors of opportunity costs and development risk (Fig. 4.3) [6].

The opportunity costs refer to the shares of turnover which cannot be attained due to a late launch on the market or failure to enter the market. The urgency with which product development requires to be accelerated is thus quantified in monetary terms. Three important development strategies can be distinguished in the portfolio of opportunity costs and development risk.

Crash program: A strong rival company is already present on the market. It is therefore necessary to concentrate on currently available capacities. Lacking know-how should be purchased by means of classical relations with suppliers. In the medium term, a decision must be reached as to whether the lacking capacities are to be developed as core competences at the company concerned.

Step-by-step innovation: A significant technological advance has been attained with the preceding model, as a result of which important shares of the market have been conquered. The aim is to further develop the product in the direction of special variants, in accordance with customers' specific requirements. Cooperation is to be carried out primarily in vertical direction, in order to open up new markets and to increase the benefit to the customer via cooperation with the buyers.

Quantum leap: The low opportunity costs indicate that the rival companies have not yet discovered the potential market. Cooperation in core areas serves to develop the necessary know-how. The sharing of risks is also an important motive here, in view of the great uncertainty which prevails.

The establishment of development partnerships on the basis of a company's development strategy is therefore suitable primarily for medium to long-term projects. In the long-term, potential development partners should provide only "peripheral competences", i.e. areas which do not belong to the company's own core competences, on a subcontracting basis. Each company willing to enter into cooperation must therefore identify its own core competences, on the basis of which product differentiation ispossible.

4.3 Characterization of the product development partnership

Three cases are distinguishable in the area of cooperation with external partners, whereby a clear trend away from the "classical" relationship with parts suppliers towards the development partnership is to be observed, regardless of the industrial branch concerned (Fig. 4.4). The proportion of development work carried out by subcontractors is increasing in terms of overall expenditure on value-creation.

The "classical" contractor-subcontractor relationship does not involve any development competence whatsoever on the part of the suppliers. Each supplier produces in accordance with the contractor's exact specifications, and thus serves essentially as an extended work bench. In the area of standard and catalogue parts, the customer practically purchases the development work "off the peg". This case, which is to be described as "in-house" development from the contractor's point of view, thus represents an extreme arrangement, and is mentioned here solely in the interests of completeness.

The contrary extreme case applies when a component or system is completely developed by a systems supplier. In this case, the contractor regards the work carried out by the supplier as a "black box". Apart from the function to be fulfilled by the system, only the installation space and the adjoining interfaces are specified. Although this alternative enables specific utilization of the systems suppliers' development competence, there is a danger that only a sub-optimal overall system will be developed. The restrictive structuring into installation spaces means that important degrees of freedom for optimal system configurations are stipulated prior to the actual commencement of development work. Time- and cost-savings due to integrated problem solving are not achieved.

Consequently, the research and development competence which is required from external companies is being acquired to an increasing extent via development partnerships. Contractor and systems supplier cooperate

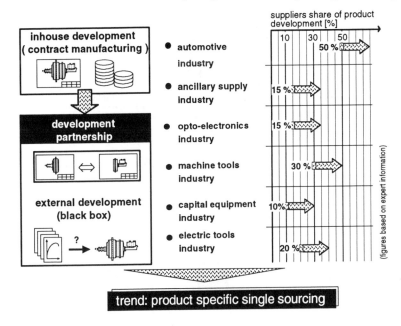

Fig. 4.4 Types of supplier involvement

closely right from the early phases. The initial basis for the configuration of the development partnership is provided by the contractor (primary manufacturer) breaking down the functions of the finished products into an organized structure, which enables the development work to be structured without restricting the scope of the various development tasks in a negative manner. Cooperation between the partners in the course of product development is then characterized by diverse contacts between the relevant employees of all the companies involved in the partnership. The openness in all technical matters which is vital to the success of such a cooperation requires each partner's contribution to be accepted and fairly rewarded. It must further be ensured that the partner companies are well matched in every respect. In addition to fundamental and strategic requirements, it is particularly important that the "cultural fit" be ensured, i.e. that the corporate cultures be suited to one another [7].

In order to verify fundamental and strategic compatibility between potential partners, it must be examined whether supply and demand for development services can be brought into alignment. Fundamental compatibility applies when the potential partner possesses all the development competences which are required for the specific problem area

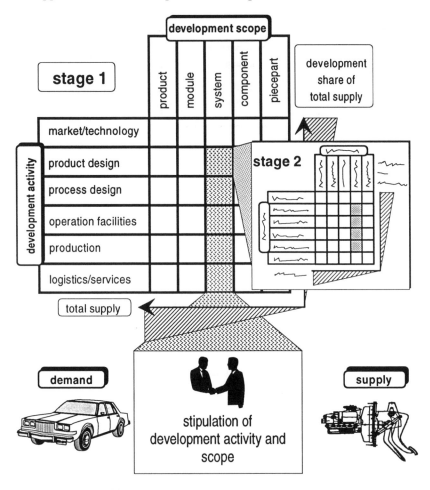

Fig. 4.5 Matching of supply and demand for development services

concerned. Strategic compatibility applies when the potential partner pursues an equivalent corporate strategy with regard to product development and does not, for example, participate in cooperative development projects as a preliminary measure to direct canvassing for new business. There must also be a mutual commitment to pursuing the aims of cost minimization, time reduction and quality improvements, with the ultimate objective of attaining an overall optimum solution, observing the same priorities in the process.

The portfolio shown in Fig. 4.5, which is based on a description of the required development work according to the criteria of "development activity" and "development scope", can be employed as a practical aid in assessing the above-stated factors. The development activities cover the complete product development process, beginning with the identification and implementation of market requirements in a specification. The proportion of pure development work relating to the activities procured from outside suppliers decreases in the direction of the logistical services for the start-up of production operations. The quantitative scope of work is characterized in terms of modules, systems, components and individual parts, corresponding to the product structure.

Fig. 4.5 presents the portfolio for the example of automobile development. With the aid of the portfolio, the company requiring development work, in this case the automobile manufacturer, specifies the development work which is to be purchased, which in the example shown concerns the complete brake system for a new automobile. The full scope of required development work must be covered by one or more potential development partners. In the example shown, a brake system manufacturer provides the complete scope of development services for this system as a stage 1 systems supplier, from product design through to just-in-time delivery of the fully assembled brake system. This does not necessarily mean that the brake systems manufacturing company actually carries out the complete scope of tendered development work on its own, however.

In a second supply stage, the brake systems manufacturer can, in turn, contract out development services, in order to acquire the required development know-how or capacities which he does not possess and to organize these services for his own purposes.

The presented portfolio can thus be used by any potential development partner in order to position himself in a supply chain and to specify the required services which are to be rendered. The closer the partner's position is to the primary manufacturer (first or second supply stage), the greater the scope of services to be rendered. This imposes high requirements on system competence and may necessitate extensive coordination measures for the lower stages of the supply chain.

The prime requirements to be observed when matching supply and demand are process integrity, the minimization of transaction costs and the avoidance of duplicated work.

In order to ensure process integrity, one partner should be responsible for a specific system in clearly defined terms. In order to minimize transaction costs, the number of contact persons is to be kept low. The partners cooperating directly with the primary manufacturer must be

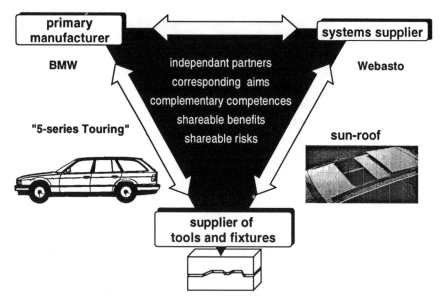

Fig. 4.6 Characteristics of a product development partnership (example: car body development)

motivated to coordinate the rest of the supply chain as system suppliers via payment for their coordination services or the possibility to increase their scope of services. As a long-term objective, efforts should be undertaken to develop trusting partnerships, in order to limit the costs of establishing new contacts for subsequent projects. Appropriate methods are to be applied to support operational cooperation between the developers and designers of the partner companies, in order to avoid the duplication of work. It is expedient to apply the methods of simultaneous engineering on an inter-company basis.

For the purposes of product development as a whole, several complementary cooperation arrangements are to be configured in this way, to form a comprehensive product development partnership. In contrast to companies with interlocked capital, the companies involved in the product development partnership remain legally and economically independent. Further essential characteristics of a product development partnership are explained below by reference to the example of a cooperative partnership in the field of automobile development (Fig. 4.6). The development partners involved here are an automobile manufacturer, a systems supplier and an equipment supplier, cooperating in the development of a new car body variant.

The fundamental features required of a successful product development partnership are corresponding aims and complementary competences. In this case, BMW wishes to offer a two-panel sun roof as an additional product feature, in order to appeal to special groups of buyers. In its capacity as systems supplier, Webasto possesses the necessary know-how and the required capacities and is interested in further expanding this know-how, for the purpose of possible application for other customers. As the susceptibility of the product "sun roof" to copying means that a competitive advantage is expected to apply only to the current series, the supplier's objectives are quite compatible with those of the contractor. BMW is interested in attaining an additional differentiation criterion for the "5-series", and it is to be expected that imitations by rival companies will follow swiftly in case of a market success. For this reason, long-term development of the appropriate know-how and the required capacities is of no interest for BMW. For the same reason, the systems supplier can rest assured that the manufacturer's wish to enter into cooperation is not based on any intention to build-up know-how with the long-term aim of acquiring independence from the supplier.

The success of operational cooperation within the product development partnership is dependent on both the benefits and risks being suitable for clear and fair division among the partners. The sharing of benefits primarily involves dividing up the attainable profits according to the respective levels of investment which are required. With regard to the sharing of risks, it is necessary to distinguish between financial risks, technical risks and the risk involved in the uncertainty of the products success on the market. Both of these problem areas must be regulated as interconnected factors, via the stipulation of appropriate pricing in the contractual arrangements.

In the example shown, a third partner is incorporated into the partnership in the form of a tool supplier. The relationship between this third partner and the systems supplier is of the same type as the described relationship between BMW and Webasto. In this case, however, the primary manufacturer is only indirectly involved with the tool maker, as Webasto's tool costs are paid in full by BMW. BMW thus bears a substantial proportion of Webasto's financial risk.

It is also quite possible for rival companies to cooperate within a product development partnership on the development of a sub-system which does not contribute to product differentiation on the market. The

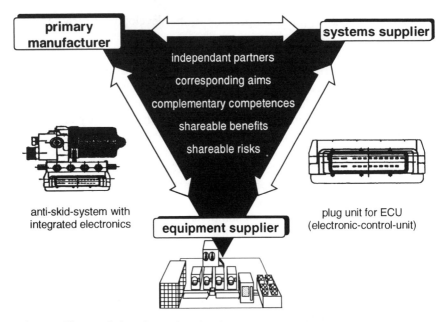

Fig. 4.7 Characteristics of a product development partnership, example: component development, (source: ITT-Automotive)

primary source of motivation here is the sharing of the attendant risks by the development partners. This form of cooperation becomes possible when potential customers in the market segments concerned fail to recognize a differentiated system developed by a rival company, as a result of which their decisions to buy remain unaffected by any such differentiation. Fig. 4.7 shows a further example illustrating the characteristics of a product development partnership. The primary manufacturer in this case is a brake systems manufacturer, who has developed an ASS system (Anti-Skid System) in cooperation with a manufacturer of electronic components. As the ASS system as a whole represents a safety component, the primary manufacturer, who is responsible for the reliability and operational safety of the unit, must also be able to monitor and influence the production process, including manufacture of the ECU unit (Electric Control Unit) by the systems supplier. This is why the supplier of the production equipment for the ECU manufacturing process is incorporated into the partnership at a very early stage of the development process in this case. Indirect control, as applies in the form of cooperation shown in Fig. 4.6, is not sufficient here.

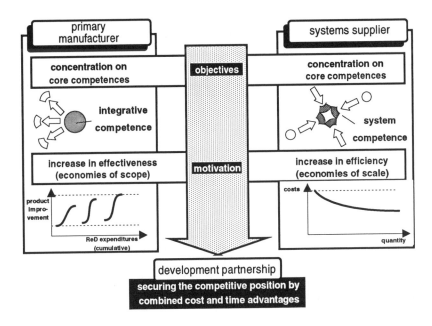

Fig. 4.8 Opportunities offered by product development partnerships

Cooperation between independent companies in the development of products enables increasing quality requirements imposed by the customer and market conditions to be met. The special potential of such a cooperation is based on the combination of cost and time advantages, the likes of which are not attainable with alternative strategies, such as company acquisitions or the further development of existing competences (Fig. 4.8). The primary manufacturer is provided with an opportunity toconcentrate on the various core competences which are available to t, whereby the ability to integrate sub-systems into one overall system is also to be regarded as a core competence. The peripheral competences required to this end are acquired by placing orders with system suppliers. This enables the company's own know-how and resources to be deployed directly for the purposes of product differentiation, thereby ensuring the effectiveness of the company's own product development operations.

The peripheral competences purchased by sub-contracting constitute core competences for the system suppliers. As the systems supplier acts as a development partner for various primary manufacturers in providing this system competence, economies of scale are attainable which ultimately

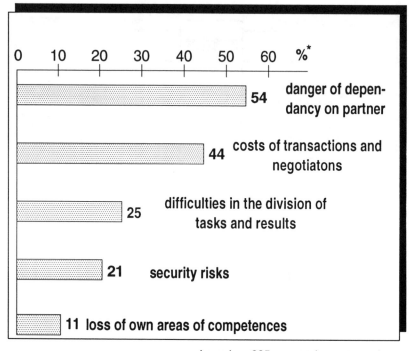

source: Rotering

based on 385 companies surveyed
(* multiple answers possible)

Fig. 4.9 Risks involved in product development partnerships

benefit every primary manufacturer. The described sharing of work by the partners enables concurrent development work to be carried out, resulting in a reduction in the overall product development time.

In spite of the obvious advantages, cooperation in the field of product development is nevertheless subject to risks, as the results of recent surveys have revealed (Fig. 4.9) [8]. The most commonly mentioned risk was the danger of becoming dependent on development partners. This line of argument can be refuted when it is assumed that the external partners are to provide only those competences which - from the manufacturer's point of view - do not belong to the required core competences. The aspect of high transaction and negotiating costs must be taken into account when

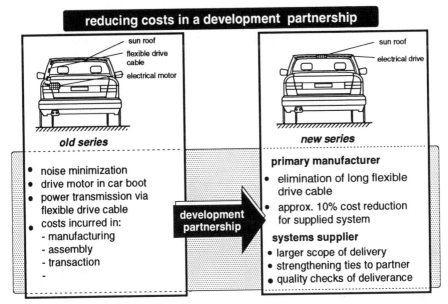

Fig. 4.10 Cost reduction in a product development partnership (based on: Webasto)

selecting the development partners. In certain circumstances, transaction costs may outweigh all the cost benefits which are to be expected from subcontracting in low-wage countries. Difficulties regarding the division of development tasks and jointly acquired results must be solved by systematic structuring of the development project to be carried out and corresponding ontractual rangements. Patents are unable to fully eliminate security risks resulting from the possibility of the unauthorized utilization of the know-how acquired in the course of the cooperation. Insofar, the loss of a company's own areas of competence is not critical, provided that the areas of competence concerned involve peripheral competences to be assigned to partners in accordance with cooperation agreements.

In order to avoid the stated risks in an effective manner, the aim of the cooperation on the part of the primary manufacturer must be long-term cooperation in the area of product development. Systems suppliers must bring about a situation involving two-way benefits ("win-win situation") for the manufacturers and themselves by actively contributing their know-how. This requires systems to be developed which provide the primary manufacturer with potential for rationalization while at the same time leading to economic benefits for the systems supplier. This provides positive motivation for the primary manufacturer to establish a long-term partnership.

Fig. 4.10 shows an example of such a situation involving two-way benefits, in which a systems supplier has taken the initiative of carrying out an improvement to a product for a long-standing customer which has resulted in significant reductions in costs. The initial measurable advantage for the systems supplier was an increase in sales resulting from an increased scope of delivery. In the long term, this measure consolidated the customer's trust, providing a favourable basis for future orders.

The case illustrated in the example concerns the modification of an existing configuration for an electrically-driven steel sun roof. Originally, this system consisted of the roof itself and the drive unit. The drive unit consisted of an electric motor, which was installed in the car boot in order to minimize noise generation, and a flexible drive cable to transmit the power to the roof. The supplier of the sun roof developed a configuration with an integrated drive in the roof itself, which satisfied the customer's requirements with regard to low noise generation. As a result, the long drive cable leading to the car boot was no longer necessary and the scope of assembly work to be carried out by the automobile manufacturer was substantially reduced. Although the systems supplier did not participate directly in this reduction in costs, the automobile manufacturer subsequently purchased the complete system from this development partner, whose scope of delivery was thus increased. The automobile manufacturer's purchasing operations were also simplified, resulting in reduced transaction costs.

4.3.1 Strategies for cooperative product development partnerships

Assuming that the primary manufacturer is the system leader, two strategies are distinguishable with regard to the configuration of a product development partnership (Fig. 4.11).

The first strategy is applied primarily to pursue the aim of minimized transaction costs referred to in the preceding section. The system leader is interested in defining the scope of development work as early as possible and sub-contracting this work out to a small number of systems suppliers. Sub-systems developed in compliance with the overall system are to be delivered ready for installation in the final assembly stage. The entire scope of responsibility for the operational sub-system is assigned to one systems supplier. The small number of development partners reduces the

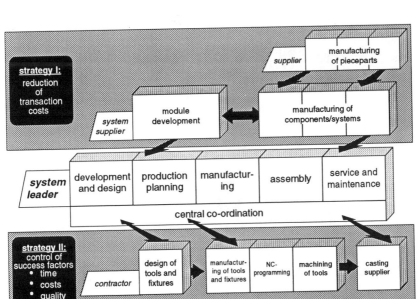

Fig. 4.11 Configuration strategies for product development partnerships

scope of coordination work for the system leader and effectively simplifies purchasing operations. The primary manufacturer's own development competences can be concentrated on the core areas contributing to product differentiation and system integration. The systems suppliers belonging to the first stage are provided with an opportunity of effecting large-volume deliveries directly to the primary manufacturer, in the form of sub-systems.

As direct contact partners for the primary manufacturer, the system suppliers take on the responsibility of coordinating the supply chain at the second and third levels, thereby strengthening their economic positions.

The second strategy focusses on direct control of the success factors of time, costs and quality. The system leader concentrates its own competences in particular on efficient organization of the product development process. The project planning is geared towards splitting up the value-creation chain, to enable as many areas of work as possible to be carried out simultaneously. On the basis of a detailed functional structure for the development work, the system leader will involve external experts for specific aspects of the development work, if these are able to work more efficiently than his own development specialists, for cost- or capacity-related reasons. At the same time, this strategy enables certain

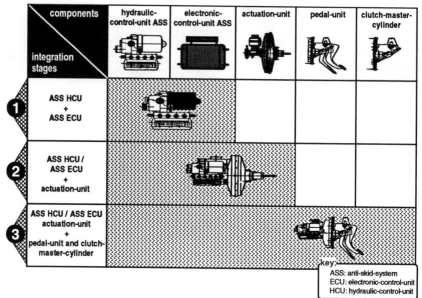

Fig. 4.12 Development of brake systems: Integration stages (source: ITT-
Automotive)

areas of the value-creation process which are subject to relatively low
qualification requirements to be transferred to low-wage countries, thereby
reducing the production costs.

The first strategy is particularly suitable for the development and
production of complex products which involve a large number of
production stages, such as is the case in the automobile sector. Current
trends in the automobile components supply industry demonstrate that the
resultant opportunities for competent systems suppliers are being
recognized and utilized [9].

Using the example of the production of ASS systems, the path from the
delivery of components to the delivery of complete modules ready for
installation can be shown from the system leader's point of view (Fig.
4.12).

In the final stage of integration, the module comprises components
from second-level suppliers, such as the pedal unit and the clutch master
cylinder, as well as the components of the system leader, such as ASS
system (cf. Fig. 4.7) and the actuation unit. The system leader - in this case
ITT-Automotive - is thus able to expand its scope of delivery and services
on a step-by-step basis, through to preliminary testing and logistical supply

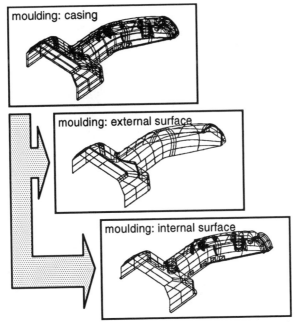

moulding: casing

moulding: external surface

moulding: internal surface

Fig. 4.13 Example of structuring the development tasks (source: Bosch)

of the modules ready for installation to the contractor's assembly line. Making use of these services helps the cooperating primary manufacturer (which may be any automobile company) - in conjunction with a drastic reduction in the scope of coordination work - to effectively reduce his transaction costs.

The second strategy is aimed in particular at controlling the success factors of time, costs and quality of value-creation. An appropriate example here is the development and production of plastic housings for electric tools. These are produced via the injection moulding process and are normally referred to as "preforms". Due to the long lead times involved in the production of injection moulding dies, the preforms normally represent the most time-critical factor in the development of electric tools. Structuring the required development work provides the basis for efficient coordination of the development partners involved in the value-creation process (Fig. 4.13).

The example of a casing, used as housing of an electrical tool, illustrates that the development task can be split into the development of the outer and inner "shell" of the housing. This structuring process is advantageous, as the outer contours of electric tools are determined to a

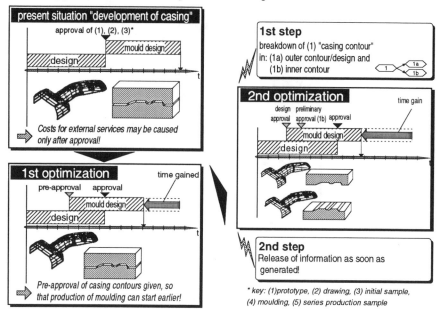

Fig. 4.14 Reduction in development time by information management within the product development partnership (source: Bosch)

very large extent by the design, in spite of their being utility goods. A design model is therefore first decided upon, and only then are the technical details of the product developed and specified.

For the product development partnership, this procedure enables the early involvement of both the tool designers and the tool manufacturers into the development process. This means that appropriate activities can be commenced before the inner contours of the housing, with the ribbing and recesses for bearings and switches, have been finally decided upon. The primary manufacturer as project leader is responsible for continuous coordination of the internal and external developers with regard to both technical progress and compliance with the planned schedules. Splitting the various development tasks in order to reduce the development time by carrying out various subtasks simultaneously bears the risk of extensive modifications. That is why the project leader should be supported by an advanced information management system (Fig. 4.14).

The most important element of such an information management system is the definition of approval procedures which enable fragmental results to be cleared for further use by means of preliminary approval, for example. In spite of this precautionary measure there is nevertheless an increased risk of modifications, as costs for external services are caused prior to overall approval being issued for the product. In the case illustrated here, however, Bosch accepts this risk, as considerably higher priority is attached to the time which is gained, enabling an earlier launch on the market.

4.4 Guidelines for product development partnerships

The presented practical examples illustrate that significant savings in production time and costs are possible by means of product development partnerships. It has also become evident, however, that no generally applicable procedure can be specified for the establishment of product development partnerships and for the operational organization of cooperative product development, due to the various branches and types of companies which may be involved. However, successfully implemented product development partnerships do enable central influencing factors to be identified to which the positive results of the cooperation are accountable.

On the basis of these influencing factors, a success profile can be established for each company, by reference to which it can be determined whether the requirements for cooperation on a partnership basis are fulfilled for each specific company and each specific case of development work. To this end, the individual influencing factors require to be weighted in accordance with the specific scope of development work (Fig. 4.15).

The essential basic prerequisite for successful cooperation is the so-called partnership potential. After determining the company's own core competences, it must first of all be assessed whether the complementary competences which are necessary for partnership-based cooperation are available (fundamental fit) A company is of interest as a development partner only when its competences effectively complement the partner's capabilities with regard to the objective of the development project. In addition, the medium- to long-term objectives of the potential partners with regard to the development project must be compatible (strategic fit).

Finally, it must be ensured that the different corporate cultures are suited to one another. The resultant cultural fit places the focus of attention on the staffs of the participating partners, both as individuals and as teams, as it is they who ultimately decide the success or failure of the development partnership.

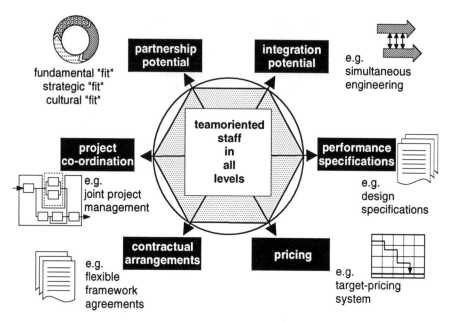

Fig. 4.15 Success profile for product development partnerships

The integrative potential describes the capabilities of a company to integrate, in terms of both time and space, its own work and external work of varying levels of detail. A good initial basis for such integration is provided, for example, when experience with simultaneous engineering (S.E.) is already available within the company. The function-related structuring of the products to be developed, which is necessary in order to optimize the time required for the development process, overcomes the departmental limits and classical areas of responsibility, which is essential to break down for the purposes of product development on a partnership basis. From an operational point of view, integrative potential also includes in particular the capacity to participate in inter-company communication. The requirements involved here range from the configuration of data interfaces to link different EDP systems through to the operation of inter-company computer networks.

For potential development partners, the specification of the scope of services relating to the development competences to be purchased is of decisive importance. A systematic, interdisciplinary specification in line with standard performance specifications is necessary for this purpose. In this connection it is particularly important to involve the members of staff responsible for the operational levels at the earliest possible stage.

The pricing system must be closely linked with the specification of the scope of services, whereby a distinction is to be made between two cases. In the case of development services which are ultimately rendered in the form of systems or components ready for installation, target price systems have proven most expedient. On the basis of a fixed upper price limit, target prices are stipulated which must be attained in time- or quantity-related stages to be established on a cooperative basis. To ensure successful and fair negotiations, this requires an exact knowledge of one's own cost structures, on the basis of which well-founded budgetary planning can then be carried out. When, on the other hand, the development work is to be purchased in the form of actual services, the transaction costs method provides a suitable aid [10].

Although the specification of the scope of services and the pricing are closely linked, it is recommendable to assign these two areas of work to different persons. While the development specialists from the operational areas who are concerned with the specification of the scope of services are required to work together with their "counter parts" in the subsequent course of development work, the detailed and often controversial price negotiations have more of a one-off character. In order to develop the trusting relationship which is required for the development work right from the outset, the development specialists should therefore provide only the technical input information, on the basis of which buyers experienced in technical matters can then carry out sound pricing. The buyer thus acquires an active role in the value-creation process, his negotiating efforts applying not only to the external procurement of certain parts, but also to defined sections of the value chain.

The contractual arrangements must regulate in particular the rights and duties of the development partners, as well as the customary conditions of purchase. Risks are to be covered by appropriate contractual provisions. Flexible general agreements are recommended, in order to reduce the transaction costs relating to the area of contractual arrangements. The establishment of long-term partnerships is also expedient, as the resultant relationships based on mutual trust then render it unnecessary to lay down every conceivable detail in writing.

The various areas of responsibility for inter-company project coordination are to be clearly defined. The organizational planning should be carried out in cooperation with the partners. This point once again underlines the fact that employees who are capable of working in teams ultimately represent an essential prerequisite for successful product

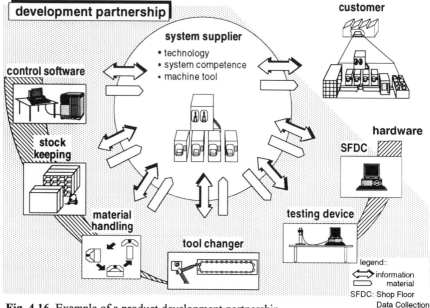

Fig. 4.16 Example of a product development partnership

development partnerships.

The presented success factors can be employed to assess the suitability of a specific company for cooperation within a partnership. The subsequently determined success profile does not enable a decision to be reached as to the level at which it is suitable to involve the company, in terms of the qualitative scope of services to be rendered. The problems involved in selecting the appropriate level for integration of a company into a product development partnership are illustrated by reference to the example of an equipment supplier (Fig. 4.16). A customer requires a flexible production cell for finishing gearboxes. In the example shown, a machine tools manufacturer wishes to act as a systems supplier for the customer, having developed the necessary system competence on the basis of the relevant production technology and machine tool know-how. All the necessary sub-systems are to be developed via subcontracting.

The division of the sub-systems into hardware and software components, which applies in all cases, provides a good starting point for the configuration of the product development partnership. Due to the diverse problems relating to interfaces it has proven effective, for example, to delegate the responsibility for the control software of all components, from the machine tool itself through to the tool-changer, to a competent

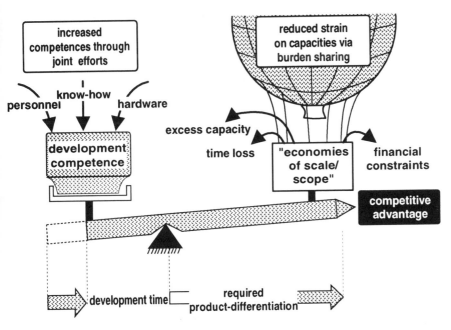

Fig. 4.17 Effects of supplier involvement in product development on the competitive position

systems supplier at the second level. This is an expedient procedure, as the sub-systems to be integrated differ primarily in the area of hardware. A similar procedure can be adopted for sub-systems which differ in the area of software and hardware being the common denominator.

The close cooperation in the supply chain right from the innovative phase enables significant cost and/or time benefits to be attained for the supplier of the overall system and for the ultimate customer. In a concrete case, the coordinated introduction of a shop floor data collection system (SFDC) and a quality assurance system, which both run on the same kind of personal computers, helped to save 40% of the required hardware. Because of the close cooperation of the individual system suppliers it became obvious that part of the SFDC-terminals could serve as data acquisition and evaluation stations for quality assurance as well.

When suitable partners wish to cooperate in a product development partnership, each individual company must first of all be assessed, therefore, in order to establish the scope of development work to be taken on by each partner. It is quite possible to apply a graduated systems supplier chain, when individual partners are able and willing to contribute not only special know-how, but also system competence. In the subsequent course of development work, it is then vital to provide all partners with the most comprehensive information possible on the overall system, in order to utilize potential synergetic effects. Via this form of cooperation, it is possible to increase development capacities while at the same time minimizing the attendant risks (Fig. 4.17).

4.5 Summary

Supplier involvement in product development represents a suitable strategy for effectively increasing a company's development competences while at the same time minimizing the attendant risks. The central objective must be product differentiation. Every company should be capable of configuring and specifying product characteristics suitable for the purpose of product differentiation on the market, while any other areas of work may be carried out by external development partners. When due account is taken of the stated factors, product development partnerships provide an effective strategy to meet the challenges arising from ever-shorter development times in conjunction with the increasing requirement for product differentiation. Within a product development partnership, each company can concentrate on its own competences, avoid wasting scarce, cost-intensive resources and thus improve its own competitive position on a lasting basis.

5

THE DESIGN WORKSTATION OF THE FUTURE - FROM INITIAL DESIGN TO NC PROGRAMMING

5.1 Introduction

Changed market conditions, such as the requirement for reduced throughput times in conjunction with increased product quality and the need for environmental compatibility, necessitate new concepts for the area of product development. The main focus of interest is on optimization of the operational processes in all areas of production. As it is the design which bears the greatest responsibility for the success or failure of a product, the designer's workplace occupies a position of central importance in systematic rationalization of the product development process. Future concepts must also take due account of other important influencing factors, however (Fig. 5.1) [1].

Current efforts relating to the optimization of product development are characterized by a move away from the strict division of respective areas of work, with the aim of overcoming the traditional boundaries between design and production via the comprehensive utilization of information technology [2]. This involves a number of advantages:
- reduced scope of processing for design data for areas preceding production, such as operations scheduling and NC programming;
- reduction in errors as a result of the internal transfer of product data

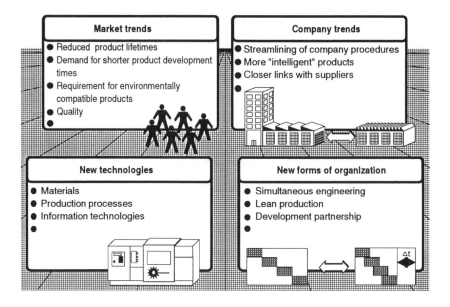

Fig. 5.1 Trends in the areas affecting product design

within one computer system, without any interface problems;

– shorter throughput times in conjunction with simultaneous engineering[1] (see glossary at the end of the chapter):

– low-redundancy storage and management of process data.

From the point of view of organizational aspects, the channels of communication between the various functional areas involved in the product development process should be improved. In the context of the overall system, the designers, operations schedulers, etc. involved in the development process are to be incorporated into an integrated information system which not only ensures all users random access to their own product data and programmes but further enables them to utilize the software of traditionally neighbouring departments (Fig. 5.2). Within this system of inter-departmental cooperation, each network user is able to set up his user interface in accordance with his individual requirements, i.e. he configures the interface in such a manner that only the system modules which are of relevance to his area of work are available at his user interface. This ensures a clearly configured user environment without isolating the individually configured workstation from the information technology system operated by the neighbouring departments.

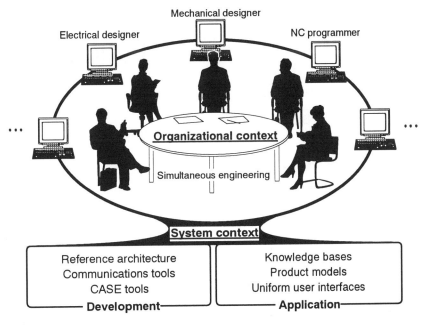

Fig. 5.2 Scenario for the design workstation of the future

The essential prerequisite for such a system is an open reference architecture comprised of various application programmes and software development tools, together with communications tools and an integrated product model covering all phases of the product's life.

With regard to the practical configuration of the software for the overall system it is to be examined which existing system modules can be used and to what extent further development of these modules is necessary.

5.2 State of the art in the field of computer-aided product development

Computer tools are currently available for carrying out at least some of the work involved in many tasks relating to product development. These tools include conventional CAD systems for the production of drawings, CAD systems for hydraulics, pneumatics and civil engineering, electrics and electronics.

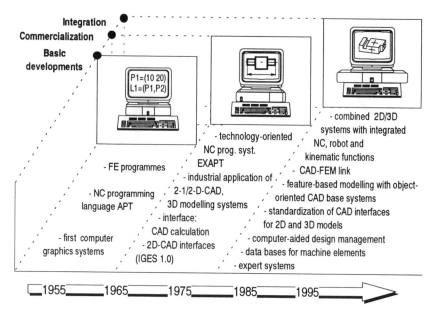

Fig. 5.3 Milestones of computer-aided engineering

Fig. 5.3 shows the stages of development which have led to the CAD/CAM tools available today [3, 4]. Finite element and design programmes for machine elements and simulation programmes for kinematic and dynamic analysis are also frequently used in the fields of development and design. Recent new tools are supplier catalogues on data carriers, company data bases and expert systems.

In general, simplified terms, a distinction can be made between three periods (Fig. 5.3). On the basis of initial concepts developed in the 1950s and 1960s, broader commercial utilization of CAD, CAP and CAM systems began at the beginning of the 1970s, whereby the appurtenant hardware and software was usually distributed throughout separate computers. Although increasing integration of the available computer tools in the areas of design and operations scheduling has been in progress since the beginning of the 1980s, a commercially available system integrating all software modules within one uniform user environment is yet to be produced.

Furthermore, the department- or activity-related application of present-day design systems results in serious obstacles which future system development will be required to overcome (Fig. 5.4).

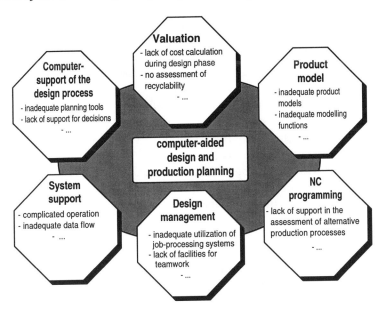

Fig. 5.4 Problematic areas in computer-aided design and production planning

At present, no uniform system of computer assistance embracing all phases of the design process is available in industrial practice. This is due firstly to the fact that hardly any product-related design logic systems are available which model the design procedure, while secondly there is a lack of context- and task-related assistance to provide the designer with alternative procedures based on his current results, when working with the computer.

Growing interest is being directed towards Valuation processes accompanying design operations, to enable the early estimation of production costs or the environmental compatibility of production and subsequent use, for example, which are not available in current systems.

A further obstacle which stands in the way of a fully-integrated design system is the lack of suitable product models. This is accountable primarily to the geometry-oriented modelling of current CAD systems, as these operate primarily on the basis of drawings. In contrast, the designer's approach to designing technical products involves a complex interrelated combination of machine elements, individual components and complete modules.

There is also a lack of computer assistance to enable the designe r to simulate alternative production measures and to compare the available alternatives with regard to production restrictions and costs. Considering the area of CAD, CAP and CAM as a whole, there is a clear lack of integrated and transparent data flow between the departments generating product data and the project management.

Finally, the system support which is available for comprehensive, inter-departmental team work lacks correctly functioning interface processors. The employed systems themselves often involve user environments which are difficult to learn and operate [5].

In the context of the above-described weaknesses and problem areas relating to the current application of computer technology in the fields of design and process planning, the changes presented in the following section are in progress.

5.3 Trends in computer-aided product development

Advanced CA systems suitable for industrial practice must be geared to the current development trends. Standardized components are being used to an ever-increasing extent, as a result of the growing number of development and production partnerships which are being established to carry out projects beyond the confines of the single company. This applies equally to the employed hardware and the system software installed on this hardware (Fig. 5.5). Market analyses confirm the trend towards the use of higher-capacity workstations and personal computers [6, 7]. On the basis of these facilities, operating systems with networking capability are being deployed which will permit so-called CAD conferences in the future. This term refers to the possibility for the designers of various departments within a company, and possibly at other companies as well, to have the same design model available in their CAD systems, in a form of conference system. Design modifications can be carried out by any participant in interactive graphic mode, and are subsequently displayed on all connected CAD screens. At the same time, it will be possible to transmit language via an acoustic link.

Object-oriented languages which enable the generation of parametric or object-oriented product modelling cores are also being employed to an ever-increasing extent. These languages provide a convenient means of generating combined 2D/3D models and quickly producing variants by simply changing the dimensions of a design parameter. The designer, operations scheduler, etc. works with graphic user interfaces and an

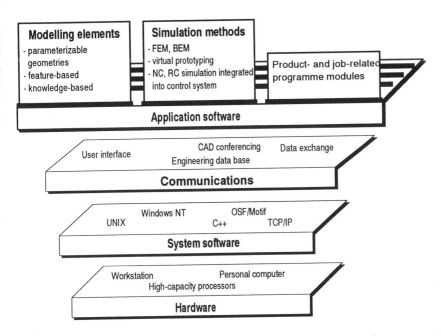

Fig. 5.5 Future components of CAx facilities

engineering data base which is accessible to all departments and which enables the exchange of data between departments. With regard to data exchange with suppliers, work is in progress to establish international standards beyond the scope of the IGES[2] and VDA/FS[3] standards, with the aim of enabling not only geometric data, but also production data, such as surface specifications and operating schedules, to be exchanged between different systems [8].

Requirement for a drastic reduction in development times increasingly minimizes the scope for practical testing. In addition to the well-established design computing systems, so-called "virtual prototyping" is acquiring increasing importance. This term refers to recent, software-assisted processes for realistic testing in the course of production and assembly of a certain module, for example. One such method of realistic testing is provided by virtual reality. Using a so-called data-glove and a pair of stereo goggles, the designer or operations scheduler is placed in an artificial 3D world for the purposes of NC or robot programming, for example. Another new process, the "digital mockup" process, enables joining processes to be examined in connection with the computer-aided

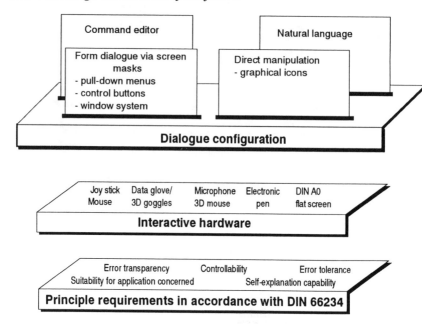

Fig. 5.6 Methods of interaction for the design workstation of the future

assembly of components or modules in CAD systems (Fig. 5.6).

The high requirements which these processes impose in terms of computing capacity, graphic resolution and memory capacity are already fulfilled by virtually all commercial computer systems. But beyond these requirements a user-friendly operator interface is equally important, in order to attain broad acceptance for the new computer tools in practical engineering work. In this context, the configuration of the interactive system acquires special importance (Fig. 5.6).

The importance of the classical form of interaction via alphanumerical commands further diminishes in the light of the fact that in parallel with the development of more sophisticated technical products the software employed in this development work is becoming more complex and extensive. Alphanumerical interaction is giving way to graphic user interfaces. The primary advantage of these interfaces is that more information can be displayed in a clear manner on one screen. Each application programme is presented in its own window, which can be displayed on the screen or removed from the display, as required. The actual system control process is event-oriented, i.e. it is the user who takes the initiative and controls the interactive process. The use of graphic functional symbols, also known as icons, the meaning of which is intuitively evident, effectively guides the user through the command structure. To date, pop-up menus, control buttons, etc. which are activated with a mouse are available for control purposes. These facilities are inadequate for applications in the field of design and operations scheduling. A higher level of user-friendliness is required with regard to internal computer configuration and handling, when work with complex geometrical models is involved. With regard to search operations for standards, repetition parts and externally purchased parts, for example, it is more expedient to communicate the search inquiry to the computer by entering keywords on the keyboard or via language processing, rather than conducting an extensive dialogue via subject characteristic lines.

In the area of geometric modelling, the facilities which are available for generating different views are in need of improvement. Whereas such operations are normally carried out today via rotary buttons or keyboard inputs, the use of a 3D mouse is one example of a cost-effective alternative for the future. The mode of functioning of such a 3D mouse involves the surface of the user's hand being placed on a ball and the movements of the hand being converted via force sensors into commands for altering the position of the viewer. A more expensive, but also more vivid and realistic alternative is provided by virtual reality. However, the extensive hardware required for the data glove and the heavy 3D goggles present an obstacle to the broad use of this technology.

Irrespective of the employed hardware, development of the above-stated interaction facilities must be carried out in accordance with the principles of ergonomic communication design, as stipulated in DIN 66234.

5.4 System architecture of the design workstation of the future

The preceding sections have described trends and developments which influence the design process as such. The area of design activities for which suitable computer-aided tools are available is constantly increasing. But, as has been shown, the scope of design work itself is also growing, as a result of which additional software tools require to be developed and combined with existing tools [9].

The stated inadequacies result in a need to provide more effective support for the design process as a whole and to fulfil design-related requirements on the deployed software more effectively [10]. A new class

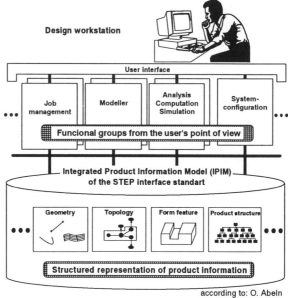

according to: O. Abeln
STEP: Standard for the Exchange of Product Model Data

Fig. 5.7 Reference model of a design system

of computer-aided design systems is thus required. These systems should be configured in the manner of a tool box, containing at least one suitable programme for each area of design activities. The creativity of designers should not be restricted by schematized procedures or awkward operation. Wherever possible, the current form of interaction between designer and computer via points and lines should be abandoned in favour of communication via function- and production-related product characteristics.

The following paragraphs describe a number of concepts and prototype developments for future software structures and software tools which are intended to enable the required integrated procedure for product development and production scheduling. Such an integrated software system is referred to here as a design workstation.

The central objective of the design workstation is to provide engineers, designers, operations schedulers and NC programmers with a broad range of computer-aided tools to apply to all the tasks encountered in development and operations scheduling. Time lost due to the need to process exchanged data and to become acquainted with unfamiliar

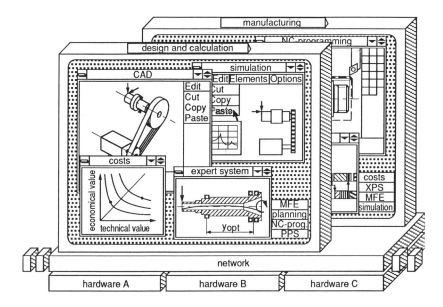

Fig. 5.8 Parallel use of computer-aided design tools

programmes is to be reduced. The simplified process of learning the ways in which a large number of individual programmes think and operate and the resultant acquisition of a comprehensive scope of knowledge from various areas are intended to produce synergetic effects which will lead to qualitative improvements in the results of design work [11].

The individual components of the design workstation are either based on existing, established software or will be incorporated into the system as newly developed modules. As the elements of the resultant integrated system are of a very heterogeneous structure, concepts and software tools will be required which are able to assist system designers in integrating additional modules, such as specific company modules, into such systems. Fig. 5.7 shows the general configuration of a design system in the form of a reference model [12].

The basis of the reference model is its integral architecture, created by dividing the design-related activities into numerous definable task areas. To each task area, e.g. geometric design, calculation or procurement of information, modules can be assigned which exchange information among each other or with a product model data base, via a communication system. The IPIM[4] product data model developed in the course of the STEP[4] development programme provides the architecture of such a data base and describes its components [13].

The proposed modular concept for a design system enables additional software modules to be incorporated and integrated into the system at any time. A special configuration module is thus employed both to implement and incorporate software modules and to configure and adapt the modules in accordance with the specific requirements of the individual user.

To enable the described design system to be used in an effective manner, the expandability, modularity and openness of the system must also be presented to the user, i.e. via the user interface (Fig. 5.8). In addition to the familiar and established modules, new interactive facilities and additional functions require to be developed, in order to ensure acceptance of the design workstation on the basis of the improved performance capabilities [14].

An important feature of user interfaces for future design systems is the use of modern window techniques. Originally implemented by Apple Macintosh, this technique has meanwhile become established on all hardware platforms of the most common classes of computer, the personal computer and the Unix workstation. The more natural method of working with graphic symbols, mouse cursor, help functions and the immediate display of results reduces the familiarization time and clearly displays the on-going work process for novices and specialists alike. The window technique further enables the most diverse programmes to be used simultaneously on the same screen and provides simple mechanisms, such as "cut-copy-paste", for the exchange of data between the systems.

The introduction of local computer networks (LAN[6]) in conjunction with the window technique is increasingly weakening the link between the hardware which executes a certain programme and the screen on which the user interface for the programme is displayed, as programme calls and even windows can be transmitted via the network to remote workstations. Similarly, the link between the user, his or her workstation and the software modules available to the user is also weakening.

Each programme integrated into the design workstation represents a specific approach to the product to be developed [15, 16]. In addition to the indisputably most important approach, which places prime importance on the geometry of the product and is implemented via conventional CAD systems, there are a large number of software modules for additional tasks which are of importance in the product development process. Management programmes for standard parts and drawings provide information on existing isolated technical solutions, for example. Expert systems serve as

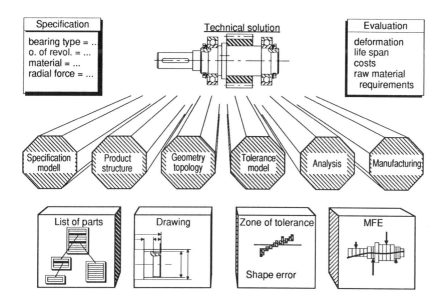

Fig. 5.9 Partial models as parts of the product model

decision-making aids in specialized areas in which measurable empirical knowledge suitable for formal presentation is available. Simulation systems are employed to verify individual aspects of a product's technical functions and its manufacture. Further design-related modules worthy of mention are cost information, calculation and NC-programming systems.

The creation of comprehensive data models for broad-ranging product information cannot be carried out by individual software suppliers or national institutions alone, and is rather to be considered as an original task for international standardization councils. The developers of the STEP product data interface have faced up to this challenge and are involved in the development of a product model incorporating several partial models, which represents the entire scope of all production information (Fig. 5.9). A further objective is to accompany a product throughout its entire life cycle, from the formulation of a list of requirements, through design, manufacture, testing, and analysis to maintenance and recycling. Each item of information is to be contained only once in the data model, and due account is to be taken of mutually dependent items of information. This data model, part of which has been completed while part remains in the development stage, is thus intended to guarantee completeness,

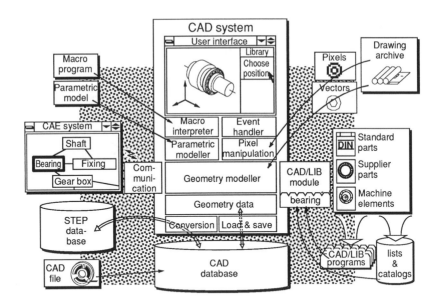

Fig. 5.10 Methods of exchanging geometry-related information with CAD systems

consistency and freedom from redundancy with regard to all information sets on the product which can be mapped within the computer.

The strictly formal structure of the data models developed in the STEP programme, based on the EXPRESS[7] data definition language, renders these models suitable for mapping in data bases and in files, as a result of which they go far beyond the original aim of providing a standard unrelated to any specific CAD system for the exchange of geometric and geometry-related data between the most diverse programmes. In addition to the two processes of data-base linking and file transfer, a number of other methods are also possible for the purposes of data exchange with CAD systems (Fig. 5.10).

The content of drawing archives, which represents a substantial part of a company's know-how, can be converted into raster graphics or simple vector graphics with the aid of scanners and displayed on the screen under conventional CAD drawings, for further processing. A satisfactory solution has yet to be found, on the other hand, to the problem of directly converting technical drawings into CAD models using intelligent algorithms.

Frequently recurring modules, components and form elements can be

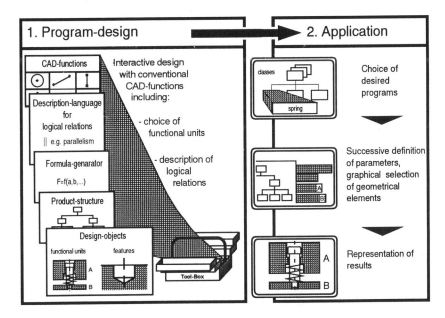

Fig. 5.11 Design and application of non-system-related program modules for specific tasks

linked to the appropriate communications modules of the CAD system in the form of macro, variant or standard parts programmes. The VDA-PS[8] or CAD-LIB[9] programming interface, which was originally developed by the Association of the Automobile Industry and is to be standardized at European level, possesses the greatest potential in this area, as it ultimately enables any external system to utilize functions available within the CAD system for its own purposes. Currently available software tools for inter-process communications and for the development of client-server architectures will be employed in the development of CAD interfaces.

Several studies relating to industrial projects show that CAD systems can be used efficiently only when the user adapts and expands their functions via applications geared to the specific requirements of the tasks on hand. To this end, more or less user-friendly methods of variant programming are available, such as the generation of command macros or specifically system-related programming languages based on FORTRAN, for the purpose of creating application programmes.

A further improvement in system support can be attained via task-related application modules and application systems produced by software companies, which can be incorporated into the CAD system concerned. The costs involved constitute a considerable disadvantage here, however.

In the light of this situation, new concepts are required, to enable the CAD user to generate and implement system functions for specific tasks without any need for a special knowledge of programming. These specific functions should relate to recurring tasks and activities, which occur in virtually all areas of design work. For such tasks, the designer should be able to map a completed design step once in interactive mode, without any knowledge of programming, so that when a similar task recurs it can be carried out to a large extent automatically, after entering the necessary parameters.

With regard to such modules, a differentiation must therefore be made between a design phase and an application phase (Fig. 5.11). In the design phase, a programme is created for a specific task, using
- available CAD functions,
- design objects,
- structural information on existing part-designs,
- a descriptive language for logical relations and
- a formula generator.

The design process is comparable to the creation of a new programme, whereby virtually all the CAD functions of the design system are available for use. Any design objects can be employed, i.e. existing part designs, functional units or technical shaping elements, known as features. A characteristic of the employed functional unit is that it is comprised of several individual components and adjoining component elements between which logical relations exist, e.g. the diameter dependency between a screw and the appurtenant borehole. These relations can be defined as the designer considers appropriate, by using structural information on the functional unit, by stipulating parameter dependencies for the individual elements with the aid of a specially developed descriptive language for logical relations and by using mathematical formulae. Possible elements of this descriptive language are, for example, parallelism, contact between objects, etc. [17].

In practical applications, such a programme module is selected by the designer, whereby the programme is executed automatically, after entering the required parameters and selecting the geometry required to carry out the functions concerned.

The created programme modules are stored in a library of the design system. Object-oriented programming methods and the attendant heredity mechanisms mean that programme modules which do not fulfil the full required scope of functions can be developed in the first step of a rapid prototyping procedure and extended at a later juncture for the purpose, for example, of further reducing the scope of interactive input operations.

An essential aspect of these programme modules is a data management concept which is not dependent on any specific system, to enable data exchange between any systems. The STEP interface standard provides a basis for such a concept. However, the specification which is currently available restricts data exchange to static data, i.e. dynamic data, which is a characteristic feature of variant programmes, and cannot be mapped. Expansion of the existing partial models of the Integrated Product Information Model (IPIM) from STEP is necessary here [18, 19].

The use of such job-related programme modules is expected to provide substantial scope for rationalization and may help CAD users from small and medium-sized companies in particular to increase their efficiency in the area of computer-aided design.

5.5 System modules for the design workstation

Fig. 5.12 shows an overview of the possible components of a future design workstation. Such a workstation comprises tools for the areas of development and design, which tend to involve a technical, function-oriented approach to products and their elements, and for operations scheduling, which is concerned with workpiece machining and the deployment of resources. At the interface between these two areas are modules for processing product-related data of such fundamental importance as geometry or the deadline schedule.

The design workstation is to be regarded not as a turnkey system, but as a framework into which existing, commercial, branch- and company-related software can be integrated. To enable the configuration of an integrated design-supporting software module which is tailored to the demands of the specific application concerned, system developers are to be provided with tools within the design workplace which allow them access to the internal representation of data and procedures. These tools are also to enable branch-related supplementation of the sets of information which are presentable in the product model and to facilitate the creation of additional software modules and integration of these into the system.

Such a design workstation, which can be configured in accordance with the specific requirements of the company concerned, enables users to support the design process from definition of the list of requirements

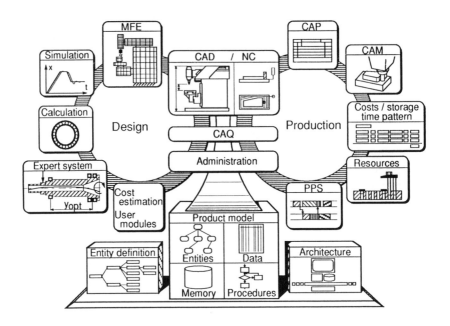

Fig. 5.12 Components of the design workstation

through to the generation of production data in the course of design operations.

5.5.1 Generation of design data

At the beginning of the design process it is necessary to provide the persons involved in development of the product with detailed information on the actual task to be carried out. Studies confirm that failure to clearly specify the task on hand leads to costly and time-consuming correction loops which would have been avoidable, had the task on hand been clarified. Often, the management responsible for a new product is not sufficiently involved in the definition process at the decisive phase of production specification. Modifications are then initiated by the management at a later stage of development, resulting in lost time and costs.

For some time now, priority has thus been attached to clarifying and specifying the requirements concerned at the beginning of the design

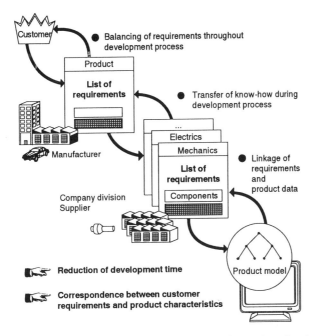

Fig. 5.13 The list of requirements as the starting point for product development

process. This involves determining the actual problem, compiling all the available information, identifying gaps in information, examining and adding external requirements as well as incorporating requirements imposed by the design department and the company as a whole [20]. In the face of the prevailing market restraints, these requirements necessitate a drastic reduction in development time via new concepts and resources obtained in cooperation with development partners.

In this context, the list of requirements is to be seen not only as the starting point for product development, but also as an inter-disciplinary information medium to accompany development measures (Fig. 5.13).

The list of requirements is not only to specify the customer's requirements in complete and precise terms - to which end close cooperation with the customer is required -, but external development partners and other areas of the company or development disciplines formerly considered to represent separate areas are also to have an opportunity to contribute their know-how to the definition of product requirements. Although a limited scope of flexibility is permissible when drawing up the list of requirements, the aim is to stipulate a final product specification at the earliest possible juncture. The stipulation and coordination of suitable interfaces between various areas of work and component areas thus represents an essential prerequisite, in particular for more complex tasks involving a large number of development partners. This requires a comprehensive knowledge both of the overall product and product-related factors and of the capacities of the respective development partners. This is the only way of ensuring that a fixed list of requirements is available at a defined time which reconciles the customer's wishes and product requirements in an optimal manner.

For all requirements which are ultimately documented in structured lists of requirements, the connections are established between these requirements and the defined characteristics of the product to be developed. This enables stipulated product characteristics to be assessed with regard to the specifications contained in the list of requirements and to identify the product characters concerned when modifications to requirements occur. This link can also be utilized to examine originally stipulated requirements, when decisive advantages are to be expected with regard to product costs, for example, via small reductions in the requirements.

Computer-aided drafting of the list of requirements is already available today in prototype form. It can therefore be expected that system modules for drafting lists of requirements in conjunction with development partners will become a standard element of multifunctional design systems in the future [14].

An essential requirement for the development of technical products, apart from the designer's experience, is a methodical approach to the design process. Consequently, efforts have long been directed at systematizing the design process by means of an appropriate methodology, thereby rendering the process more transparent and efficient.

A methodical procedure is attained by dividing the design process into successive stages, known as design phases. The aim is to assign items of work to the individual phases, to specify the points at which decisions are required and to describe the results of the work to be carried out in each phase [21]. The individual stages are carried out in succession, whereby the result of each stage is examined after completion. Should the result be found unsatisfactory, no clearance will be provided for the next stage, and the unsatisfactory stage will be repeated. The design process can thus be characterized in terms of an iterative procedure [22].

VDI guideline 2222, "Design methodology - the development of concepts for technical products" contains an initial attempt to standardize

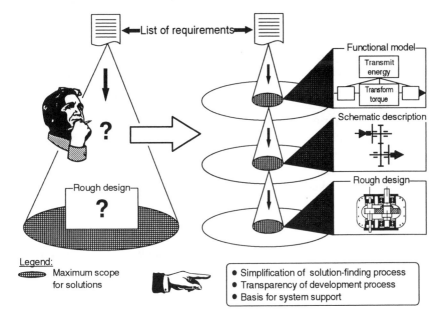

Fig. 5.14 Function modelling and schematic description as the link between the list of requirements and the rough design

the procedure for creating new products. A distinction is made between the four general phases of "planning", "concept development", "rough design" and "final design". Guideline VDI 2221 specifies further sub-divisions for the purpose of possible application of the resultant system to the computer-aided design process, whereby the solutions obtained in one step of a phase define the problem for the next step (Fig. 5.14).

The step-by-step procedure clarifies the process for establishing the appropriate solution and makes it suitable for transfer to the computer-aided design process. This systematic approach is thus based on numerous systems developed in the recent past [14, 23 - 25].

From the point of view of product development, geometrical representation of the product constitutes a very important and vivid form of representation which inspires the powers of imagination. But the overall physico-technical reality of the product embraces a wealth of additional aspects which cannot presently be mapped onto internal computer algorithms and data schemes with sufficient accuracy or completeness. Developments in the field of knowledge-based systems have shown that the human powers of imagination remain the most complete and adaptable vehicle for determining and configuring all the aspects of a product, as the imagination has such fascinating capabilities as perception and the capacity to learn at its disposal.

In most cases, however, the theoretical and practical knowledge of human beings and, in particular, the knowledge of designers is clearly linked to definable objects and situations. For example, the electrical connection data for an electric motor is linked to its geometric form and its order number from the manufacturer to form an overall mental picture, the various parts of which are to be found in separate places and in different software tools. The mental picture of the structural configuration of a machine is also linked to the design-related objects. The task for a motor within a machine is essentially described in full, for example, by specifying the objects connected to the motor in mechanical, electrical, air-condition terms, etc., and can be incorporated, for example, into a structural design model, a kinematic model or an assembly model, depending on the configuration of the computer programme concerned.

In the rough design of products, it is therefore expedient to utilize the mental concept of the object to be designed as a reference standard for the creation of internal computer data models and design tools based on these models [26]. The division of a product into objects forming the subjects of individual stages of design work can furthermore be incorporated into a communication mechanism between computer-aided design tools in the design, configuration, calculation, simulation and functional analysis phases (Fig. 5.15). Data exchange between the individual systems then takes place at the level of design elements, while the transmission of simple geometrical elements, for example, remains hidden from the user. At the centre of such an integration concept is a module which is required only to know the types of the individual design elements employed in the product, which manages all the display forms for each individual element in the connected systems and which determines the fundamental physico-technical aspects of the product. The system communications configuration thus reflects the designer's way of thinking, which has been developed during the designer's training and practical experience, at the level of components and machine elements.

The procedure presented in concise terms in the following scenario could apply in the case of the configuration and mechanical design of the main drive of a machine tool:

After establishing an initial rough design with the aid of the CAD system on the basis of the specifications contained in the list of requirements with regard to geometry, dynamic performance and interface

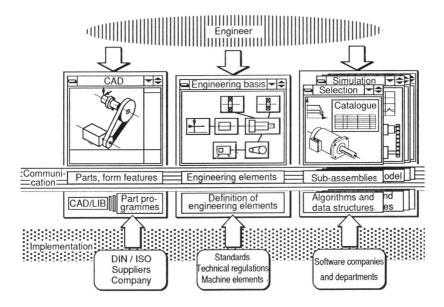

Fig. 5.15 The engineering basis, an integrative module of the design workstation

of the drive, the designer activates a programme for the preliminary dimensioning of electric main drives. The motor selected in this programme is copied into an intermediate design file ("Copy"). As additional analysis and simulation operations may be necessary at a later stage the selected motor, which is now clearly identified by specifying its manufacturer and order number, is removed from the intermediate design file and inserted into the technical engineering basis for the drive to be designed ("Paste"). Without any need for a detailed knowledge of the spatial configuration of the other design elements of the drive, the engineering basis is completed by selecting and linking bearings, spindle, belt drive, attachment elements, brake, machining point, etc. Analysis of the flux of forces and a knowledge of the rotational speeds is required for precise specification and dimensioning of the suitable type of bearing for the spindle bearing, to which end an expert system is available to provide technical assistance. To enable intermediate checking of the drive's three-dimensional form and the more detailed configuration of internally produced components, the standard and externally purchased parts selected up to this point are transferred to the CAD system with the aid of the Copy-Paste process. This same mechanism is employed to create a

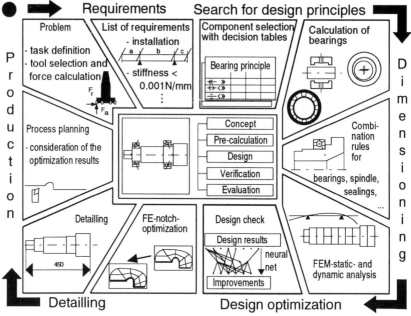

Fig. 5.16 Knowledge-based spindle-bearing design system

dynamic model of the drive in a simulation system. Instead of the variant geometries of the individual components and machine elements, the appropriate dynamic model incorporating mass, damping and stiffness is generated automatically or by the designer.

The above described process does not present the designer with a rigid procedure; rather, the designer is able to freely decide at any time which step is to be carried out, which programme is to be used and what result is required. In completing his or her work, the designer has gained time due to the improved data-exchange mechnisms and developed higher-quality solutions to the problems on hand by carrying out design work and simulation operations in parallel.

Fig. 5.16 presents a knowledge-based design system for spindle-bearing systems as an example of the application of an integral design system.

On the basis of a list of requirements to be drawn up by the designer in interactive mode, such a system enables the complete computer-aided design and design assessment of spindle-bearing components in machine tools. After the geometrical parameters and machining forces have been specified in the CAD system, a qualitative design model is first of all

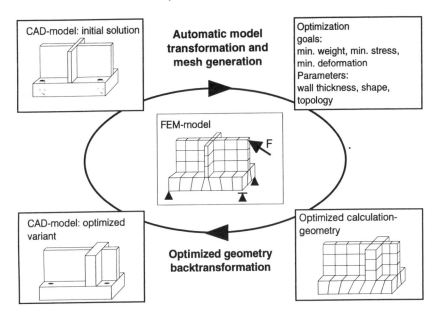

Fig. 5.17 Integration of CAD and FEM via the bidirectional exchange of geometry

generated on the basis of additional user questions. For this purpose, the design components such as bearings, seal and lubrication are selected according to the prevailing requirements, using decision tables. The appropriate bearing design programmes are activated in accordance with the system proposal for the bearing principle generated by the system via interactive communication with the designer. The results obtained by these programmes are incorporated in the first design, which is carried out on the basis of stored design rules. The design model is then transformed automatically into an FE bar model, for the purposes of structural and dynamic analysis. On the basis of the results of the calculations and the geometric design data a characteristic number model is produced, which serves as the input for a neural network for the purposes of assessing and improving the design. The results of this computer-aided analytical process can be used by the designer or substituted by the designer's own proposals, in order to initiate improvement of the design [27].

Finally, the detailed spindle geometry can be incorporated into a programme system for optimizing notch geometries, in the area of shaft shoulders subject to high levels of strain, for example. The results of the optimization process can be fed back into the initial geometric model, for the purposes of subsequent operations scheduling and NC programming.

The design and configuration of geometrically complex components, such as machine bases, require user-friendly programme interfaces which facilitate use of the FEM[10] or BEM[11] design programmes necessary for these purposes (Fig. 5.17). Although geometric data is processed by both CAD programmes and FE and BE programmes, geometric models which require processing cannot be transferred directly from the CAD programme to the FEM or BEM programme. Whereas an accurate, detailed model of the component to be designed is processed in the CAD system, a simplified component with a low level of detail provides the basis for BEM or FEM computing operations. The geometric information is reduced in the course of generating the model. At present, this conversion from CAD to calculation models requires to be carried out in interactive mode. Similarly, the modified geometric models which represent the results of the analysis and optimization processes cannot be fed directly into the CAD system; rather, the CAD models must be adapted to the modified FE geometry by hand.

A suitable modelling process therefore requires to be developed which enables the geometric data of relevance to the computing processes to be removed from the CAD model automatically. Also, the data which is not contained in the FEM/BEM model must be placed in intermediate storage.

An FEM modelling process which meets the above-described requirements can be derived from the CSG (constructive solid geometry) modelling process, by means of which CAD models consisting of simple elements such as right parallelepipeds, cylinders or cones linked via boolean operations are developed. The CSG data records are filed as binary tree structures, containing geometric and topological data and information on the sequences and types of logic operations carried out. As identical component geometries can be represented by different CSG tree structures, the CSG modelling process enables the geometric information of relevance to the FEM computing processes to be filed in a CSG sub-tree and utilized as a partial geometry for a calculation process. An FE network is then to be placed fully automatically on this partial geometry. Equally, with the aid of the employed CSG data structure the partial geometry modified after the calculation process can be reunited with the geometric details which were not included in the computing process, in the course of CAD remodelling [28].

In addition to configuring machine elements, in the area of component design the designer is often confronted with optimization problems, for which a large number of methods and processes have been developed.

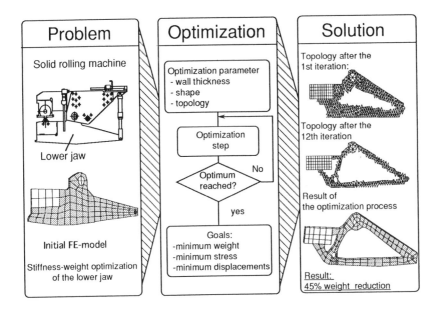

Fig. 5.18 Automatic component optimization in mechanical design

Parameter optimization occupies a position of particular importance in the application of optimization processes in conjunction with the finite element method (FEM) (Fig. 5.18).

In order to apply parameter optimization processes, so-called "mathematical modelling" requires to be carried out [29]. This term describes the process of transforming a problem to be solved into a model to which special optimization algorithms can be applied. The "mathematical modelling" process first of all requires the specification of an optimization target in the form of a target function which is dependent on the parameters to be optimized. In addition, the requirements to be fulfilled are to be incorporated into the "mathematical optimization model" in the form of restrictions.

In order to determine the optimal solution, optimization algorithms are employed, most of which operate by means of iterative processes. In each iteration, these algorithms calculate at least once the functions describing the problem and their derivatives in accordance with the optimization parameters. In the course of optimization calculations employing the finite element method, target and restrictive functions are calculated by means of finite element analysis. The overwhelming majority of currently employed optimization algorithms also require the results of derivative calculations, which are referred to as sensitivity analyses.

Finite element optimization calculations require high-capacity programme systems, which are generally configured as shown in the middle of Fig. 5.18. The finite element model is generated on the basis of the available design documents, with the aid of FE preprocessors operating in interactive graphic mode. Definition of the optimization programme is also carried out in interactive mode. In addition to specifying a target function, this also involves the definition of optimization parameters and restrictions. Once a convergence criterion defined prior to beginning the optimization process is fulfilled, analysis of the optimization process can be carried out using analysis programmes which operate in interactive graphic mode.

A characteristic common to most of the optimization processes known today is that they require an initial design as a basis. In developing these initial designs, all questions relating to the topology of the component, such as those concerning external design and internal structure, require to be answered by the designer on a purely intuitive basis. Topology optimization processes aimed at the automatic configuration of component structures with optimal mechanical properties promise to improve this situation.

The strength of topology optimization processes is that they enable an initial, complete description of the geometry of mechanical structures on the basis of only a small number of specified factors, such as the points of application of force, the installation points and the prevailing loads.

The potential for structural improvement represented by the application of topology optimization is illustrated by the result of a stiffness-weight optimization process for the lower jaw of a solid rolling machine. The optimal component topology shown in the diagram results in a 45% reduction in the weight of the component, while maintaining the same shape at the machining point as on the initial structure.

5.5.2 Generation of production data in the course of design work

Before operations scheduling and NC programming can be carried out, the equipment designer is required to design the appropriate devices on the basis of the detailed and possible optimized component geometry. An expert system provides assistance in carrying out this development work. Such a system is able both to generate a proposal for the elements of the

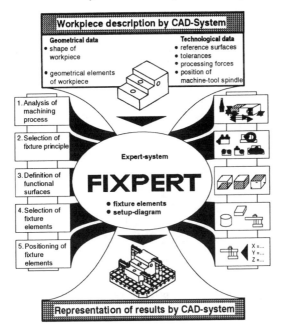

Fig. 5.19 Linking of CAD and expert system for the configuration of modular devices

devices to be employed and to configure these elements in relation to the component.

Such a system further enables the designer not only to configure a modular device, but also to optimize the workpiece which he designs with regard to use of the available modular elements.

In this connection, an expert system has been linked to a CAD system at the Chair for Industrial Engineering of RWTH Aachen. The area of application for the developed expert systems and the interlinked system known as FIXPERT (Fixture Expert) is in the design of modular devices (Fig. 5.19).

Modular devices comprised of standardised elements are employed in order to position and secure workpieces in machine tools. A conventional, standard CAD system which enables the three-dimensional, volume-oriented representation of workpieces and devices was employed in the development of this system [30].

The aim of the development work was to attain a close link between the expert system and the CAD system, so as to enable graphic support to be provided throughout the entire design process. This was intended to

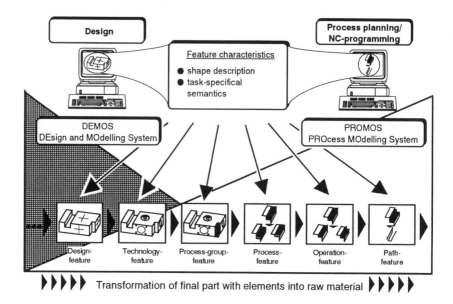

Fig. 5.20 Process modelling with the "PROMOS" system

enable the designer to trace and monitor the design process, thereby also helping to increase the level of acceptance for the expert system. Control of the design process, the search for solutions to the problems on hand and processing of the results are carried out with the aid of the expert system. As a supplementary facility, the CAD system is employed primarily for storing, processing and displaying the geometric data and as the basis for the integrated user interface of the overall system.

The starting point for the equipment design process involves transferring the detailed geometric and technological workpiece data from the CAD system to the expert system. The expert system then derives a design proposal for a suitable device on the basis of this data. Intermediate results of the design process are displayed with the aid of the CAD system, thereby enabling the system operator to intervene in the solution-finding process in order to request alternative proposals, for example. This consultation of the expert system results in stipulation of the selected device elements and their configuration in relation to the workpiece. This data is then transferred back to the CAD system, where it is utilized to visualize the device with the fitted workpiece.

In addition to improving the human-machine interface and enhancing the transparency of the displayed knowledge, it is to be assumed that the successful integration of expert systems with CAD systems in particular will have a decisive effect on the commercial success of application systems based on expert systems technology. The above-described integrated system represents an initial step in this direction.

The PROMOS (Process Modelling System) knowledge-based system is employed to support process planning for NC operations in the course of operations scheduling (Fig. 5.20).

The feature-oriented input information is generated by a CAD system and subsequently undergoes further processing by the PROMOS system. The initial step involves completing the scope of input information by adding technological data. The starting or rough-part geometry is then generated. Further planning steps involve planning the sequence of the machining operations, assistance in the selection of tools and determining the appropriate cutting strategy. The planning process results in the specification of individual operational sequences, the tools required per operation and the overall sequence of individual operations. On the basis of the individual operational sequences operational features are generated, for which the NC traversing paths are then stipulated, and this data is subsequently fed into a CAD system. For the first time, this procedure enables tasks which were traditionally carried out as independent measures to be considered as one integrated task for the purposes of operations scheduling and NC programming [31, 32].

In the future, the STEP-based product model will enable increased integration of design and NC programming. The product data for a simple shaft illustrates the situation concerned here. With the aid of the STEP-based product model, design- and NC-oriented approaches to one and the same product can be pursued and organized in a consistent manner. The integrating element is the product shape, which is based on both approaches (Fig. 5.21) [33].

It will also be possible to carry out modifications with regard to both CAD and NC applications which initiate corresponding, consistent modifications in the other application system. Fig. 5.21 shows this situation for a borehole. Should the NC programmer decide, on the basis of the available tools, to modify the diameter of the borehole belonging to the design concerned, modifications to the CAD design will be initiated when the appropriate tool is selected. The same applies to any NC programme which may already have been created - any alteration of the operating equipment would initiate a modification of the NC programme data. Final implementation of the modification should be carried out by the person

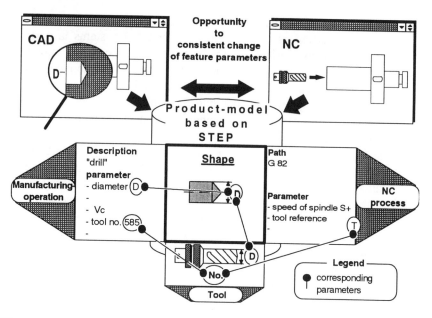

Fig. 5.21 Connections between design data and NC data

responsible for the area concerned, however. The necessary iterative loops can thus be carried out with a minimum scope of work, in accordance with the principles of simultaneous engineering. To this end, however, it is essential for the data and their complex interrelationships to be mapped in a suitable data base.

5.6 Summary and outlook

The present organizational forms of manufacturing companies have resulted from a process of increasing specialization and work-sharing. Current developments, on the other hand, are characterized by efforts to overcome the traditional division of systems into the areas of product development and production. This situation results in the need to develop and supply a new generation of design systems. Such systems must go beyond the scope of the systems which are currently available, to provide continuous and integrated support for all design phases, from drafting of the list of requirements through to operations scheduling and NC programming, and they must be geared to the specific requirements of each individual user.

In the long term, an integrative concept on the basis of specialized system-to-system link-ups will be unable to satisfy the requirements for flexibility, expandability and inter-departmental application. Future generations of systems will be required to possess an integral system architecture on the basis of an integrated product model which represents various approaches to the product. Only when such a system architecture is available will it be possible to operate various application modules in parallel and in various departments, employing a uniform user interface, and to freely exchange information between the application modules. Such a system configuration will enable the designer to generate a provisional operational plan during the design phase, for example, and to assess the product design with regard to its production characteristics.

The presented application systems based on the required, expandable system architecture are currently at the prototype stage. These prototypes demonstrate the feasibility of the proposed multifunctional design environment on the basis of a STEP-based product model. Before the entire scope of functions of a design workstation can be implemented, important concepts and system modules require to be developed in various quarters (Fig. 5.22).

Systems suppliers are required to gear their future system developments to the appropriate standards which are already available today. In particular, these include the standards for system-independent data archiving specified in STEP.

Beyond this, agreements are also required on the design of user interfaces and division of the systems into clearly structured and compatible systems, in order to enable the flexible combination of components from various suppliers, if necessary. Of particular importance is the development of system functions relating to configuration of the system, which are to ensure the configuration of user interfaces in accordance with the user's specific requirements and optimal system support.

New ideas and developments from research establishments are awaited for the purpose of further development of the systems. A problem which still remains largely unsolved concerns system support for the teamwork which is required by a modern organizational concept such as simultaneous engineering. To this end, data management concepts require to be developed which enable the parallel processing of a product by several different departments throughout various development phases, ensuring consistent data files at all times. A further area in which research work is

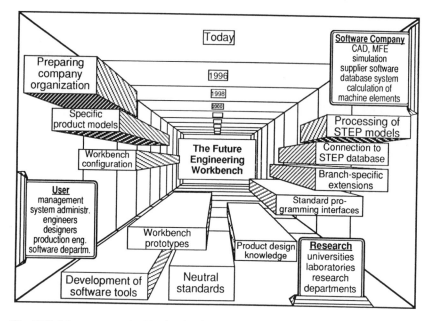

Fig. 5.22 Measures required in the development the future design workstation

required concerns the development of methods for evolving ideas, for the systematic planning of new products and for the assessment of product concepts with regard to diverse and sometimes contradictory criteria, such as low fitting costs for integral parts as against the higher production costs. The integration of quality assurance methods into the design process is a further subject for research work.

The users of the systems also play a significant role, as their requirements, particularly those relating to long term data archiving and the safe exchange of data, influence system developments. The users, in turn, are required to establish the organizational basis required for interdepartmental and parallel product development in accordance with the principles of simultaneous engineering.

Strict application of the discussed concepts in the form of efficient facilities is essential, if innovative and competitive products are to continue to be produced in the future.

5.7 Glossary

1) Simultaneous engineering:
parallel development of product and appurtenant means of production

2) IGES:
Initial Graphics Exchange Specification, format for the exchange
of plotting information

3) VDA-FS:
Verband der Automobileindustrie - Flächenschnittstelle,

4) STEP:
Standard for the Exchange of Product Model Data ISO CD 10303,
study groups TC184/SC4/WG4+5

5) IPIM:
Integrated Product Information Model, product data model in STEP

6) LAN:
Local Area Network

7) EXPRESS:
Data definition language in STEP

8) VDA-PS:
Verband der Automobilindustrie - programming interface, now DIN
V.66304: Format for the exchange of data on standard parts

9) CAD-LIB:
CAD Programming Interface for Parts Library
CEN/LCL/IT/WG CAD-LIB N87 Standardized programming
for CAD systems

10) FEM:
Finite element method

11) BEM:
Boundary element method

6

INNOVATIVE SOFTWARE TECHNOLOGIES FOR QUALITY MANAGEMENT

6.1 Requirements relating to the application of data processing in quality management

6.1.1 General situation and objectives

The tasks involved in quality management are becoming increasingly more complex. This applies above all to product development, where new, preventive methods have been introduced to an increasing extent in recent years [1]. In accordance with the principles of integral strategies, such as Total Quality Management (TQM), the implementation of these methods requires the cooperation of all the involved company levels and departments. Inter-departmental cooperation on a teamwork basis is necessary, for example, for carrying out failure mode and effects analysis (FMEA) and Quality Function Deployment (QFD).

Whenever the application of software technologies in quality management is discussed, a large number of objectives are soon encountered (Fig. 6.1) which quality management is normally required to pursue and for which it is assumed in particular that the deployment of data processing will provide at least a certain degree of support.

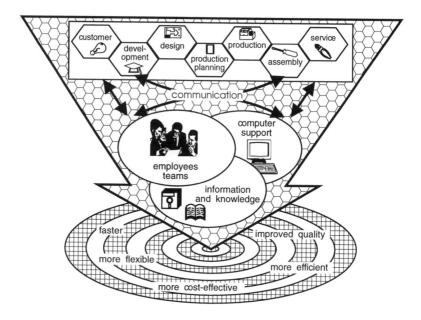

Fig. 6.1 General situation and objectives for the deployment of data processing for quality management

The situation is influenced to a decisive degree by the primarily departmental approach of the staff involved in work relating to quality management. The comprehensive character of quality management results in complex communicative relations between departments and employees, whereby communications, i.e. the exchange of information, are carried out both between specific persons and at a non-personal level.

The use of data processing to support individual employees or groups of employees in the form of teams leads to the establishment of additional communicative relations between the quality management data-processing systems and the employees (man-machine interface) and/or with data-processing systems of other company departments. Overall, a highly interlinked communications structure is produced, embracing the entire company.

The basis for all these systems is provided by the utilization of *information*, which has recently acquired central importance as the so-called *4th resource*. In this context, the improved utilization of the resource *information* can ultimately be defined as the central objective

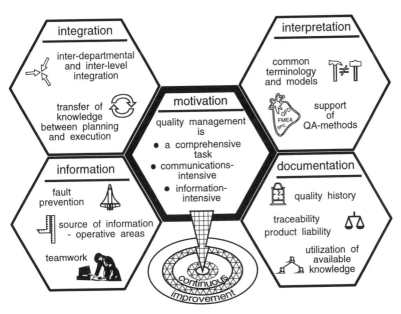

Fig. 6.2 Motivation for data processing in quality management

pursued in the application of data processing in quality management.

More detailed examination of the resultant requirements relating to the use of data processing in quality management (Fig. 6.2) reveals four main areas. Motivation forms the core of all these areas, due to the fact that quality management is a comprehensive, communications- and information-intensive task.

It is directly evident that data processing, if it is to support quality management, must enable inter-departmental and inter-level integration, whereby the prime emphasis is on the transfer of knowledge between planning and execution.

A second, less obvious aspect concerns the interpretation or more precise interpretability of the exchanged knowledge. The primary requirements here are the establishment of common terminologies and models, and also support for the various quality management methods.

The information itself must be suitable for avoiding faults, both in a preventative and in a retrospective mode. This gives rise to the need to incorporate the operative areas into the flows of information as an essential source of information. As information will remain related to specific persons in the foreseeable future, data processing must be suitable for

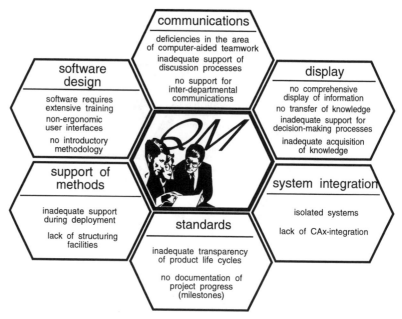

Fig. 6.3 Deficiencies of data-processing systems for quality management (QM)

applications in teamwork operations.

The scope of requirements is rounded off by the area of documentation. This is where the greatest advantages of data processing traditionally lie. The area of quality management is subject to special requirements, however, not least of all on account of the situation with regard to liability. Applications in this area include the maintenance of a quality history in accordance with DIN ISO 9004 and the possibility of maintaining complete records in connection with product liability. In each case, the aim is to ensure the broadest possible availability of the existing knowledge.

6.1.2 Deficiencies of conventional data-processing systems

The software technologies in use today possess a number of deficiencies with regard to the above-stated requirements (Fig. 6.3). The primary weaknesses are to be found in the areas of the configuration, ergonomics and social compatibility of the systems.

With regard to the configuration of the systems, aspects relating to the support of teamwork have failed to be considered in many cases. The implemented systems provide only inadequate support for discussion processes. There is also a lack of interfaces for inter-departmental communications; the exchange of data is rendered difficult by the fact that virtually all CAQ systems are designed as isolated systems. This precludes integration into a company-wide network - also for the purpose of linking the systems to other CAx components.

Another large area in which improvements to the existing data-processing systems are urgently required concerns the ergonomics of the deployed software and hardware. Existing data-processing systems often lack user-friendliness and require extensive training. The need for on-line help, an adequate degree of fault tolerance, transparent structures and a uniform user interface is often ignored. The deployed hardware is often too expensive and requires an excessive level of maintenance. Mainframe systems, which are still in use for MRP and CAQ systems, are unable to meet today's requirements for ergonomic workstations. In future, a high level of priority will therefore be attached to the design of the man-machine interface.

When evolving software concepts it must never be forgotten that the systems are operated by humans, and that only humans are able to deploy the systems in an effective manner. An innovative data-processing concept must therefore also be designed in such a manner as to ensure that the staff who are required to work with the system are able to understand and control it. This means that, on the one hand, it must promote and recognize the operator's competence. On the other hand - and this again concerns the aspect of support for teamwork - the system must not isolate operators and must leave adequate scope for decisions. The operator must maintain a clear understanding of the project in which he is currently involved at all times. The operator's own work must always remain clearly identifiable in the context of the level of progress attained on the project, and the value of this work must be evident.

6.1.3 Innovative software technologies as tools for quality management

The importance of humans within a quality management system is also reflected in the integral strategy of Total Quality Management (TQM) [2, 3]. The methods, such as failure mode and effects analysis (FMEA), QFD, DOE, can only be tools of this strategy (Fig. 6.4). A *quality-based approach* by all members of staff is intended to enable a certain degree of

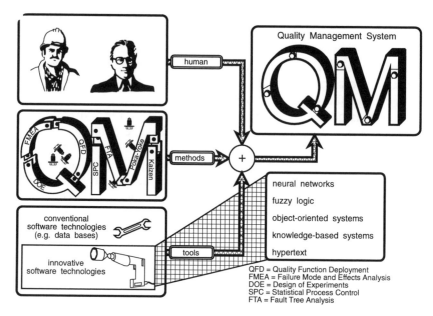

Fig. 6.4 Innovative software technologies as tools for quality management

independence from the stated methods. The example of the quality circle illustrates the promising nature of this approach [4].

Another approach to improving quality management systems is aimed at improving the interface between man and method, i.e. the tools employed to apply the methods. This aim is to be attained by simplifying the methods. This reduces the scope of personnel required to carry out the methods, while the benefits increase - not least of all due to a higher level of acceptance.

In many applications, adequate results can be obtained via the use of conventional software technologies. It is nevertheless evident that the potential of these conventional technologies is now exhausted in most areas. Weak points are evident at the interfaces between the methods and with regard to the suitability of the technologies for application by humans (user interface). "Every data-processing system, from the simple accounting programme through to control systems for production plants, can be used to its full potential only when it renders the objects and operations of that part of the real world with which it is concerned accessible to the user in a natural and clear manner." [5]

In certain areas of application, innovative software technologies can help to overcome these deficiencies. The innovative character of the application of these software technologies is discussed in the following section.

It is not intended to provide a detailed explanation of the software technologies themselves. The basic principles (Fig. 6.5) are therefore referred to only briefly.

(a) Neural networks

The quality characteristics of a product often manifest themselves in a significant form, e.g. in the curve shape of a measurement signal pattern [6]. While the qualitative analysis of such patterns, i.e. the assignment of individual quality characteristics to the respective patterns, does not pose a significant problem for humans, the automated identification and interpretation of patterns with the aid of an EDP system is a complex task [7, 10]. The application of neural networks represents one approach to carrying out this task. Neural networks are mathematical models which acquire knowledge relating to a given problem by means of a thought and decision-making process similar to that employed by the human brain, i.e. in the course of a learning phase the neural network acquires the knowledge required to solve a given problem with the aid of a representative sample from the problem area concerned.

Such a neural network is comprised of a large number of so-called neurons, which may be interlinked in various topologies. The knowledge acquired during the learning process is distributed at the junctions between the various neurons, in the form of so-called *weights*. After completion of the learning process, this knowledge can also be applied to patterns which did not explicitly belong to the learning sample [8, 9]. This capability is referred to as the *generalizing capacity* of neural networks.

The interfaces between a neural network and data input and output are provided by so-called input and output neurons. The specific assignment of a quality characteristic to a given pattern is thus effected via the so-called activations of the output neurons after the pattern has been applied to the input neurons, whereby calculation of the output activations can be carried out by the individual neurons of a network in parallel.

(b) Hypertext and hypermedia

Hypertext processes offer new methods of storing and accessing complex information. In hypertext systems, a document consists of a number of information units (*nodes*), which can be linked via associative links (*hyperlinks*). A complex *organization* of items of information can then be treated

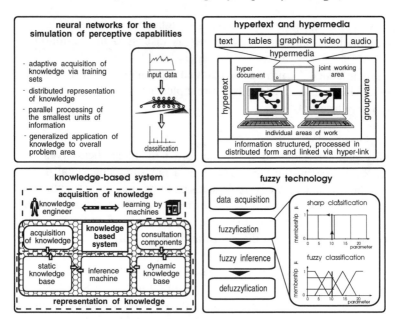

Fig. 6.5 Basic principles of selected software technologies

as a *hyperdocument*, which can be displayed in a structure as a *hypergraph*, by means of the nodes and hyper-links (11).

Hypertext technology also provides a concept for linking the objects on the user interface directly with the internal objects in the data base, by means of mechanized links. This concept supports concepts such as *Multi-Windows* and *WYSIWYG* (What You See Is What You Get). The use of hypertext and the closely related hypermedia and Computer-Supported Cooperative Work (CSCW) technologies results in the following advantages:

– **Hypertext:**

A hypertext system supports the user with regard to the following two aspects: The user can firstly enter information as an *author* and secondly access this information as a *reader*. *Writing hypertext* [11] involves creating hyperdocuments, i.e. acquiring and organizing the necessary information and, when necessary, manipulating the information structure. In contrast to the reading of conventional documents, *reading hypertexts* involves a non-sequential process which enables the user to access the stored information according to his or her own subjective criteria. These capabilities are able to facilitate the development and utilization of information systems substantially.

– Hypermedia:

In addition to text, a hyperdocument may also include tables, graphics, optical data and acoustic signals. The techniques employed to integrate comprehensive displays of information are referred to as *hypermedia* [12] or *multimedia*. The hypermedia techniques facilitate the development of intuitive man/machine interfaces and utilization of the information systems.

– Groupware and computer-supported cooperative work (CSCW)

Recent work has further developed hypertext systems in the new research area of *Groupware* or *CSCW* [13, 14]. Using the hypertext concept, it is possible to structure complex items of information into *individual work areas*, to process these items of information in distributed form and then to integrate them in a joint work area, using hyperlinks. New concepts from this field of research overcome the restrictions of the classical processes for multi-user operation and directly support cooperation between several users on the computer to solve complex problems.

(c) Expert and knowledge-based systems

Utilization of the empirical knowledge available at a company in developing large quality control circuits is of decisive importance. Conventional systems are subject to substantial deficiencies with regard to the acquisition, storage (*representation*) and utilization (*consultation*) of this expert knowledge.

Knowledge-based systems or expert systems [15, 16, 17, 18, 23] enable expert knowledge which can be described via language or rules to be mapped in computers. In addition to assisting the user, knowledge-based systems also enable the transparent and traceable documentation of expert knowledge, which subsequently facilitates the initial training of new staff, for example.

Current knowledge-based systems are generally free of the restrictions of earlier systems with regard to the acquisition of knowledge, which necessitated the deployment of so-called knowledge engineers. The knowledge acquisition process is now carried out automatically, i.e. while the user carries out a method with the aid of the computer, the system inserts the entered data into the knowledge base.

This procedure requires adequate structuring of the problems under examination. The system is able to process the items of knowledge which have been stored in a structured manner by applying rules, which are also stored in the knowledge base. When it is possible to represent a problem in such a manner as to render specific aspects of the knowledge contained in the problem, such as the similarity of technical processes, explicit for the system, by means of extensive structuring and a suitable set of rules, the user can be provided with effective support in carrying out his work, in the form of proposals and warnings.

(d) Fuzzy logic

When analyzing complex states characterized by a large number of parameters (e.g. in technical systems), the use of boolean logic often fails to provide satisfactory results. Boolean classification on the basis of sharp, digital values fails in such systems, the complexity of which means that classification of the states according to *true* and *false* is not adequate. Humans also think in terms such as *quite round, almost round* or *almost square* when assessing situations. The incorporation of these fuzzy values extends and enhances the informational value and enables a decision-making process which is more suited to the situation concerned. Fuzzy logic [19, 20, 21] incorporates indefinite values into the decision-making process. Classification based on fuzzy logic is thus able to evaluate complex states with a substantially higher degree of reliability. To date, the primary areas of application for fuzzy logic have been control engineering and process control engineering. Fuzzy logic is able to improve the decision-making process in expert systems. Highly efficient, adaptive classification systems can be created via the combined application of fuzzy logic and neural networks.

(e) Object-oriented techniques

According to Stroustrup [22], object-oriented programming (OOP) is a "programming technique - a paradigm for writing good programmes for a specific class of problems". Although the principles of OOP are not bound to any specific programming language, they are supported particularly effectively by certain languages. They represent the need for new approaches to software development, with the aim of rendering the programme creation process more transparent, more efficient, less susceptible to error and more economical.

The object-oriented techniques comprise on the one hand object-oriented modelling and programming and, on the other hand, object-oriented data bases. The application of these techniques enables simplified implementation, modification and expansion in particular when developing inter-company networks.

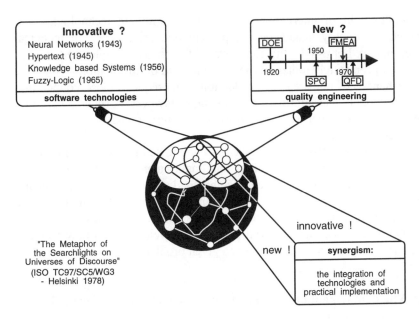

Fig. 6.6 The integration of software technologies and QM methods

6.1.4 Are these techniques and systems really new?

Looking back over the history of technical development, the techniques and systems mentioned here do not appear new (Fig. 6.6) [9, 11, 23, 24, 25]. Software developments which have been in existence for over 3 decades remain "innovative" primarily on account of one reason - namely, the complexity of the increasingly developed theories relating to the technologies. By means of appropriate, application-oriented development it should, however, be possible not only to make these technologies appear more comprehensible, but also to broaden their scope of application.

In the area of production technologies, including quality management systems, many of the concepts employed in Japan have become new bywords for success. This is due not only to the level but also to *the pace* of technical development in Japan. It is interesting that significant concepts which were originally developed some time ago are now recurring in different guises. Why should these "old" concepts cause such a stir in their new form? Perhaps it is the consequence of the many technological elements which previously existed separately being fused together and developed into new concepts [26].

In the field of international competition, software technologies and methods of quality management represent two prime areas of technical development. Let us take the above-stated factors as examples and attempt to uncompromisingly utilize our lead in the field of information technology to the benefit of our areas of production! This requires close cooperation between quality and information technicians, however. A schema based on the Helsinki concept [27] specifies a common focus and a common language for this cooperation.

The joint objective to be pursued in the course of cooperation between the areas of information technology and quality management is to integrate the innovative software technologies and the methods of quality management and to develop a comprehensive system for implementing these integrated technologies and methods for the purposes of practical production engineering. This will enable the development of new methods of establishing innovative solutions to given problems.

6.2 The deployment of innovative software technologies for quality management

A large number of systems employing innovative software technologies are already in practical use. The most well-known are the so-called expert systems, which operate primarily by means of rule-based inference, but which also employ fuzzy logic, neural networks, etc. This section presents examples of technologies from the area of quality management, some of which are already in industrial operation. This is intended above all to demonstrate that the advantageous use of such technologies in everyday industrial operations is already feasible and expedient. The first system which is discussed has already been deployed in daily practice for some time. The special problem of knowledge acquisition is considered by reference to this system. Two examples are then presented which have already been tested in field trials and are on the brink of practical deployment.

6.2.1 The DAX expert system

The DAX expert system is presented below as an example of a system incorporating innovative software technology which is already in operation (Fig. 6.7). The system is a joint development of the Mercedes Benz

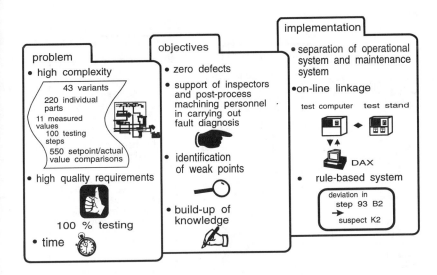

source: Mercedes Benz and FBK

Fig. 6.7 The DAX expert system - problems, objectives, implementation

company and the Chair of Production Engineering and Industrial Organization at the University of Karlsruhe (FBK) [28, 29]. It is employed at Mercedes Benz AG for diagnosing automatic circuit boards. Automatic circuit boards are highly complex components for controlling switching operations in automatic gearboxes.

The specific task to be solved here concerned automation of the diagnosis process for this highly complex component in order to fulfil high quality requirements. For this purpose, each circuit board undergoes a 100% testing and inspection process, whereby it is tested in various operational situations on a special test stand, in the course of which a large volume of test data is produced. Manual analysis of the resultant complex test records thus constitutes a special problem.

A primary objective here which involved the application of a new technology was to provide support for inspectors and post-process machining personnel in carrying out fault analysis. Weak-point analysis of the component was also to be carried out and the resultant information made available for development of the next generation of the product. This aspect is of particular importance in view of the increasing emphasis on reducing the time which elapses prior to launching new products on the

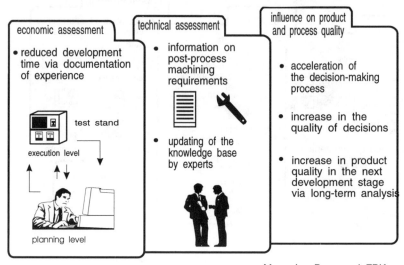

source: Mercedes Benz and FBK

Fig. 6.8 The DAX expert system - assessment of deployment of the system

market (*time to market*).

The stated objectives were attained by employing a rule-based expert system linked directly to the computer controlling the test stand. A rule-based concept was selected which links pressure deviations as symptoms and circuit board faults as diagnoses. This concept enabled the implementation of a fast and efficient system.

Assessment of the DAX system

The following paragraphs provide an assessment of application of the DAX expert system (Fig. 6.8), focussing in particular on the economic and technical aspects of deployment of the system and the direct influence of this technology on product quality.

Apart from the more long-term, indirect economic benefits which result, for example, from a long-term improvement in process quality as a result of deploying the DAX system, the reduction in the development time for the next generation of circuit boards in particular represents a competitive advantage with far-reaching economic consequences. Short-term savings on a scale in the region of one half man-year per year were also attained due to the possibility of carrying out specific post-process machining on defective parts. By comparison, the financial investment of less than one man-year for development of the system appears justified.

Technical advantages of employing the system in the course of production operations include the simplified post-process machining of defective products on the basis of appropriate information provided by the system, and direct updating by the expert of the knowledge base which represents the core of the system.

The system helps to ensure product quality in the area of development. The long-term analysis of quality-related data which has become possible with the aid of the expert system, for use in development work on the next product generation, has already led to an increase in product quality.

In the short term, application of the expert system improved the quality of decisions reached in assessing product quality, particularly with regard to the objectivity and reproducibility of these decisions, as well as accelerating the decision-making process.

6.2.2 Different knowledge-acquisition processes

A substantial proportion of the scope of work and investment involved in the development of automated diagnostic systems, such as the presented DAX system, is accountable to the acquisition of the expert knowledge required for the diagnostic process.

The indirect acquisition of knowledge represents a particularly problematic method in this context (Fig. 6.9). This involves the expert concerned passing on the necessary problem-related knowledge to a so-called knowledge engineer, who then processes this knowledge in a manner suitable for the diagnostic system. In addition to the very high level of labour costs resulting from deployment of the knowledge engineer, the success of this method depends to a very substantial degree on the willingness to cooperate which prevails between the knowledge engineer and the expert.

A new approach, which moves away from the previously employed method of knowledge acquisition, is also under development for the DAX system presented here.

An initial step towards reducing the scope of work and expenditure involved in knowledge acquisition can be provided by the direct acquisition of knowledge by the expert. However, this direct method imposes very high requirements on the employed techniques. A suitable example here is provided by the expressive capacities of formalism to

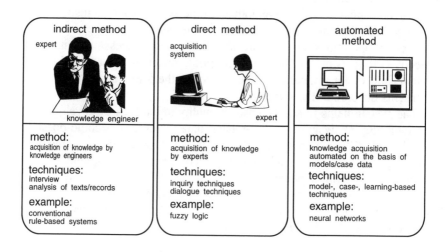

Fig. 6.9 Knowledge acquisition methods

represent expert knowledge in a form suitable for automated processing. The capacities of rule-based concepts in particular, based on established boolean algebra, are soon exhausted here. Systems based on fuzzy logic, which are also capable of expressing indefinite conditions, are more in keeping with the natural manner of expression of an expert.

The final step now involves the fully automated acquisition of knowledge. For this purpose, techniques are employed which learn the expert knowledge required for the diagnosis process automatically, on the basis of case and/or model data. An example of such technologies with a learning capability is provided by neural networks. In addition to the neural networks, which belong to the so-called sub-symbolic processing methods, concepts with learning capabilities also exist in the area of symbolic data processing, which includes rule-based processes.

6.2.3 Knowledge-based failure mode and effects analysis (FMEA)

Whereas the presented DAX expert system represents a system in practical use, the following example concerns an application which is currently still under development, namely, the application of knowledge-based methods to perform failure mode and effects analysis (FMEA). The results of the

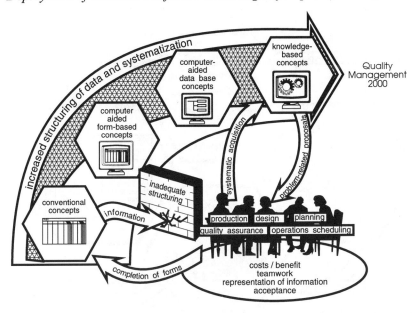

Fig. 6.10 Development options for failure mode and effects analysis (FMEA)

research work in this area are currently being implemented, which means that such systems will become available in the near future [30].

Failure mode and effects analysis (FMEA) is a preventative quality method involving the examination of design and production drafts for possible weak points in the planning stage [31]. Such weak points can then be eliminated in good time, by initiating suitable measures. The inter-departmental and inter-level nature of the problems under examination requires the cooperation of an inter-disciplinary team in carrying out the failure mode and effects analysis. The labour, organizational and coordination costs involved in this teamwork make the acquired knowledge expensive. It must be ensured that the benefits of the FMEA justify this expenditure. The benefits of FMEA are, firstly, the ability to identify weak points in the examined production process or component. Secondly, expert knowledge is documented in the course of the FMEA. In addition to displaying production processes and describing the functions of components, the FMEA above all results in the filing of knowledge on causalities, such as the propagation of faults throughout the component and the overall system which results from impermissible deviations.

The procedure employed to date in the documentation of knowledge

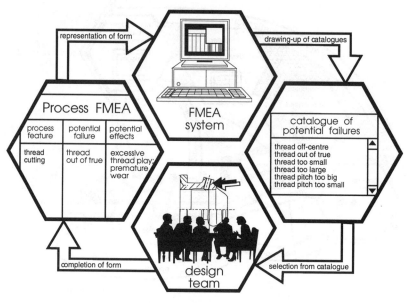

Fig. 6.11 Computer-aided failure mode and effects analysis (FMEA) by means of a form-based concept

does not permit this knowledge to be utilized in an adequate manner, however (Fig. 6.10). The reason for this is the manner in which the knowledge is acquired and represented by means of standard forms, on which the information is filed in an unstructured manner.

A potential solution to this problem is provided by the use of knowledge-based systems. In such systems, the complex causalities discussed in the course of a meeting of the FMEA team are recorded in a systematic manner and stored in a knowledge base, thereby enabling selective recourse to this knowledge.

Various criteria are to be taken into account when assessing the different concepts. Economic viability is certainly a prime factor here. The high level of labour costs involved in carrying out the FMEA accounts for the major part of the costs involved. Firstly, therefore, everything possible must be done to provide optimal support for the teamwork and - where possible - to prepare the teamwork; secondly, a method of displaying information is required which facilitates the utilization of stored FMEA knowledge.

Form-based systems for FMEA support the analytical process by

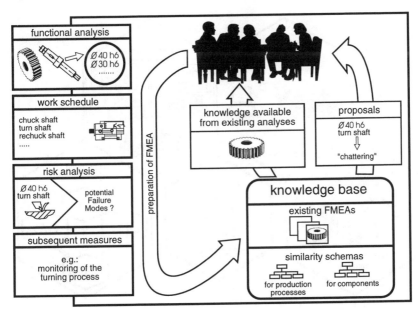

Fig. 6.12 Computer-aided FMEA by means of a knowledge-based concept

enabling the user to complete the form displayed on the screen (Fig. 6.11). A large number of users have been taught how to complete the analysis forms in the course of an appropriately extensive training programme. The entered information is stored in catalogues, whereby a separate catalogue is organized for each column of the form (potential faults, causes of faults, measures, etc.). In performing FMEA, the user can call up these catalogues and utilize the data which they contain. There are no facilities, however, for restricting the data to be displayed to a specific problem area, such as thread cutting.

The lack of structuring for the data on the form and the lack of clearly defined terminology mean that a team of experts working on a specific problem cannot access the relevant information in a *selective* manner. This results in a poor cost/benefit ratio.

Knowledge-based concepts provide potential for more effective utilization of the available knowledge, as is illustrated by the following example (Fig. 6.12): A company at which primarily rotationally symmetric parts (shafts, gearwheels, etc.) are produced has received an order for the production of a complex gear shaft; the customer imposes the very highest quality requirements and has stipulated fault-free execution of the order as

Fig. 6.13 Advantages of knowledge-based systems in FMEA

a precondition for further orders. As FMEA has been applied consistently in the past, it now appears appropriate to carry out an FMEA for the gear shaft, thereby utilizing the existing knowledge. The development team can be supported in an effective manner via the selective utilization of the analytical knowledge stored in the system in accordance with the specific context concerned. If the team is to examine the context of *finish-turning of the gear shaft*, the system will select data from the knowledge base which has been entered in relation to the finish-turning of similar parts. If no data relating to this context is available in the knowledge basis, the context will be generalized and data will be selected which relates, for example, to metal-cutting production processes.

The use of knowledge-based systems offers further advantages, beyond the facility for generating proposals (Fig. 6.13).

Structuring of the data and the attendant systematic acquisition of knowledge result in simplified operation. The reason for this is that the user is not required to understand the basic structures in order to complete the form, but is provided with extensive system guidance in FMEA. The system solicits data by displaying a specific context, such as a work schedule, and requesting the user to enter items of information, such as potential faults which may occur in a specific operation. The user is not required to learn how to operate such programmes or to become acquainted with the methodology on which the analysis form is based; in carrying out the analysis work he can concentrate fully on contextual questions. The majority of the knowledge-based systems for FMEA [30, 32] are implemented under a graphic user interface. This enables simple and intuitive learning of the operating process, and the degree of acceptance in the use of such systems is markedly higher.

There is no substitute for the creativity of an interdisciplinary team, but improved preparation of FMEA can optimize this teamwork, thereby reducing personnel costs. Structuring the data in a knowledge-based analysis system enables the analytical work to be divided into individual steps and individual areas of work to be removed from the analytical process. These areas of work can then be carried out separately by the responsible experts in the departments concerned. An analytical process which has been prepared in this way can then be discussed, expanded and finally passed in a few team meetings.

This aspect of supporting teamwork is of primary interest in the following example, which concerns the increasingly more important area of simultaneous product and process development.

6.2.4 Supporting simultaneous engineering

In future, the simultaneous execution of product and process development, which is known as simultaneous engineering (SE), will help to reduce development times and prevent errors at the planning stage [33].

Whereas a characteristic feature of conventional procedures for product and process development work is that the activities involved are carried out in sequence, at departmental level and in isolation from one another, *simultaneous engineering* involves the parallel and interdisciplinary execution of these activities. This requires cooperative, joint planning, scheduling and improvement of the stated activities, from market analysis through product design to the selection of production equipment.

The effective deployment of simultaneous engineering requires a smooth flow of information between the company departments involved in the development project concerned. It is therefore expedient to carry out simultaneous engineering with the aid of computer technology. However, computer-assisted simultaneous engineering can only be implemented when concepts have been developed to fulfil the following requirements:

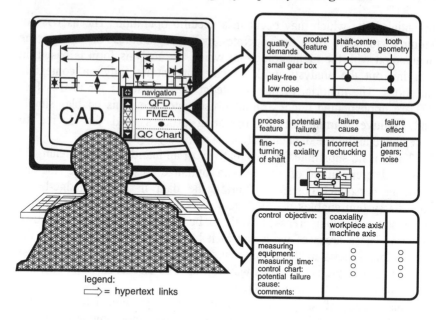

Fig. 6.14 An example of the use of hypertext

- Uniform data models for displaying the product characteristics at the various phases of development.
- A system of information management to organize the information arising in the various phases and amendments to this information.
- Communications capabilities to support the simultaneous exchange of information between various disciplines.

Potential concepts for fulfilling these requirements are provided by hypertext processes and CSCW (Computer-Supported Cooperative Works). In contrast to conventional documents generated with the aid of data-processing facilities, hypertext documents incorporate non-sequential structuring of the information which they contain. This form of structuring is attained via so-called hypertext links. Hypertext links are comparable to cross-references in a dictionary which connect different subject areas. This enables the mapping of multi-dimensional connections between customer requirements, product characteristics and process characteristics. Hypertext also enables substantially more efficient and more flexible accessing of complex information, as the user is able to access the desired information in a simple and fast manner, by following a few clear instructions. The management of data-editing operations can also be

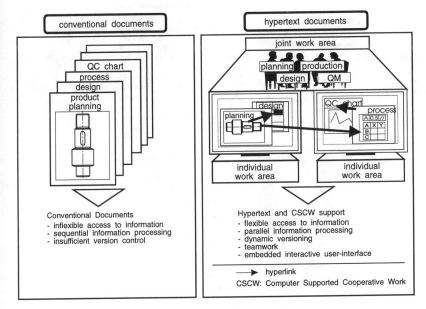

Fig. 6.15 Support of simultaneous engineering via hypertext components

simplified substantially via hypertext.

The possibilities opened up by the use of hypertext processes in simultaneous engineering are illustrated by the following example (Fig. 6.14).

The constructional design of a gear shaft is to be examined, in the course of a design review, for example. When working with a CAD system, the use of hypertext enables all the available information on this component to be displayed and called up. After selecting a characteristic, cross-links are displayed in an overview (navigation). The user can then branch off directly into the appropriate system on the basis of this overview. Via a simple procedure, the user is able to call up the QFD for the characteristic under examination, in order to view the customer requirements relating to the characteristic, for example.

The linking capability of hypertext systems has provided the basis for the further development of processes in the area of CSCW (Fig. 6.15). Various types of links, e.g. links according to aspects of time and area of work, can be further developed with the aid of CSCW processes. A primary area of development work here concerns the utilization of standardized links to compile the specialist knowledge of several persons.

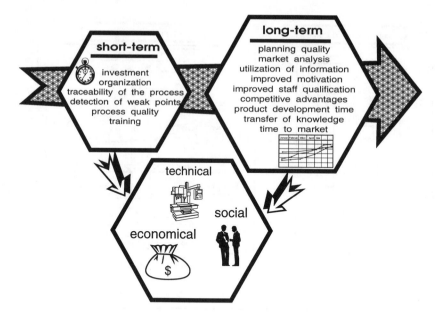

Fig. 6.16 Technical, economical and social effects of employing innovative software
technologies

New concepts from the area enable the simultaneous exchange of
information on identical topics and provide a good basis for the
implementation of team-oriented information processing, such as is
required in the area of quality management in particular.

6.3 The deployment of innovative software technologies -
a competitive advantage

The deployment of software technologies at a company results in a large
number of changes affecting virtually all areas of the company. Not only
the productive departments, but also the organizational and social areas of
a company are affected to a substantial degree. A fundamental distinction
can be made between short- and long-term effects (Fig. 6.16).

The initial effect on the short-term side is the necessary investment.
The development and procurement of software, training and introductory
measures involve costs which impose a substantial strain on the corporate
budget, while the only direct advantages concern the transparency of the

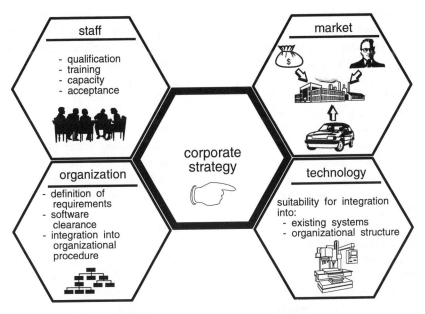

Fig. 6.17 The deployment of innovative software technologies - a challenge

process and the identification of weak points. Any investment must be repaid within an economically feasible period. This means that above all the long-term aspects, i.e. cost reduction in conjunction with increased productivity, play a major role. The examples of applications which have been presented show that the deployment of innovative software technologies can help to achieve this aim.

As demonstrated by the example of the DAX expert system, the planning quality and planning cycle time of subsequent product generations can be improved and shortened respectively via the structured processing of specialist knowledge. It is also possible to integrate the remachining of defective parts into the production process in a substantially more efficient manner. The hypertext facilities enable the corporate information system to be restructured. The non-sequential storage of information enables substantially more flexible and rational access to data. The available knowledge, which is essentially extremely complex and difficult to handle, can be deployed quickly and expediently for the purposes of product and/or process planning by means of a knowledge-based system for analyzing the probability and influence of faults.

The advantages are not restricted to the improvement of internal production and organizational processes, but also include improvements in the situation of the company concerned with regard to the market in which it is required to assert itself. Improved planning quality, enhanced capabilities to gear product development to customer requirements, shorter development times, more stable processes and lower quality costs create competitive advantages which, in the long term, are able to secure the existence of a company [34].

The stated effects show that the deployment of innovative software technologies represents a great challenge to the companies concerned (Fig. 6.17). These challenges can only be met by an appropriate corporate strategy which takes due account of the effects on staff, organization and technology and thus supports the effective implementation of the new systems.

The areas most affected are staff and corporate organization. These areas thus represent the greatest challenges.

The introduction of a new software system usually involves increased personnel requirements; these additional capacities have to be made available by the company. Altered channels of information and data flows must also be taken into account and appropriate restructuring measures may be required.

The most vital objective of a corporate strategy must therefore be to ensure acceptance of the software among the users. A system which is not accepted by the staff can never be deployed in an effective and profitable manner. The support of the staff can be attained by involving them in the configuration and development of the software and by extensive training.

An essential task for the company's organizational system prior to introducing the technologies under review here is to precisely define the requirements which apply to the software to be introduced. This is a crucial task, as the technologies under consideration here will hardly allow recourse to standardised software packages in the area of quality management. The corporate organizational system is thus further required to clear the software concepts developed for the individual problem areas involved. A point which should be incorporated into the planning procedure at an early stage is integration of the systems into the corporate organizational procedure.

Deployment of the technologies under review here also presents a great challenge to systems suppliers. Of particular importance in this respect is the integrative capacity required of the offered systems, in order to avoid isolated processes. The systems must furthermore possess sufficient flexibility to enable them to be incorporated into existing corporate structures without any problems whatsoever.

7

RAPID PROTOTYPING: THE WAY AHEAD

7.1 Time as a competitive factor

New products are generally required to perform more sophisticated functions and are thus usually more complex, batch sizes are becoming smaller while the number of variants increases, and the lifetimes of products are becoming ever shorter [1, 2]. Product developers are thus compelled to carry out the required development work in a shorter period of time, in order to meet the planned deadline for launching the product on the market. The economic importance of observing this deadline is clearly demonstrated by the lost revenue which may result from delayed marketing. In the example presented below, a drastic increase in development costs in order to meet the deadline is the more economical alternative (Fig. 7.1).

A further, equally fundamental problem becomes evident when analyzing the course of information and communication within the development department of a company more closely. During early product development phases, the management is compelled to make decisions on the basis of unreliable planning data, thus introducing a high risk of misconception. Each manager will therefore delay a decision for as long as possible, thus involuntarily increasing the scope of necessary modification required whenever a false decision needs to be corrected, on account of the advanced stage of planning and implementation. The consequences of this problem are demonstrated by studies which conclude that avoidable

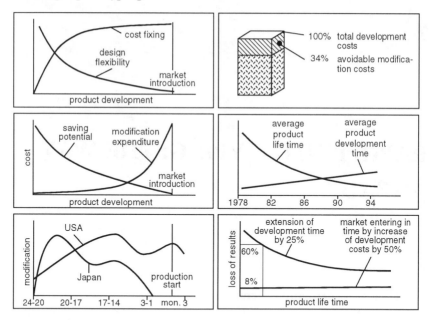

Fig. 7.1 Present situation in product development (based on: ASI, IAO, IBM,
Sullivan, Siemens)

modification work accounts for approximately one third of total
development expenditure.

Products which are not fully developed at the time of launching them
on the market are no rare occurrence. They are accepted and tolerated as a
consequence of the pressure of time, when the customer is unable
topurchase a more fully developed product at the same cost. But this is the
point where the advantages of the entrepreneurial philosophy which
prevails in the Japanese automobile industry become evident: the number
of design modifications during early development phases is substantially
higher among Japanese companies than at American or European
companies. Various alternatives and/or iterations may be tested thoroughly
without causing high costs, while a high level of design flexibility may be
maintained at the same time - an essential requirement for the
implementation of product modifications and the attainment of high
product quality. In turn, such design flexibility requires a quick and
comprehensive spread of information about the level of development of the
product concerned; this necessitates the fast availability of specimens of
the product.

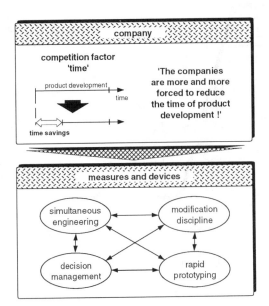

Fig. 7.2 Time as a competitive factor

Along with product quality and production costs, the competitive factor "time" in the sense of "time to market" becomes increasingly important (Fig. 7.2).

This situation should induce companies not only to undertake appropriate organizational arrangements, but also to adopt a quite different approach to planning procedures and the decision-making structures involved. In this context, "simultaneous engineering" - interdisciplinary and simultaneous cooperation between the corporate departments involved in the development process - is regarded as one of the most important means of reducing development times [1, 2, 3, 4, 5]. The parallel implementation of planning procedures requires additional structures to enable the synchronization of simultaneous operations. To this end, a decision-making management structure needs to be established by defining schedules and milestones which stipulate binding deadlines by which the various departments must complete the concepts for the following planning stages.

In the interests of ensuring product quality, a binding decision is also necessary for subsequent areas which prevents later and thus expensive modifications. The enforcement of a system of "modification discipline" in the departments and among the individuals concerned, by means of fixed

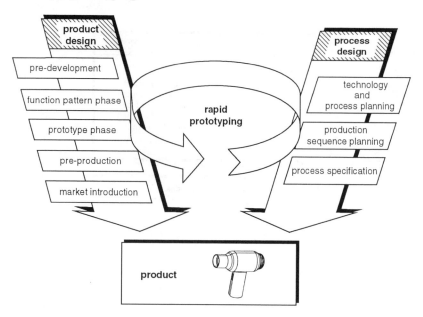

Fig. 7.3 Synchronized product and process design

stipulations regarding both the duration of a development stage and the number of iterative cycles up to product optimization, provides a further means of preventing unnecessary product modifications.

Rapid prototyping opens up new areas of potential, which may be utilized to shorten the planning phases or to improve product characteristics, depending on the specific requirements of the case concerned. The fast availability of specimens offers previously unknown possibilities to the designer: complex components may be manufactured within days, whereas the production using conventional methods would take at least several weeks [6, 7, 8].

The fast availability of models for the purpose of checking the design, for presentation at meetings or for test installation contributes towards design improvements and provides sounder criteria for finding decisions more quickly. There is, however, also a danger of too many alternatives being tested, which may have a negative effect on management of the decision-making process and discipline in the area of modifications.

Models and specimens are employed during the development process to test the product design and to configure the production process for subsequent series production (Fig. 7.3). They are information carriers and thus provide the link which enables synchronization of the parallel product and process design procedures. The fast availability of models and specimens determines the speed of information exchange between the various areas and thus has a decisive influence on the product development time.

7.2 Prototype requirements in product development

The designations used for the various forms of prototypes, models and specimen which are required in the course of product development vary considerably at the individual companies and in the respective branches of industry. It is therefore intended to establish a generally valid identification of different product development phases and to classify the various forms of prototypes.

An analysis of the product development cycle reveals that prototypes are required in all phases of development - from the original idea for a product until the launch on the market. For the branch of industry producing consumer and capital goods, the product development cycle can be divided into six phases (Fig. 7.4). The prototypes employed in the respective development phases possess highly varied features with regard to quantity, material properties and geometrical, optical, haptic and functional requirements (Fig. 7.5). The fields of application for prototypes in the area of product and process planning are correspondingly diverse (Fig. 7.6).

In the pre-design phase, design models and geometrical prototypes are often employed, and these are generally produced as single items. Only approximate dimensional accuracy is required for the **design models**. They are, however, subject to very high visual and haptic requirements. As the functional requirements are of secondary importance, such models are often produced out of typical model materials. The models are useful for studying design and ergonomics and for initial market analysis work.

In the case of **geometrical prototypes**, on the other hand, the visual and haptic component characteristics and also the functional performance are of secondary importance; the focus is rather on the high requirements regarding dimensional and shape accuracy as well as tolerance of form and position (surface flatness, parallelism ...). The geometrical prototypes do not necessarily have to be produced out of the material which is to be used later for the components in series production; model materials are usually employed here, too. This form of prototype is utilized primarily in the area

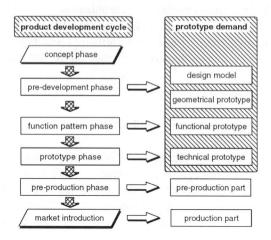

Fig. 7.4 Need for prototypes in product development

of process planning. Typical fields of application are the development of product concepts, verification of the suitability of products for manufacturing and installation, and the rough planning of production and assembly operations, where prototypes are required not least of all as a means of communication.

In the functional testing phase, 2 to 5 **functional prototypes** are generally employed with the aim of checking and optimizing the operational and functional principle. At this stage in the development process, the emphasis is on analyzing the functions of individual product components and modules. Functional prototypes are employed in the area of process planning for the purpose of planning systems, processes, production sequences, assembly operations and equipment. External appearance and dimensional tolerances are of secondary importance, as long as they do not impair the functions to be checked. The requirements imposed on the prototypes with regard to the mechanical (strength, elasticity, hardness, ...), thermal and chemical stability of the component are limited to those needed for the purpose of functional testing.

In the following development phase, **technical prototypes** are produced in larger numbers (between 3 and 20, depending on the application concerned). These technical prototypes are to correspond as closely as possible to the final product, with regard to the material used and the final production process. For a detailed analysis of the function, the mechanical, thermal and chemical permanent rating of the product as well

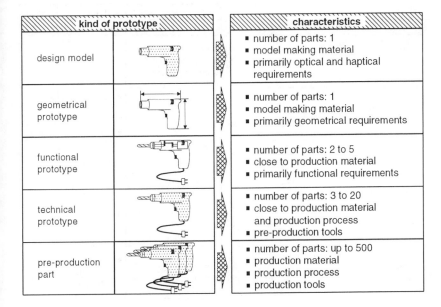

kind of prototype		characteristics
design model		■ number of parts: 1 ■ model making material ■ primarily optical and haptical requirements
geometrical prototype		■ number of parts: 1 ■ model making material ■ primarily geometrical requirements
functional prototype		■ number of parts: 2 to 5 ■ close to production material ■ primarily functional requirements
technical prototype		■ number of parts: 3 to 20 ■ close to production material and production process ■ pre-production tools
pre-production part		■ number of parts: up to 500 ■ production material ■ production process ■ production tools

Fig. 7.5 Characteristics of prototypes

as the acceptance of the clients and the producibility, deep-drawn, injection-moulded and die-cast components are produced using pre-production tools. In some cases, these products are delivered to prepared clients acting as beta-users, so that the results of these first tests may be used for optimizing the design.

Prior to launching the product on the market, up to 500 **pre-production components** are produced during the pre-production phase, depending on the branch and product concerned. These components are made out of the material to be used for series production, applying the series-production tools and processes. Pre-production components are required in the area of product planning for intensive product and market tests. Production operations are started up in this phase; the required process parameters are determined and optimized. For the last time, minimal modifications are carried out to improve the final product. Any extensive modifications to the design of product components will result in very high subsequent costs at this stage of product development.

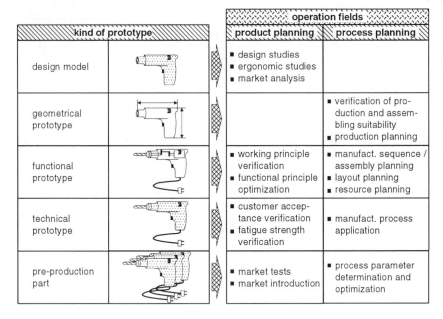

kind of prototype		operation fields	
		product planning	process planning
design model		■ design studies ■ ergonomic studies ■ market analysis	
geometrical prototype			■ verification of production and assembling suitability ■ production planning
functional prototype		■ working principle verification ■ functional principle optimization	■ manufact. sequence / assembly planning ■ layout planning ■ resource planning
technical prototype		■ customer acceptance verification ■ fatigue strength verification	■ manufact. process application
pre-production part		■ market tests ■ market introduction	■ process parameter determination and optimization

Fig. 7.6 Fields of application for prototypes

7.3 Methods of prototype production

The production of models and prototypes for the purpose of product development is currently carried out by means of conventional production processes, sometimes in combination with subsequent casting processes. The processes applied include, in particular, NC-milling, copy-milling, turning and grinding, together with manual joining and laminating processes (Fig. 7.7). The above stated methods of prototype and model construction involve high production costs, and considering the small batch sizes and the frequent modifications to the product specimen they are thus the prime reason why prototype production accounts for a large proportion of the costs and time involved in the product development process. It is not least of all on account of this situation that the production of models and prototypes is often omitted from product development processes.

The introduction of CAD/CAM technology principally offers the possibility, to produce models and specimen directly on the basis of design data. This approach is consequently pursued by new types of production processes known by the names of "Rapid Prototyping", "Desktop Manufacturing", "Solid Freeform Manufacturing", "Layer Manufacturing",

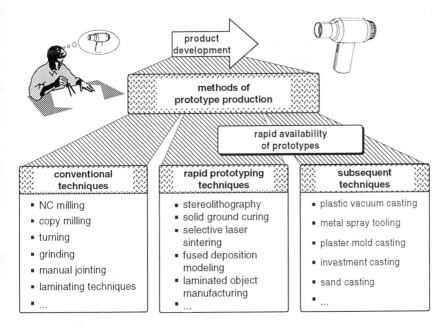

Fig. 7.7 Prototyping technologies

etc.

With these processes, component production is carried out in the briefest of periods, without any moulds or tools. A common feature of all these technologies is that the workpiece is formed not by the removal of material - as is the case with conventional cutting processes - but by adding material, or via a phase transition of a material from liquid or pulverized to solid state. A further common feature is that for the purpose of NC data generation the component geometry produced in the CAD system is first of all cut into closely adjoining planes ("slices"). On the basis of the edge contours obtained by the slicing process, the workpiece is then built up in layers in the course of the actual production process ("layer technique").

The first process of this type was developed around 6 years ago, in the form of Stereolithography (STL). Since then, additional processes, such as Solid Ground Curing (SGC), Selective Laser Sintering (SLS), Fused Deposition Modelling (FDM) and Laminated Object Manufacturing (LOM), have become available. Further processes, such as 3D-Printing (3DP) and Ballistic Particle Manufacturing (BPM), are already at advanced stages of development [11, 13, 19].

In combination with the Rapid Prototyping process, subsequent processes, in particular the conventional casting processes, the plastic vacuum casting process and the metal spray tooling process, offer interesting potential for the fast production of plastic prototypes in large quantities, or of metallic prototypes.

7.3.1 Rapid Prototyping processes and subsequent techniques

On the basis of the requirements applying to the specified categories of prototypes, the following pages present possible production processes and production sequences. The currently available Rapid Prototyping processes and new subsequent processes are presented, together with new concepts for processes which enable the direct manufacturing of metallic prototypes.

The production of a geometrical prototype for flow pattern and installation tests is illustrated by the example of a water-channel bend for an automobile engine. These prototypes, which are manufactured in serial production operations by means of the sand casting process, were subjected primarily to geometrical requirements and, to a limited degree, functional requirements, with regard to thermal stability and strength. Considering the relatively low strength requirements, it was possible to produce the prototype by means of the **Stereolithography** process, which provides the possibility to manufacture plastic prototypes (Fig. 7.8).

Using Stereolithography, the component geometry is created by hardening a photopolymer on a layer-by-layer basis with the aid of an ultraviolet laser (photopolymerization). The 3D geometry described in the CAD system by free-form surfaces is first of all approximated via triangulation and converted into a standardized format for Rapid Prototyping processes (STL format) to simplify mathematical processing. A supporting structure is then attached, which ensures subsequent detachment of the component from the support platform and supports and fixes the component in position during the construction process. For the example discussed, three hours were required to produce the supporting structure.

The STL data for the component and the supporting structure then undergoes further processing in a separate computing operation which slices the 3D geometry into individual cross-sections of defined height. The customary layer height is 0.1 to 0.2 mm. The data for the individual cutting planes is used to control an XY scanner unit, which guides the laser

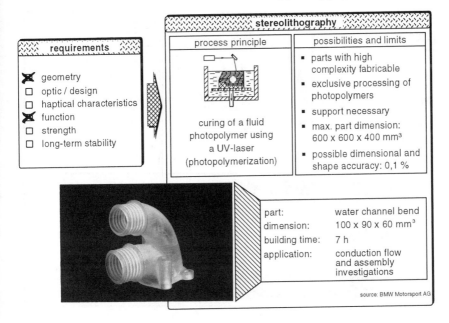

Fig. 7.8 Possible applications for Stereolithography - geometric prototypes

beam over the surface of the liquid photopolymer bath in accordance with the calculated cutting areas. The component is successively built up on a support platform which is located directly under the surface of the bath at the beginning of processing. The three-dimensional component geometry is produced via layer-by-layer hardening of the liquid photopolymer and subsequent lowering of the support platform [6, 9, 10, 11, 12, 13, 14, 15].

The actual construction process is followed by post-processes. The supporting structure must be removed and the component cleaned to remove unhardened photopolymer. The component is then completely hardened under ultraviolet light in a post-process cross-linking cabinet. Surface finishing may be required on functional surfaces and/or on surfaces subject to increased requirements with regard to external appearance.

The water-channel bend, as a possible application, which took only 7 hours to produce, clearly illustrates the possibilities of the stereolithography process. The attainable dimensional and geometrical accuracy is influenced to a substantial degree by the process control and parameters. Dimensional inaccuracies primarily result from process-induced material shrinkage and subsequent deformation of the component. The maximum level of dimensional and geometrical accuracy

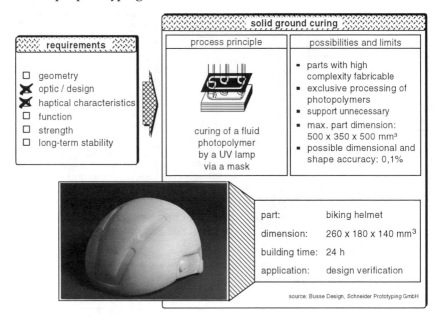

Fig. 7.9 Possible applications for the solid ground curing process - design models

which is currently attainable is around 0.1% of the component size.

Stereolithography is suitable only for processing photopolymers (acrylic, vinyl and epoxy resins) with varying material properties.

For many products, the design is crucial to subsequent success on the market, so that the possbilities for design concerning the desired function of the product is strongly influenced by the specifications for the outer appearance. At the beginning of the development process, models are therefore required for design and ergonomic studies as well as market analyses. Thus, the visual and haptic characteristics during the manufacturing of such models are in the foreground. One possible alternative to milling the model - illustrated here using the example of a bicycle helmet - is the **Solid Ground Curing** process, which is related to the stereolithography process. The advantages of this process over milling are the simpler generation process for the NC data and the possibility to produce the component without any retooling operations (Fig. 7.9).

The solid ground curing process is also derived from the principle of photopolymerization. In contrast to Stereolithography, which involves exposing the surface of a layer to light on a point-by-point basis using a laser, solid ground curing is based on an exposure of the entire surface via

via a mask using an ultraviolet lamp. On the basis of the description of the component geometry in STL format, the component geometry is built up in two separate cycles. A negative mask is first of all produced in an ionographic process, and this mask serves as the lithographic structure for the exposure process. At the same time, a thin film of liquid photopolymer is applied to a bearing plate for actual build-up of the component geometry. After exposure, the non-hardened photopolymer is drawn off and replaced by liquid wax, which is face-milled to a defined layer thickness of normally 0.15 mm after cooling. The construction cycle then begins anew, with application of the liquid photopolymer and the production of a new mask [11, 13, 17).

Using this process, several components (generally between 5 and 10, depending on the size) can be produced in a cube of 500 x 350 x 500 mm3. Embedding the components in wax eliminates the need for the supporting structure which is characteristic of the stereolithography process. Separate post-hardening is also unnecessary, as the mask exposure process ensures complete hardening of the photopolymer. The components are released from the wax block in a subsequent cleaning process, using citric acid. Components produced by means of the solid ground curing process also possess a dimensional and geometrical accuracy of approx. 0.1% of the component size. Solid ground curing is suitable for processing photopolymers only. The total production time for the bicycle helmet shown in Fig. 7.9 came to 32 hours: 24 hours of construction time, 5 hours of cleaning and 3 hours of manual post-processing (surface finishing).

The use of geometrical prototypes is generally sufficient for the purpose of limited functional tests which do not involve any special requirements with regard to the materials. Geometrical prototypes are also suitable, for example, for testing the installation characteristics and the draw-in mechanism during development of a code card reading device, when various prototype cases are required for installation of the internal components. Conventional prototype production for the complex case cover incorporating intricate structures (Fig. 7.10) would have taken several days and would have restricted the number of iterative cycles to be carried out for optimization because of both time- and cost-related reasons. With the **Laminated Object Manufacturing** technology, on the other hand, involving a production time of only 10 hours per part, it was possible to carry out all the necessary variant tests.

With this process, the component geometry is created by cementing individual paper foils on top of each other and then cutting these foils along the contour lines with the aid of a laser. The control data for the laser

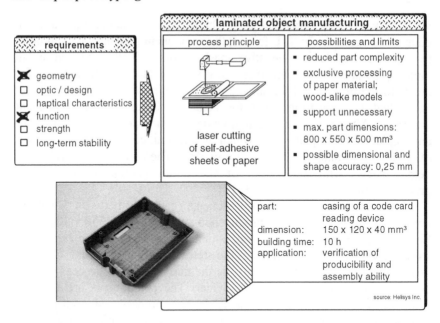

Fig. 7.10 Possible applications using laminated object manufacturing - geometric prototypes

is calculated for each layer on the basis of the 3D CAD design data provided in STL format. The individual foils are deposited on a support platform movable in vertical direction and subsequently compressed by a roller. The laser moves along the contour lines of the component in accordance with the generated control data and cuts out the geometry of the current layer. Exact focussing of the laser beam and control of the laser power ensure that in each operation only the last layer is cut out. The areas which are not part of the workpiece are cut into rectangles to facilitate removal at a later stage. Through cementing the individual paper layers together one on top of the other a wood-like, three-dimensional model is produced.

After completion of the component geometry the areas which do not belong to the workpiece must be removed and manual finishing of the part is carried out in accordance with the surface requirements. Paper foils with material thicknesses between 0.05 and 0.5 mm can be processed using the laminated object manufacturing method. The attainable levels of dimensional and geometrical accuracy are in the 0.25 mm range [11, 13].

In contrast to the previous examples, the development of components

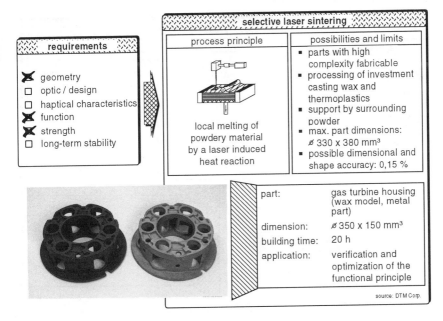

Fig. 7.11 Applications for selective laser sintering - technical prototypes

subject to thermal and mechanical strain requires functional prototypes to be produced out of materials similar to those to be employed for serial production already during early development phases, to provide the possibility of testing performance characteristics. The increased stability requirements exclude the utilization of model or substitute materials for the production of the prototype in such cases. Thus, for the example application of a gas turbine case presented below, the attention was primarily focused on the functional aspects, since the performance characteristics of the turbine were to be tested. The **Selective Laser Sintering** process shown in Fig. 7.11 was used for the production of a wax model of the turbine case within 20 hours. On the basis of this wax model, it was possible to produce the metal turbine case by means of conventional investment casting processes (model melting processes).

The principle of the selective laser sintering process is based on the local welding or melting of powder-form materials under the effect of laser-induced heat. As in the case of the previous processes, the control data for the CO_2 laser is generated directly on the basis of the 3D CAD geometry. The starting material is applied to a support platform on a layer-by-layer basis under an inert atmosphere, with the aid of a levelling roller.

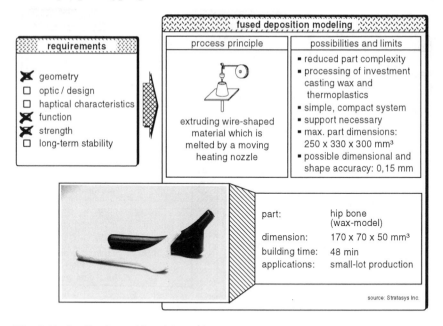

Fig. 7.12 Applications of fused deposition modeling for small lot sizes

The powder is preheated to a temperature just below it´s melting point, using infrared light. The laser beam, which is controlled via a scanner unit, sinters/melts the powder at the areas belonging to the component structure. The surrounding powder supports the component. The supporting structure typical of the stereolithography process is therefore unnecessary here. The component geometry is created layer-wise, by continuously lowering the support platform [16, 17, 18, 19, 20].

After completing the final layer, the component is removed from the working area and undergoes post-process treatment in accordance with the requirements of the specific application concerned. The non-melted powder can be used for future construction processes. The manufacturing time for the case shown in Fig. 7.11 was 20 hours. The attainable dimensional and geometrical accuracy stood at around 0.15% of the component dimensions. Principally, all powder-form materials which can be melted or softened by heat are suitable for use with the selective laser sintering process. At present, PVC, polycarbonates, ABS/SAN, polyamide and mould wax are employed. A significant advantage of the selective laser sintering method is the ability to process wax materials in order to produce premasters for the investment casting process.

The example from the field of medical equipment shown below in Fig. 7.12 illustrates the possibilities provided by the combined application of Rapid Prototyping processes and subsequent casting processes to produce functional components and also presents the special method of 3D data generation. On the basis of a hip joint bone (shown at the front in the photograph in Fig. 7.12), the geometry was first of all digitized and ported into a CAD system. The preprocessed 3D data record for the component geometry formed the input information for the **Fused Deposition Modeling** process, with which a three-dimensional wax model of the hip joint bone (shown at the rear in the photograph in Fig. 7.12) was produced within 48 minutes.

Using the fused deposition modeling process, the geometry is built up by extruding and layering a wire-shaped material which is melted by a moving heating nozzle. The raw material, which is wound on a spool, is fed to the heating nozzle, which is controlled by a plotter mechanism, and heated to a temperature just above the melting point. The melted material is then extruded onto the support platform or the previously produced layer, whereby the gap between the tip of the nozzle and the substrate causes a flattening of the material's round cross-section. The layer thickness ranges between 0.025 mm and 1.25 mm, depending on the type of application concerned, and the minimal wall thickness is 0.22 mm. After completion of a layer, the support platform is lowered by the set layer thickness and the next layer is applied. Supporting structures made of cardboard, polystyrene or similar materials may be required to support projecting parts of the component [13, 17, 20].

The dimensional and geometrical accuracy attainable with this process is approximately 0.15 mm; suitable materials for processing are thermoplastics and mould waxes for the production of components for the investment casting process. In the application presented here, the fused deposition modelling wax model was employed to produce a cast wax titanium hip joint prosthesis.

Technical prototypes are subject to high requirements with regard to stress-resistance and long-term stability. In order to obtain reliable information on the stress-resistance of the series-manufactured product, the technical prototypes must be as similar as possible to the final product with regard to the material employed and the applied production processes. Hollow steel moulds are currently produced in industrial practice for the production of technical plastic prototypes or small series of 20 to 50 units; the production of these moulds usually requires 4 to 8 weeks.

The use of rapid prototyping processes in combination with

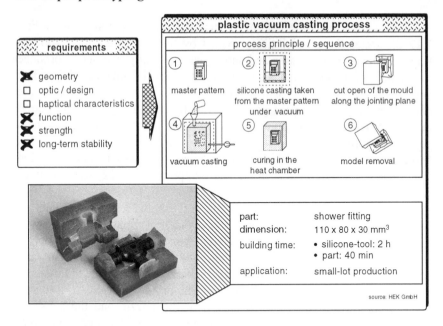

Fig. 7.13 Possible applications for plastic vacuum casting - technical prototypes

subsequent casting processes provides substantial potential for reducing product development times also for this type of application. Such a production sequence is illustrated below by reference to the example of a shower fitting (Fig. 7.13). First of all, a model of the shower fitting was produced using the stereolithography process; the production time for this model was 3 hours. The resulting component then served as the master model for the **Plastic Vacuum Casting** process, with which the original was duplicated in accordance with the required number of copies.The initial measure required for the plastic vacuum casting process is to provide the master model with sprues and air gates. The model is then fixed in a rectangular moulding box and cast in silicone rubber in a vacuum chamber. After hardening in a heating chamber, the silicone mould is cut open along the parting plane and the master model is removed. For the subsequent component production process, the mould is joined together and filled under vacuum. The range of 2 component resins suitable for filling the mould is highly diverse with regard to the mechanical properties and the colour of the material. After completion of the casting process, the mould is removed from the casting chamber and cured in a heating chamber; post-processing finishes the production process of the component

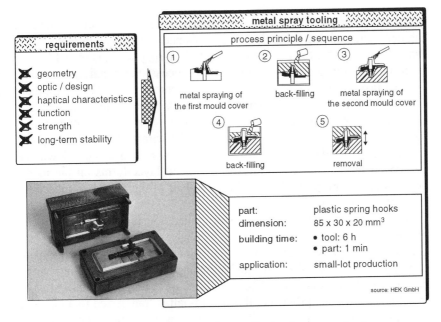

Fig. 7.14 Applications of metal spray tooling application for pre-production part

[17, 18].

This process makes a very close reproduction of the master model possible. Production of the intricate areas of the component incorporating undercuts presents no problems whatsoever, due to the simple mould removal process (elastic moulds). The production time for the silicone die was 2 hours, and a further 40 minutes have to be allowed for the production of each component. It was thus possible to produce 20 shower fittings in only 3 days, on the basis of the CAD design data. By selecting an appropriate material, the material properties of the final series component were approximated sufficiently so that the prototypes could be used for testing stress-resistance and long-term stability. Whether the application of such a production sequence provides a suitable means of acquiring reliable data on the final series-produced component in other cases depends to a decisive degree on the influence which the series production process has on the development of the required product characteristics. Similar restrictions also apply to the following example.

When larger quantities (50 to 1000) of technical prototypes, pre-production series or small batches of workpieces are required in plastic, the **Metal Spray Tooling** method may be applied to manufacture experimental or production dies. The principle behind the process and the characteristics of this technology are explained below by reference to the example of a die for the production of plastic spring hooks (Fig. 7.14).

Similar to the plastic vacuum casting process, the metal spray tooling process first of all requires a master model, onto which a layer of a metal alloy with a low melting point is applied using a metal spraying gun, similarly to the spray-painting process. The application of voltage to two metal wires produces an arc, which melts the wire material. The liquified material is diffused into fine particles using compressed air and blown onto the master model. After solidification, a metal layer forms on the model, and this layer can be used to produce one half of the mould in subsequent steps of the process.

Production of the die begins with embedding the master model in a plasticine compound along the previously stipulated parting plane. Undercuts are to be taken into account by appropriate partitioning of the mould and the use of inserts and drawing dies. After spraying with a parting agent, the first half of the mould is sprayed with metal. After the metal layer has been produced, this mould-half is transferred to a moulding box and lined with a low-melting alloy or a synthetic resin compound containing aluminium chips. The moulding box is then turned over, the embedding compound is removed and the second half of the mould is produced according to the described procedure. The use of standard moulding boxes makes it possible to join the two halves of the mould with a perfect fit [17, 18].

Typical materials employed for the metal spray tooling process are alloys consisting of bismuth, tin and zinc. In principle, other materials, such as steel, aluminium, bronze and copper, may also be used for this process, but they are rarely suitable for producing dies, on account of the higher melting temperatures and the resultin deformation. The attainable levels of dimensional and geometrical accuracy correspond to those of the master model. Restrictions to the scope of applications for this technology result firstly from the geometric characteristics of the model, e. g. grooves with a width-to-depth ratio of less than 1 to 5 cannot be produced via this process, due to the restricted accessibility and the danger of beading. Secondly, the thermal characteristics of the metal-sprayed dies differ from those of conventional materials for injection moulding dies, resulting in different geometrical and mechanical characteristics on the component.

The costs for production of the injection moulding die shown in Fig. 7.14 utilizing the metal spray tooling process were at around DM 1,500; the production time was 6 hours.

7.3.2 New approaches to the direct manufacturing of metallic parts

Due to the limited mechanical, thermal and chemical properties of the materials which are used in connection with the rapid prototyping processes currently employed in industry (polymers, waxes, nylon, paper, etc.), the components which are produced serve solely as visual models, sample parts, or can be employed as premasters for subsequent process sequences. The comprehensive functional testing of components which is often required by users is thus possible only to a limited extent with the prototypes which are produced in this manner.

In order to attain a higher degree of correspondence between the prototype and the subsequent series-produced component, the materials employed in production of the prototype should be as similar as possible to those employed in series production. Based on rapid prototyping models, the production of prototypes in materials similar to those to be employed in series production is presently limited to subsequent casting processes. One of the goals of the continued development of rapid prototyping technologies is the manufacturing of prototypes and prototype tools with enhanced material properties. Efforts are focused on metallic and ceramic workpieces as these can currently only be manufactured using conventional manufacturing methods which restrict their geometrical characteristics. For this reason, intensive work is currently in progress on the development of existing and new rapid prototyping processes which are to provide the possibility to directly produce metallic component and tool geometries.

For example, experiments were conducted for the laser sintering of metallic and ceramic materials. To date, however, these tests were limited to experiments with polymer-enclosed ceramic and metallic powders. Laser sintering of these powders only melts the polymer around the core, which serves as a binding agent for the subsequent sintering process. The component acquires its final strength during subsequent hardening in a sintering furnace. In contrast to this procedure, ceramic and metallic components will be directly producible in the future, employing the technology of high-temperature laser sintering and reactive laser sintering [20, 21, 22, 23].

A technology related to laser sintering is laser generation; Fig. 7.15 shows the principle behind this process and machining examples. In principle, laser generation is similar to a coating process, in which a powder-form filler material is melted with the aid of a laser beam and bonded with the substrate. Three-dimensional structures are built up by placing individual layers next to and on top of one another, on a layer-by-

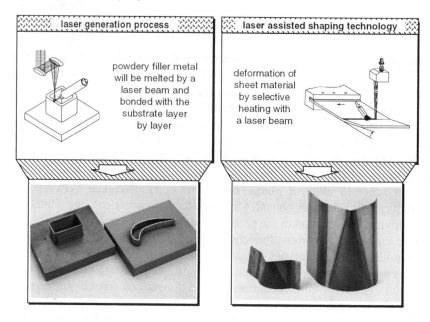

Fig. 7.15 Prototyping technologies for direct manufacturing of metallic components
(based on: IPT)

layer basis. The use of a laser beam provides a good basis for the reproducible production of intricate structures, as the dimensions of the focal point can be precisely set and the output intensity may be dosed very reliably.

With regard to the direct production of metallic components, work is currently in progress on processes known as "3D Welding" and "Shape Welding", which operate according to the principle of laser generation. In these processes, a welding electrode is guided by a robot on a plane-by-plane basis, relative to that part of the workpiece which has already been produced, thereby generating the workpiece geometry in a build-up welding process. Since the attainable levels of dimensional and geometrical accuracy for this process are on a scale of +/- 1 mm, the produced components undergo subsequent turning or milling [17]. The scope of programming required for both the laser generation and the build-up welding process is substantially higher than that required by the available rapid prototyping technologies.

For prototype production or the production of individual components or small batches of components in sheet metal, laser-beam bending provides an interesting alternative to the customary processes of bending with an elastic mould, shot-peen forming and precision forging (Fig. 7.15). Laser-beam bending involves the specific inducement of thermal stress in the surface layer of the component via partial heating with a laser beam and subsequent cooling. The induced stress exceeds the yield point of the material and consequently causes defined, permanent deformation [24, 25].

7.4 Prerequisites and procedures for the implementation of rapid prototyping

Fig. 7.16 summarizes the conditions which need to be fulfilled to render the industrial application of rapid prototyping feasible and efficient. At present, the economical application of rapid prototyping processes is possible only for products for which time is a critical factor and for complex geometries, on account of the high investment and operating costs which are involved. Furthermore, it should be noted that without appropriate integration into the flow of operational data and information and without 3D CAD design, much of the utilizable potential will be consumed by additional expenditure, thereby casting doubt on the economical application of these technologies. The introduction of simultaneous engineering structures, planning procedures adapted to these structures and an appropriate system of management for decision-making processes are further prerequisites for the successful introduction of Rapid Prototyping.

Thus, the definition of a systematic procedure for the introduction of rapid prototyping is an important prerequisite for economic success. Such a procedure divides the various stages involved into an analytical phase, a technological planning phase, an integrational planning phase, an introductory phase and a utilization phase (Fig. 7.17).

The first phase covers an analysis of the product range, the development process and the requirements related to the prototypes in the course of product development. The results are the basis for the next phase to establish the requirements for the technologies which are to be applied. An analysis of the situation regarding data flow and technology during this phase provides further information on accompanying measures which are required for the introduction of this technology in the area of organization.

The fast development of rapid prototyping technology and the lack of knowledge with regard to it's potential and limitations renders selection of the appropriate technology and assessment operations more difficult. To

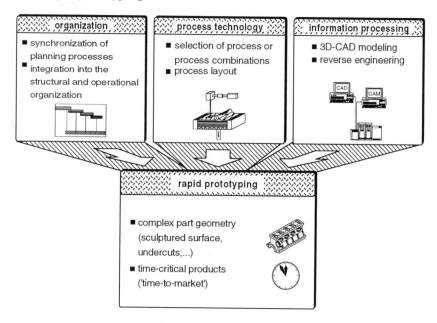

Fig. 7.16 Prerequisites for the efficient application of rapid prototyping

make a full exploitation of the available potential possible, this technological information must be systematically recorded and structured for a detailed evaluation. The comparison of prototype requirements against process capabilities provides the basis for a selection of the appropriate technology. In the course of this selection process, the individual processes must not be considered as isolated solutions; rather, possible combinations of processes must be incorporated into the selection procedure.After establishing the suitable technologies, the make-or-buy decision must be reached in the course of a plant and investment planning procedure. The difficult aspect of this planning phase is without doubt the correct estimation and assessment of the attainable time reductions and quality advantages over conventional production operations. Since the currently available knowledge is mostly based on isolated, uncomprehensive experience, this problem may easily lead to misinterpretations on account of the missing information concerning the boundary conditions when applying Rapid Prototyping technologies.

The aim of integration planning is to ensure the optimum integration of rapid prototyping into the company's organizational and operational structure. In accordance with the required kinds of prototypes and the

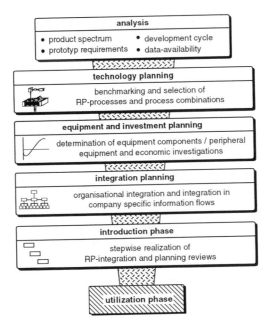

Fig. 7.17 Procedure for the introduction of rapid prototyping

individual demands, the possible methods of organizational integration have to be examined. Incorporation into the design stage, the specimen construction stage or the tool construction stage, or the establishment of an independent rapid prototyping department would be some of the intelligent integrations. The corporate development and planning procedures are to be adapted accordingly. The data technology required for data generation and data transfer needs to be provided.

A step-by-step concept should be developed for the introduction of rapid prototyping, to enable successive utilization of the potential offered by rapid prototyping and adaption of the organizational structure to the requirements of rapid prototyping technology at specified stages. At companies whose design operations have already been switched to 3D CAD and whose organizational system incorporates parallel structures within the development procedures, the integration of rapid prototyping is a simple matter and can thus be carried out more swiftly. Equally, such companies are able to implement and make profitable use of the advantages which the fast production of models and specimen provides with regard to reduced development times and product optimization.

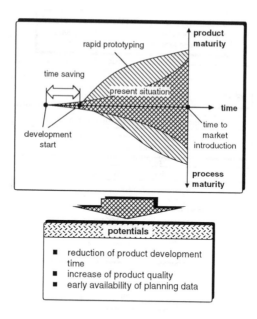

Fig. 7.18 Potentials of rapid prototyping

7.5 Conclusion

The application of rapid prototyping offers the possibility to reduce the time required to produce prototypes and has additional extensive strategic effects.

The clearance of a product for development can take place at a later juncture through the application of an accelerated prototype production process, which means that market developments and customer requirements may be converted more quickly into products, or a more quick and direct reaction to changing market requirements becomes possible. The fast availability of models and prototypes results in an advanced level of product development in early phases of the development process, leading to improved availability of planning data for production operations. As a result of this intensification of early development phases, modification costs are lower and the products are fully developed when launched onto the market (Fig. 7.18).

Therefore, rapid prototyping as a method of reducing product development times consists of substantially more than the application of the appropriate technologies. rapid prototyping is an important element for use in synchronizing parallel product and process design phases, but does not eliminate the need for management discipline in the area of decision-making.

With regard to the future of rapid prototyping, it is to be expected that new materials, further developments in systems engineering and an advanced information technology will substantially expand the potential and scope of the processes. In the future, it will be possible to employ rapid prototyping processes for the production of complex functional components as individual items and in small batches, in addition to the applications in the field of prototype production. The virtually unlimited complexity of the producible geometries provide the engineer with new possibilities for designing components which could not be produced using the conventional production processes. rapid prototyping will therefore become a firmly established element of tomorrow's production engineering, not only in the area of product development.

Part Three

PRODUCTION: RESOURCE-ORIENTED ORGANISATION OF PRODUCTION PROCESSES

8   Resource-oriented design of production systems
9   The key of efficient process and manufacturing sequences - technological development and innovation
10 Superior tools as a basis for tomorrow`s manufacturing
11 Integrated quality inspections

8

RESOURCE-ORIENTED DESIGN OF PRODUCTION SYSTEMS

8.1 Introduction and definition of the problem area

The export-oriented branches of German industry are currently suffering considerably as a result of the economic weakness of the traditional export markets. The economic depression has led to the lowest level of incoming orders for over 10 years [1]. The machine tools industry, which holds a share of over 20% of the world market, is the most severely affected branch. It is forecast that this branch will not participate in the worldwide recovery which is expected for 1994. The order situation is also very poor in the automobile industry, resulting, among other things, in extensive redundancies [2].

This means that the requirements facing companies, which were already high, are now further intensified by national and international competition (Fig. 8.1).

The customers place orders which must be completed within very short delivery times, while at the same time requiring very low prices. Political changes, such as the opening-up of Eastern Europe to free enterprize, or changing social values, also have a decisive influence on the situation in which companies find themselves.

The majority of the requirements facing companies and the resultant

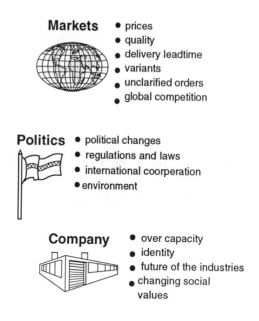

Markets
- prices
- quality
- delivery leadtime
- variants
- unclarified orders
- global competition

Politics
- political changes
- regulations and laws
- international coorperation
- environment

Company
- over capacity
- identity
- future of the industries
- changing social values

Fig. 8.1 Demands on companies

problems are not new, but the approaches to solving these problems which were adopted in the past failed to attain the objective of adapted organizational structures at the companies. The challenge now, therefore, is to induce the companies to adopt a new approach, by means of which the necessary basis can be established for successful operations in the various segments of the market. To this end, new concepts and tools are required.

The effects of the current situation are particularly severe after the years of economic boom. The strong increase in turnover was accompanied by the powerful growth of companies. No adaption of the organizational structures was carried out during this phase, however [3]. This results in the following situation for companies' production systems.

Customers' requirements for special technical configurations have led to increased complexity in the areas of operations scheduling and production structure, both for one-off and small batch production and for the series production of complex products (Fig. 8.2) [4]. In this context, the term "production structure" refers to the type, quantity and organisation of the operational equipment. For small batch production, the sequencing is

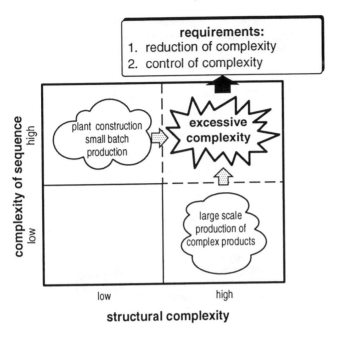

Fig. 8.2 Complexity portfolio for production operations

naturally highly complex, as each order involves specific requirements imposed by the customer concerned, affecting virtually all areas of the company. A trend towards growing structural complexity is to be observed here, as the customers' requirements can no longer be satisfied with conventional facilities. Requirements for improved quality and complex technical configurations lead to complex production and assembly facilities and subsequently impose higher demands on the organizational basis [5, 6]. The large-scale production of complex products, such as applies in the automobile industry, for example, naturally involves a high degree of structural complexity, as a certain level of automation is essential for economical production on a large scale. The necessary technical systems are highly complex and require a very extensive scope of control facilities. At the same time, the complexity of the production sequences is increasing, as the scope of variants to be produced has broadened substantially in recent years [3]. This has resulted in a substantial increase in the complexity of operations scheduling, both for product development and for the production process.

This course of development has thus placed the companies of virtually

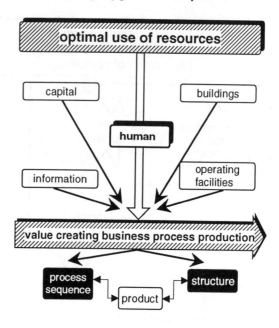

Fig. 8.3 Objectives of production system design

all branches in a situation in which an excessive level of complexity in the execution of orders renders the economical production of customer orders virtually impossible [8]. The present state of the economy further aggravates the situation, as the production systems which are currently in operation are unable to satisfy customer requirements. The objective to be pursued in the production design must therefore be as follows:

In order to acquire control over the existing complexity, it must be reduced via appropriate design and organizational measures!

8.2 Primary objectives

Diverse objectives regarding the production design can be derived from the requirement for reduced complexity. It is of decisive importance, however, that the configurational and organizational measures to be implemented should relate to the overall business process production, in order to attain an inter-disciplinary effect for the company. The optimization of individual areas or the solving of local problems no longer represent appropriate

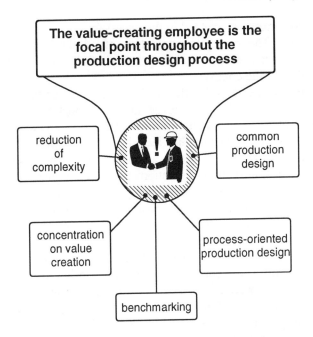

Fig. 8.4 Core elements of resource-oriented production system design

approaches. The resource-oriented production design must therefore pursue the following objectives (Fig. 8.3):
- production operations must be organized in such a manner as to ensure the optimal use of company resources and
- the organizational measures must be aimed at increased value creation.

Originally, priority was attached to optimal utilization of the invested capital and the operating facilities. This led to the local optimization of individual areas and ran counter to the overall optimum of an integral system of organization for production operations.

Efforts to implement integral EDP-aided information modelling systems in the course of CIM projects have also failed to produce the expected success [7]. In future, prime importance must be attached to the most flexible resource - humans. Effective staff involvement, assigning responsible areas of work to appropriate members of staff, offers companies the greatest potential for adapting quickly to the dynamic requirements of the market. In order to attain this objective, the process sequences involved in the production process must be optimized with regard to their value-creating potential, and human resources, above all the operative staff, must be involved in the necessary configuration process at an early stage. The complete integration of all those involved in the production process is essential, if the potential of all available resources is to be utilizable by the company.

In this paper, the term "production" is employed for those areas of a company which are directly involved in the production process. For a series manufacturer with a non-customer-related development and design process, the area of production thus includes the distribution, manufacture, assembly, logistics and shipping. For a company producing one-off batches and small batches, design and operations planning are also to be included in the area of production, as the major part of an order requires to be designed in accordance with the customer's specifications, before the product can be manufactured. As the indirectly involved areas of a company harbour particular potential for the configuration of a resource-oriented production system, these should also be included in any assessment of this problem area [8, 9]. The design parameters for a resource-oriented production system are the product, operational organization and the structure of production operations, based on the type and scope of available operating facilities. This paper is concerned with examining the configuration parameters "operational organization" and "production structure".

Shown below are the core elements of the resource-oriented configuration of production systems which are necessary in order to attain the above-stated objectives when carrying out production design measures (Fig. 8.4).

It is not intended here to consider the complete scope of all possible production design measures, but rather to concentrate on the essential measures which are required to attain the above-stated objectives. The main areas to be considered are staff integration, process orientation within the context of production design, the assessment of process sequences via indices, the increase in value creation and the reduction of complexity.

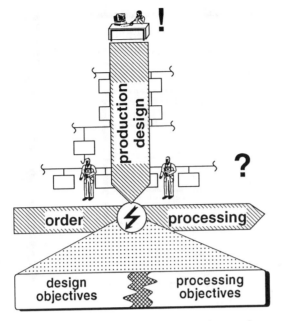

Fig. 8.5 Problems involved in production design procedures today

8.3 Elements of resource-oriented production design

8.3.1 Common production design

An essential precondition for the purposes of production configuration is staff integration, i.e. planning and operative staff must carry out the planning process in cooperation from the outset. Examination of the procedure which is commonly adopted today for production configuration measures reveals a substantial discrepancy between the objectives of the design process and the objectives pursued in the execution of customers' orders (Fig. 8.5).

The planning objectives are stipulated by the company management. Studies reveal the most commonly specified objectives here to be reduced leadtimes and reduced costs [10]. The selected measures by which to pursue these objectives are implemented at the company via the hierarchic levels extending down to the clerk or manual worker. As a result, the measures to be implemented often fail to obtain complete acceptance at the operative level of the hierarchic structure, as the aims of the operative

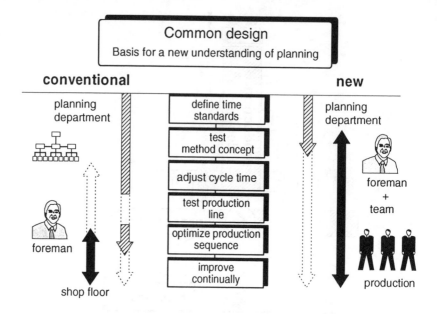

Fig. 8.6 Common production system design - example (based on Volkswagen)

staff, such as job content or the scope of responsibility assigned to operative employees, are not fulfilled. This jeopardizes the success of a measure. The discrepancy between process-oriented objectives and operative objectives in the production structure must be overcome by involving the operative staff in the production design process [11].

In industry today, planning teams are comprised of planning and operative staff (Fig. 8.6). Whereas the foremen and operative staff were previously involved only in the final phase of the planning process, these employees now actively participate right from the early planning stages.

The advantages of this new approach lie not only in the motivation of the operative staff, but also in the early utilization of the detailed knowledge which these staff have acquired in the course of their activities at the company. This enables the prevention of planning changes due to inadequate feasibility assessment.

A further aspect to be observed in establishing planning teams involves horizontal integration of the various company departments (Fig. 8.7). Prior to commencing production design measures, staff from all the departments concerned are informed of the task on hand at a joint meeting. This ensures that any conflicting aims can be reconciled at an early stage.

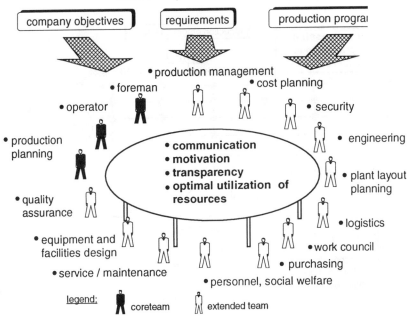

Fig. 8.7 Project team for establishing islands - example (based on MTU)

Actual implementation of the design measures involves only a so-called core team, however, in order to prevent the implementation process from being slowed down by coordination problems. The respective members of the extended team are called in to assist in appropriate areas as required.

The integration of planning and operative staff from all the company departments concerned means that organizational and technical competence are incorporated into the configuration process at an early phase of planning. This establishes the necessary basis, among other things, for the process-oriented configuration of a production process.

8.3.2 Process-oriented configuration of production systems

The objective pursued in establishing resource-oriented production operations, namely, the faster and more cost-effective processing of orders, requires companies to adopt a completely different approach to production systems. Design measures must no longer be based on internal functions at the company. Rather, the requirement for reduced throughput times means that the measures must be geared towards orders and the business

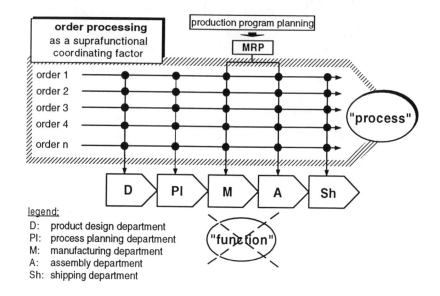

Fig. 8.8 Process-oriented production

processes which are required to process orders (Fig. 8.8) [3, 9].

By linking the deployment of resources (e.g. man, production equipment, EDP) with he process, the process-oriented approach to production configuration measures enables the specific optimization of resources and assessment of the overall potential of the measures to be undertaken. This is an important starting point to enable integration of the planning processes for operational organization and the production structure.

This means that an analysis of the business processes employed at the company and the corresponding deployment of resources must first of all be carried out in the course of production configuration measures. This is the only way of tapping the sometimes substantial levels of potential lost due to weak points, e.g. in the flow of information for the order process [12].

In the light of this situation, a process- and element-oriented model has been developed at the WZL of the Technical University of Aachen (RWTH). The appurtenant descriptive language is composed of 14 elements, the so-called process elements. This enables all the business processes employed at a company to be displayed and made transparent

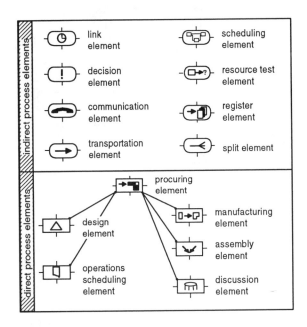

Fig. 8.9 Descriptive language for process analysis

(Fig. 8.9). A distinction is made between direct and indirect process elements. The direct elements describe business processes which contribute directly towards value creation, such as manufacture or assembly operations or the production of a drawing.

The indirect elements represent business processes which either contribute only indirectly to value creation (e.g. transportation, recording) or which consume value (e.g. storage).

This tool and the establishment of a team prior to commencing the design process provide the necessary basis on which a process analysis procedure can be carried out at the company. On the basis of interviews, the company processes are first of all represented by the elements in a so-called process sequence plan. The appropriate resources must then be assigned to each process. The order-related consumption of resources can then be determined by establishing the throughput time per process. In this way, the know-how of the operative staff with regard to operational procedures, the existing structures and the process-related resources can be made transparent (Fig. 8.10). The process sequence plan is the basis for providing all those involved in the configuration process with a uniform

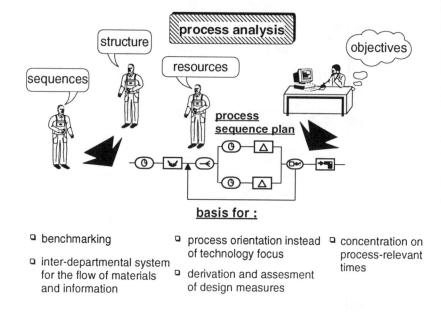

Fig. 8.10 Process analysis - The basis for resource-oriented production design

picture of existing process sequences. This is also referred to as providing a common mental picture of the operational procedures employed at the company.

On the basis of the process sequence plan, the planning staff can introduce their planning objectives, in order to identify the processes or process sequences which require reorganization in order to attain the specified objectives. The appropriate measures are planned together with the staff concerned and the attainable potential assessed.

The results are incorporated into the process sequence plan, to enable them to be assessed. On the basis of this assessment it is possible to determine the degree of success attained via the individual measures, with regard to reductions in the overall throughput time or the lowering of order costs, for example. Only such measures as reveal a measurable degree of success in the context of overall assessment of the system are to be implemented [13].

The recording of detailed data throughout the entire process sequence, such as the product manufacturing sequence, enables the planning team to carry out integrated structural and organizational planning. The most important items of data required for this purpose are shown in the process

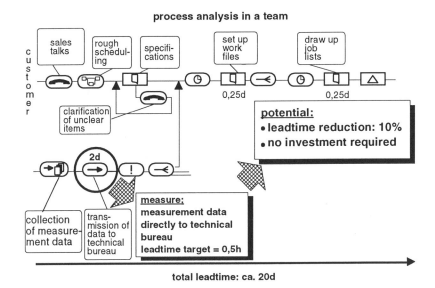

Fig. 8.11 Production design via process analysis - example (based on Krone)

sequence plan. For the purposes of planning a production island, for example, the appropriate production processes can be established and integrated to form an autonomous unit. This enables the attainable potential of such an island and the effects on the overall throughput time for orders to be estimated prior to commencing configuration measures.

Further operations, such as the establishment of group technology or detailed planning, can be continuously referred to the process sequence plan. After completing the planning stage, highly detailed assessment of the attainable order-related lowering of costs or reduction of the throughput time is possible.

The general application of this method in practice often very quickly reveals potential for rationalization, which can sometimes be tapped without any need for high levels of investment. The analysis of a certain process sequence by an motor industry supplier, for example, revealed potential for a 10% reduction in the throughput time, which was utilized by minor organizational changes and without any investment (Fig. 8.11). The analysis work and the established measures were jointly carried out by management personnel and operative staff.On the basis of a process analysis stage, the acquired data, such as process-related throughput times

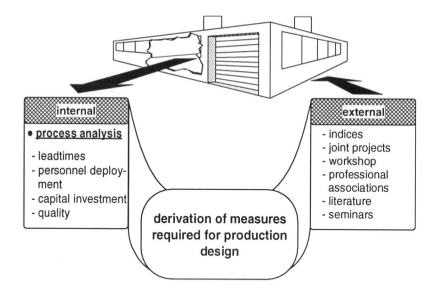

Fig. 8.12 Positioning - process-oriented benchmarking

or resource consumption, can be aggregated into indices. These indices serve as yardsticks for comparative assessment of the processes under review. The data obtained in this way provides the basis for the benchmarking process, by means of which the company's efficiency is compared against that of other companies.

8.3.3 Benchmarking

With the aid of the established indices, the company is able to determine its position in relation to its competitors (benchmarking) [14, 15]. The comparison is carried out with the company's own branch or with a different branch. To this end, the appropriate literature is evaluated and the primary areas requiring improvement measures are established (Fig. 8.12).

As an example of such a positioning process, the efficiency with which orders are executed was investigated at a company with small batch production of complex products with the aid of process-oriented benchmarking, and compared with the efficiency of an automobile manufacturer (Fig. 8.13).

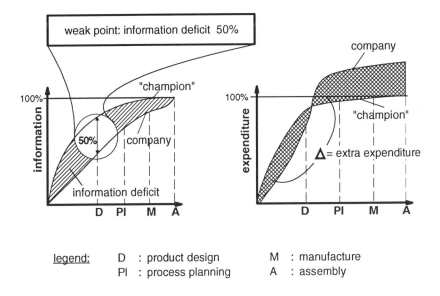

legend: D : product design   M : manufacture
         PI : process planning   A : assembly

Fig. 8.13 Process-oriented benchmarking - indices for establishing information deficits

The significant difference between the two branches is that the automobile manufacturer effects a substantially higher level of investment in product development, to ensure thorough preliminary planning of all the areas of work involved in subsequent production operations. The first company, on the other hand, is almost always confronted with orders which are not fully clarified, i.e. in the course of the production process the customer frequently submits requirements for modifications which must be taken into account. That company is thus required to cope with a latent information deficit in the course of executing the order; in the example shown, the company possesses only 50 % of the information which is available to the automobile manufacturer at this time. As a result of this information deficit, the scope of planning which the company is able to carry out at this stage of planning is less than the scope available to the automobile manufacturer, and planning is consequently carried at a lower level of detail. Due to this lack of specific detail, the modifications required at a late stage in the production process result in a substantial increase in the cost of manufacturing the product.

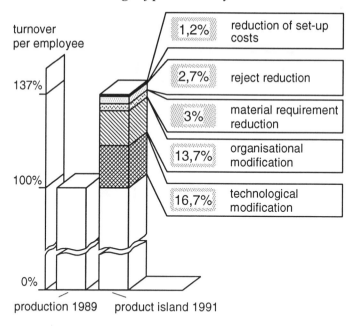

Fig. 8.14 Infices for thr product island "hydraulic engine bearing" - example (based on Boge AG)

For the plant construction company, this position imposes a need to reconfigure its own order processing operations in such a manner as to minimize the possibility of extensive information deficits towards the end of the product development process.

Not only in the indirect areas indices are suitable for the purpose of establishing a company's competitive position. They can also be used in the areas of manufacture and assembly as indications of possible weak points in the production process. An example of such an application is provided by production indices which are determined in connection with the restructuring of a previously function-oriented workshop structure into product islands (Fig. 8.14). All the production processes which are of relevance to the product concerned are brought together in these product islands [16]; the island operates autonomously. This restructuring measure enables substantial potential to be tapped. Quantifiable savings were attained, such as a strong increase in the turnover per employee, a marked reduction in the number of set-up operations required per month, in the volume of tied-up capital and in reject costs. At the same time, quantifiable advantages were also attained, such as a reduction in the order throughput

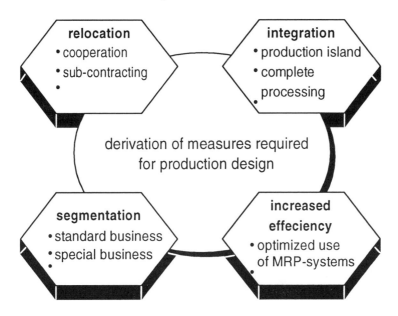

Fig. 8.15 Derivation of strategic measures

time and an increase in the maximum available capacity. The production indices determined by this process enable other companies to assess the potential of such restructuring measures, after establishing their own positions. On the basis of the benchmarking process, the need for measures to modify the processes, which can be displayed with the aid of the process sequence plan, can be determined (see Section 8.3.2). To eliminate the deficits, the production processes, i.e. the process sequences and the process elements, are reconfigured.

The process plan is employed here both to determine the required measures (interpretation) and to document these measures. Four basic measures can be undertaken for the purposes of production configuration (Fig. 8.15). The common characteristic of all these measures is general optimization of the company; detailed optimization measures are carried out subsequently.

The first possible measure is relocation. This involves removing processes or entire process sequences from the process plan and transferring them to other companies. This can be carried out via cooperation or sub-contracting, for example. These measures help the company to concentrate on its core area of activity, thereby reducing the

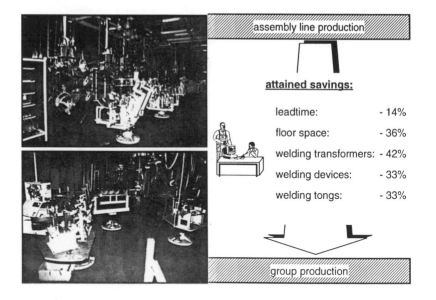

Fig. 8.16 Concentration on value creation - example (based on BMW AG)

level of complexity. This produces effects in accordance with the aim of attaining an overall optimum.

A characteristic of the other measures is that they establish a link between the requirements of operational organization and the production structure. Segmentation, e.g. via the separation of standard and special operations, enables process sequences which are identical for products with varying characteristics to be separated according to these product characteristics and sub-divided into individual process sequences. The segmented process sequences can then be reconfigured in accordance with the specific requirements concerned.

Integration involves the amalgamation of individual processes. This measure is applied in cases in which the process plan reveals a clear division between value-creating and organizational activities. An example of integration is the establishment of production islands.

The third measure involves increasing efficiency. This measure is to be selected whenever the process plan indicates deficits, the causes of which are to be found in an individual process, e.g. a production process.

The expedient measures for configuration of the production system depend on the specific requirements of the company concerned, and are

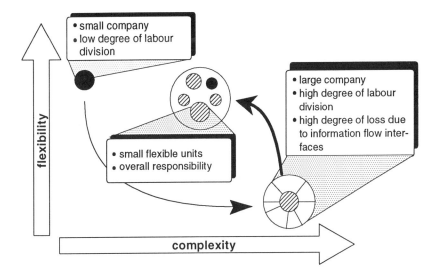

Fig. 8.17 Resource-optimized production design - segmentation

selected in accordance with the objectives set in the area of order processing. Process analysis enables the production processes to be identified and their efficiency to be assessed, providing the necessary basis for expedient deployment of the company's resources.

8.3.4 Effects on the value-creation process

The success of manufacturing companies is dependent to a decisive degree on the production processes which involve a high level of value-creation being identified and the company resources being utilized for these processes. This requires analysis of the production processes, the identification of value-creating and non-value creating processes, and appropriate joint configuration of the production system by operative and planning staff on the basis of this information.

The extensive elimination of non-value-creating processes and optimization of the value-creating processes often lead to the discovery of substantial additional potential. In the product manufacturing process, for example, down-times are minimized and a high level of resource

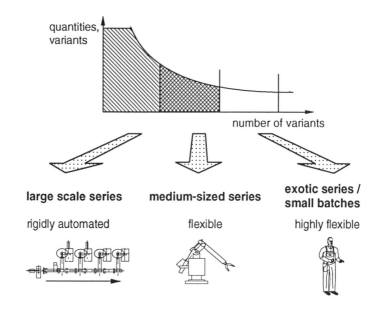

Fig. 8.18 Segmentation of assembly operations - example (based on Volkswagen AG)

consumption is avoided in carrying out the production process. For example, an automobile manufacturer's production process for the component "side frame" was restructured from an assembly line production process into a group production process (Fig.8.16).

This restructuring process was the result of intensive cooperation between the operative staff and the production planning department. The operative staff's know-how with regard to the specific requirements of production operations resulted, on the basis of the value-creating processes, in the identification of a need for a substantially modified production procedure, which in turn required extensive alteration of the production structure. Configuration of the production facilities in the shape of a U revealed considerable advantages over the previous linear structure.

The establishment of this expedient production structure resulted in substantial savings in the required resources. The required floor space was reduced, as was the required quantity of production facilities and devices. The quantity output was increased by reducing the leadtimes.

The optimized configuration presented here was possible only because all the production processes had been analyzed with regard to their value-creation potential. The combination of operative know-how and planning

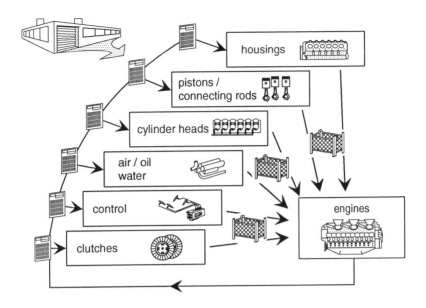

Fig. 8.19 Establishment of product islands - example (based on Pielstick)

know-how was essential, to enable identification of the areas of potential which were subsequently exploited. Staff motivation was enhanced considerably by the appropriate configuration of the production system in accordance with the requirements of the procedures involved.

Concentration on the value-creation process enables the company's resources to be deployed in an expedient manner, thereby establishing the essential basis for reducing the level of complexity at the company.

8.3.5 Reducing the level of complexity

The complexity of the procedures and structures at companies has developed over the course of time [4].

The handicraft business initially developed into small companies, which subsequently grew into medium-sized companies before finally attaining their current stature of large companies.

In the course of this development process, the advantages of the clear organizational system and overall responsibility which were characteristic of the smaller production structures were gradually lost, as the growing

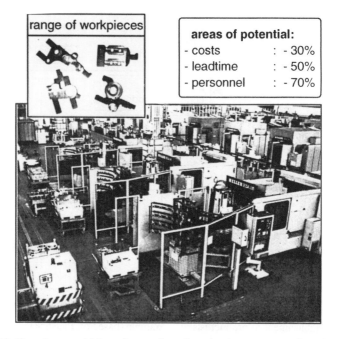

range of workpieces

areas of potential:
- costs   : - 30%
- leadtime   : - 50%
- personnel   : - 70%

Fig. 8.20 Complete machining - integration of production processes (based on HDM)

size of the companies resulted in a continual increase in the division of labour and more and more posts were involved in planning and organizing the activities to be carried out.

A remedy to this situation by way of the resource-optimized configuration of production systems is provided by segmentation [18]. This measure pursues the aim of combining the advantages of small and medium-sized structures with those of large companies (Fig. 8.17). To this end, small, flexible units are created which possess the authority and capabilities of small companies. Ideally, these units should pursue a common, general company objective, without any mutual conflict.

Fig. 8.18 shows an example of segmentation as applied for large-scale assembly operations. An automobile manufacturer required to assemble large quantities of complex products involving a high level of variant diversity at the lowest possible cost (combination of economies of scale and economies of scope).

To this end, the assembly process was segmented according to quantity- and variant-oriented technologies. Classification of the range of products identified a total of three product groups with different

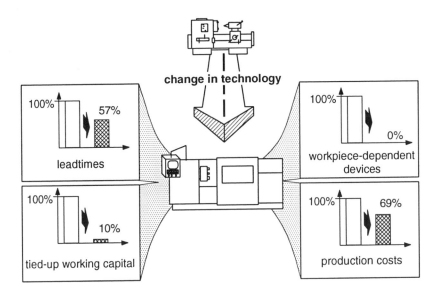

Fig. 8.21 Increase of efficiency in machining operations - example (based on Barmag)

characteristics regarding quantity and variants. Around 65 per cent of the available capacity was required for models which were produced at a constant quantity and with the same range of variants. For these models, 4 lines for large-scale production were set up, equipped with highly automated single-purpose assembly facilities. The two other product groups involved higher requirements regarding flexibility than the first product group, whereby it was possible to distinguish between high and very high flexibility requirements. Consequently, one assembly line was designed with flexible assembly facilities and one with highly flexible, manual assembly facilities. This method of segmentation enabled a suitable assembly structure to be created for each product group, as a result of which it was possible to carry out the production of all the products at overall minimum costs. This type of segmentation corresponds to a division into standard and special operations.

Another method of reducing complexity is the formation of product islands [16]. Production operations at a manufacturer of large diesel engines were divided into product islands which were able to operate as independent profit centres (Fig. 8.19). The final assembly stage (engines product island) acts in the capacity of a customer to the preceding product islands which, in turn, are suppliers to the final assembly stage. These product islands are able to operate independently of one another without pursuing conflicting objectives. The final assembly stage places orders with the preceding islands, thereby setting a control chain in motion. The preceding islands decide independently whether to accept or decline orders; they also keep and organize their own stocks of raw materials and finished goods. This form of organization ensures that the product manufacturing process is carried out within the company under market conditions, as the engines product island is also free to purchase its components on the market.

The complexity of production processes is also reduced by the establishment of production islands for machining the workpieces of a part family [19, 20]. At a printing press manufacturing company, a production line for machining a family of parts was configured in such a manner as to enable the workpieces to be fully machined in one operation on sufficiently flexible machining facilities (Fig. 8.20).

Intensive cooperation between planning and operative staff was essential in order to evolve this production configuration measure. The result was a workpiece holding / workpiece changing system which provided the basis for complete machining. The individual machines of this production line are positioned next to one another in a row and the workpieces are supplied to and removed from the machines via a driverless transportation system. The positive effects of this measure are manifested in a marked reduction in costs, personnel requirements and throughput time. Staff motivation was enhanced by the overall responsibility for therange of workpieces to be machined.

A final method of reducing the level of complexity involves making the processes flexible and increasing their efficiency [21]. An example of such a measure involves replacing single-purpose machines with more flexible, more efficient CNC machines (Fig. 21).

In this case, the increase in efficiency is attained via the use of modern, high-performance cutting tools, which leads to a substantial reduction in machining times and reduces the total retention time for the workpiece on the machine. The use of CNC technology increases the flexibility of machining operations. The order execution process is not altered as a result of this measure.

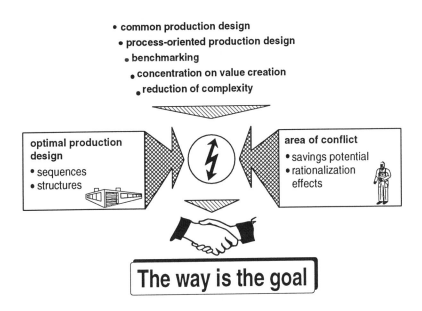

Fig. 8.22 Measures for process-oriented production design

8.4 Summary and outlook

Today, manufacturing companies are confronted with diverse challenges which can be met only by adopting appropriate restructuring measures. The design measures for the production carried out in the past, which concentrated on specific aspects of the production sequence process, resulted in the optimization of specific details, with only limited positive effects on the overall success of a company. New methods and facilities must therefore be configured in such a manner as to enable optimal utilization of the areas of value-creation potential which exist at a company for the purposes of product manufacture. This requires a radical change of approach, away from the function-oriented system which is customary today and towards a process-oriented system. This altered approach enables the analysis of the product manufacturing process to form the basis of all measures to design production. It provides the necessary information for reducing complexity and orienting design measures to the value-creation process. This ensures optimal utilization of the resources available to the company.

Utilization of the know-how available from the staff provides the essential basis for acquiring an understanding of the production processes and for identifying and exploiting the available areas of potential for restructuring purposes. It is particularly important for companies to involve their operative staff in the production configuration process. However, this places the staff in a conflicting situation, with the areas of potential for savings on the one side and the rationalization effects on the other. The central task for the future will be to overcome this conflicting situation, as otherwise it will not be possible to identify the areas of potential available at a company.

A changed situation of mutual understanding is therefore necessary when selecting measures for the resource-oriented production design (Fig. 8.22).

Production must be subject to an ongoing development process, to enable continual improvement via the input of creativity and the capacity to cope with conflicting objectives. In this way, the requirements of the order-processing processes and structures can be reconciled to establish an optimum in the long term.

9

THE KEY OF EFFICIENT PROCESS AND MANUFACTURING SEQUENCES - TECHNOLOGICAL DEVELOPMENT AND INNOVATION

9.1 Introduction

The present situation is characterized by extremely intense international competition accompanied by a decline in economic activity. In order to maintain their positions on and shares of the market, companies are required to reduce costs on an unprecedented scale. As rationalization investments are subjected to particularly critical scrutiny in such phases, the identification and utilization of existing resources is of particular importance. This is more likely to be achieved, the more the specialist knowledge of employees is utilized on an integrated, inter-departmental basis. The aim must be to accept production, and subsequently the technologies and processes, without compromise as the focus of company operations, and to gear all activities to further optimizing this area (Fig. 9.1).

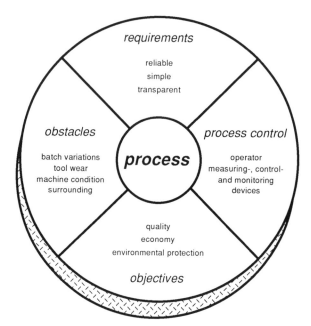

Fig. 9.1 Process and environment

9.2 Basic requirements for cost-minimized production

9.2.1 Rise of process reliability

An essential prerequisite for cost-effective production is control of the technologies employed. Individual processes and process sequences must be configured in such a way as to ensure that they run with a maximum level of reliability, irrespective of whether small or large batches are being produced. This requirement can be fulfilled only when the actions and interactions which occur in the course of the process are analyzed and understood. Only then is adequate reliability ensured with regard to selection of the machinery, tools, machining data and auxiliary materials required for the process.

The qualifications of the staff and the responsibilities which they are assigned must be in keeping with the requirements of sophisticated processes and technologies. This applies not only to the staff involved in planning activities, but also to the operative staff in particular.

Operations which are unstable and susceptible to faults ultimately indicate a lack of understanding of the processes concerned. Such deficits may have various causes. They are relatively simple to overcome when they are based on a lack of knowledge of the available information and inadequate communications within the company. Greater problems arise when phenomena occur, the causes and principles of which are unknown. In such cases, the mechanisms of action characterizing the process must first of all be analyzed, otherwise it will be virtually impossible to modify the process in a purposeful and effective manner. The use of monitoring facilities is unsuitable, as an understanding of the actions and interactions which occur in the course of the process is required, in order to develop and operate such facilities.

All processes generate signals which can be used to analyze the process status and to monitor and control the process, provided that suitable sensors are available for detecting the signals and suitable algorithms are available for signal processing. These signals relate to the forces, the structure-borne noise and the power consumption of the machines. Data on tool wear, the traverse paths of tools and machine components, the geometrical dimensions of the workpieces, the surface characteristics and the characteristics of the surface layers of workpieces, such as internal stress or microstructural changes, can also be used to monitor and control processes. Systems are already in operation today which measure the internal stress of workpieces in production operations and utilize the results for the purpose of process monitoring [1].

Modern production plants provide little scope for the machine operator to observe the process operations directly and to respond to deviations from the set parameters. Systems for visualizing processes have been developed to solve this problem (Fig. 9.2). Information on the current state of the process is displayed in graphic form, enabling the machine operator to take corrective action immediately in the event of changes to signal patterns, the exceeding of set limits, etc.

When processes are understood sufficiently to enable not only signal detection but also the specification of algorithms for processing signals and generating machining variables, process monitoring can be carried out in the form of a control circuit. The example of spark flash identification for the internal circular grinding process (Fig. 9.2) shows that time savings of around 90% are possible by automating this stage of the production process.

Despite the advanced level of sensor development and the advanced facilities which are available for process monitoring, it will not be possible

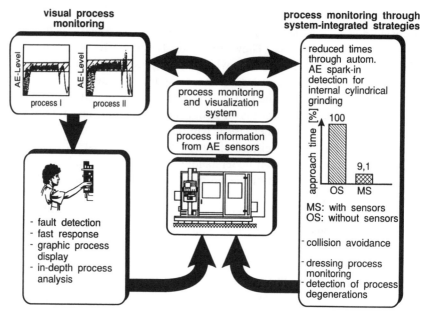

Fig. 9.2 Process monitoring strategies (source: Dittel, Schaudt)

to ensure the reliability of processes via technical facilities alone in the foreseeable future. Manual intervention is certain to remain necessary. But the initiating factor for corrections of processes must not be the notification that workpieces have been rejected.

When no automatic process monitoring facilities are available, the specialist know-how of staff must be incorporated into process configuration and process control to a greater extent than is presently the case. In this context, measures for increasing the staff's qualifications are to be considered. There is a current tendency to revise the division of labour, which is in extensive use at some companies, and to assign employees more demanding work involving greater responsibility than previously. The increased integration of technically trained staff into the production process is expected to result in increased flexibility with regard to responses to deviations from "normal" production operations. As already mentioned, the introduction of such structures into production operations requires an understanding of the actions and interactions which occur in the course of the processes concerned.

9.2.2 Reduction of the number of production stages

Newly developed and enhanced machines, tools and materials have increased the chances of optimizing production sequences, in addition to individual processes. The scope of applications for production processes has expanded continuously. Today, hardened steel materials can be turned, milled, drilled, broached and reamed. Grinding technology has been developed to such an extent that removal rates comparable to those attainable via cutting processes with geometrically defined cutters are now possible, without any impermissible effects on the surface layers [2-8].

Effective use of the components, machine, tool and material by a team of qualified staff from the areas of development, production and quality assurance enables many production processes and production sequences to be improved or reconfigured in such a manner as to enable more cost-effective production operations. This is illustrated by the following examples.

Component tolerances are based not only on the functions of the component, but also on the processes and the production sequence through which the component passes in the course of manufacture (Fig. 9.3). The traditional production sequence of rough-machining and finish-machining frequently concludes with a grinding process. This process enables high quality standards to be attained with regard to dimensions, geometry and surface characteristics. The availability of even more wear-resistant cutting materials, such as cermet, enables finish-grinding to be substituted by turning, whereby the stipulated quality standards IT5 and IT6, which impose high requirements on the configuration of turning processes, are complied with [9].

In studies concerning the substitution of production processes it is generally assumed that at least the same degree of precision will be attained with the new process as with the substituted process. In most cases, there is no critical examination of whether the given component form and accuracy result from the functions of the component or from the processes applied to manufacture the component. The persistence on present shapes and tolerances often prevents innovative and cost-effective alternative production methods.

There will be a further increase in the application of forming processes for the near net shape of components. Savings in materials are not the sole aim pursued in the application of near-net- shape technologies. A more important aim is to restrict the allowances to such an extent that only one operation is necessary to produce functional surfaces with narrow

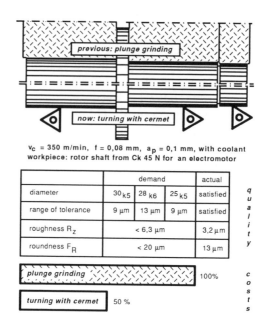

Fig. 9.3 Finishing by turning instead of grinding (source: Hertel)

tolerance ranges. Rough cutting to produce defined starting conditions for finishing is being substituted more and more by higher levels of precision in casting and forming processes. Such developments lead to drastic reductions in the scope of finishing operations, as illustrated by the example of a cast crankshaft for mid-range automobiles. In the course of time, improvements in casting technology have led to smaller allowances. As a result, the volume requiring machining has fallen from 2.2 kg to 1.2 kg.

The reorganization of production sequences is often carried out as a result of problems regarding the integration of heat treatment processes into the production line. Thermochemical heat treatment processes, such as case hardening, nitriding or boriding, can often be eliminated by switching to full-hardened materials. The study concerning a new production sequence for a drive shaft from the field of diesel injection shown in Fig. 9.4 presents the substitution of case-hardening steel by ball-bearing steel, together with the appropriate hardening processes. The use of this material is intended to enable a switch from a double-stage cold forming process to a single-stage semi-cold forming process. The residual heat from the

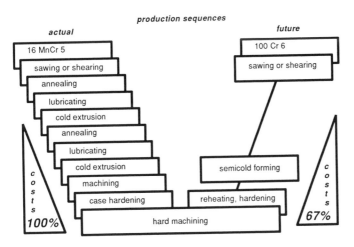

Fig. 9.4 The production sequence as a potential area of savings (source: Bosch)

forming process means that only reheating to hardening temperature is then necessary. The two annealing and surface treatment stages which are required prior to the two cold forming processes are no longer necessary. Overall, a switch to the new production process involving far few individual stages is expected to save one third of the production costs.

9.2.3 Rise of batch sizes

In past years, production plants have been structured and configured so as to enable them to respond ever more quickly to market requirements. This applies with regard to the reduction of innovation times and the splitting of products into a large number of variants. The wealth of variants is definitely due in part to the varying licensing requirements which apply in the different export countries. The remaining variants arise as a result of the need to cover certain price, performance and configuration areas as comprehensively as possible. A trend towards offering variants for niches in the market is also clearly observable, particularly for products in the consumer goods sector.

An increase in the number of variants generally results in diminishing batch sizes. This has consequences for investments in production plants, the configuration and implementation of processes and manufacturing costs.

An investment study at a large company produced the following result for the manufacture of a certain product: the installation of plants which are able to produce only a few variants, but in large numbers, requires an investment level of 100%. When around half of all possible variants are to be produced, the investment requirement increases to 135%. Full flexibility with regard to the possible variants requires an investment level of 185%. Cost disadvantages are unavoidable, unless an adequate increase in net output can be attained with the fully equipped plant.

The area of production is subject to similar factors. The application of processes involving high tool and set-up costs and production operations on linked plants is feasible only at a low level of flexibility, as they become economically viable only when large quantities are produced. Resetting the plants or individual components reduces the machine time. Losses are incurred during start-up of the processes. Overall, the need for more flexibility resulting from more diverse variants also diminishes the level of process reliability.

The presentation of this problem area is not intended to induce companies to restrict production operations to a single standard look in future. It is rather intended to prompt renewed consideration of which market segments can be covered in an economically expedient manner, whether the basis for make or buy strategies requires revising or what possibilities are available to enable the product to be split into variants at the end of the manufacturing process. In general, the question arises as to whether the cost-effective production of a limited number of variants is not more likely to lead to economic success than covering the broadest possible range.

9.2.4 Prevention or reduction of disposal costs

The costs of disposing of production waste have increased several times over in recent years, as is illustrated by two examples from quite different areas of production engineering.

Studies carried out by a mass-producer revealed that the costs per workpiece for installation and operation of the cooling lubricant supply system and for disposal of the used materials and substances were 2 to 3 times higher than the tool costs per workpiece [10].

An initial measure undertaken by the cooling lubricant manufacturer to reduce such high costs was to remove the particularly undesirable chlorine

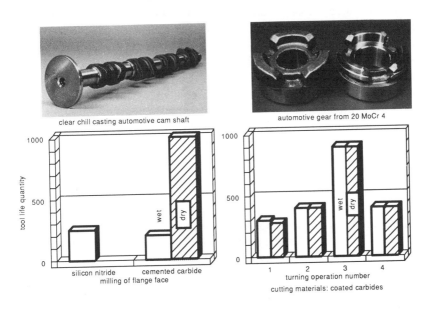

Fig. 9.5 Substitution of wet machining by dry machining (source: Mercedes Benz)

from the cooling lubricant. Other developments are aimed at producing bio-degradable cooling lubricants. In spite of these developments, a more radical approach is gradually being adopted at many production plants, in the form of a switch to dry machining operations.

This approach harbours substantial potential for cost reduction. Cutting materials such as carbides, cermets, ceramics, PCBN and diamond are essentially suitable for dry operations. Dry machining offers technological advantages over wet machining in the form of reduced thermal shock for interrupted cutting operations. Today, the lubricating effects of coolants can be replaced, where necessary, by coating the tools with hard materials. Dry machining may result in problems due to heating up of machine and tool and in poor chip removal. Examples from practical applications show that in some cases the switch from wet to dry machining can be carried out without any additional measures (Fig. 9.5). In the case of milling, the use of dry machining actually increases the tool life.

High costs for the disposal of waste are also incurred in the area of non-traditional production processes, as is illustrated by the example of residues from spark erosion processes. The disposal of one tonne of removed material from spark erosion processes employing hydrocarbon dielectrics today costs more than DM 2000.-. Disposal of the same quantity costs only DM 180.-, when erosion is carried out using an aqueous dielectric. In this case, the residues are processed in the erosion plant itself, at a cost of DM 270.-., resulting in an overall cost benefit of more than DM 1450.- per tonne for erosion processes employing aqueous dielectrics. Furthermore, the metallic content of these residues can be recycled in the furnace.

These two examples from totally different processes show that the development and application of future technologies involves severe changes in cost structures. At present, it is not possible to foresee the extent to which the costs from the area of the environment will affect process design, production sequences or tool development. It remains certain, however, that there will be a vast increase in these costs. Potential which is already available for avoiding or reducing such expenditure must therefore be exploited much more quickly and rigorously in future than has been the case to date, to avoid a sudden loss of the basis for economic production.

9.3 Available resources

A large number of approaches are available for optimizing production processes and production sequences with the aim of reducing costs. There is often no need whatsoever to carry out extensive studies before attaining this aim. On the contrary, the main problem today is usually selecting the right alternative from the broad and increasingly complicated range of available cutting materials, tools and production processes.

The essential prerequisite for selecting the correct alternative and successfully implementing new tools, processes and production sequences to solve new problems in the area of production is an understanding of the actions and interactions which occur in the course of the production processes. Only when such an understanding is available can failures be avoided which are all too often explained in terms of the apparently inadequate quality or suitability of the components deployed in the process.

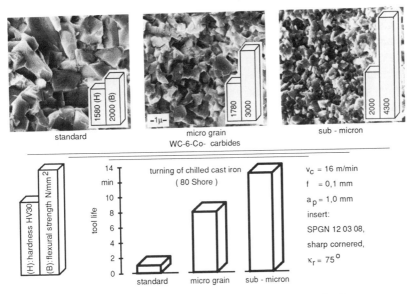

Fig. 9.6 Ultra-fine-grain carbide - Course of development and application
(source: Krupp Widia)

9.3.1 New and improved tools

Ever since tools have been produced, attempts have been made to improve their performance characteristics. The aim remains an improvement in their resistance to wear combined with increased reliability. Both of these factors are essential preconditions for any further increases in the reliability and economic efficiency of processes. This area of potential can only be exploited if the tools are deployed in accordance with their specific range of characteristics, as is illustrated by the following examples.

(a) Ultra-fine-grain carbide

In this instance, the term ultra-fine-grain carbide refers to a WC-Co alloy with a WC crystallite size of 0.5 μm (Fig. 9.6). This carbide possesses an unprecedented combination of hardness (2000 HV30) and flexural strength (4300 N/mm²). Such a combination of properties leads to markedly improved performance capabilities, as illustrated by the example of the turning of chilled cast iron. Ultra-fine-grain carbides also offer advantages over fine-grain materials for applications involving primarily abrasive wear, as in the case of the drilling of printed circuit boards [11-14].

(b) Coated carbides

The aim pursued in increasing the service properties of coated carbides is a further improvement in their ductility. This property is required in particular when machining with interrupted cutting operations. Improved ductility generally means a higher level of production safety for the user [13-19].

The high-temperature CVD coating process (HT-CVD), which operates at temperatures around 1000°C and was for a long time unrivalled, is now only one of several commercially applied methods for depositing hard materials on cutting tools. The HT-CVD process faces increasing competition from processes which operate at substantially lower temperatures, such as

- medium-temperature CVD coating (MT-CVD, t = 700-900°C)
- plasma CVD coating (PCVD, t = 500°C)
- PVD coatings (t = 500°C).

An advantage of HT-CVD coating is the possibility of depositing comparatively thick coatings (10...12 µm) on geometrically complex tools at low cost. Carbide substrates with surface zones free of mixed crystals are now being employed to an increasing extent to improve the ductility of tools coated via this process. The hard coatings usually consist of several layers. The primary hard materials employed here are TiN, Ti(C,N) and Al_2O_3 [15-18].

The coating at low temperatures in modified CVD processes (MT-CVD, PCVD) results in cutting tools with properties which are well suited to the stress induced by interrupted cutting operations. The improved ductility can also be utilized for working at low cutting speeds. It results from the reduced thermal stress on the substrate during the coating process and from the reduced tensile stress in the coating. Coatings which can be deposited by means of this process are TiC, TiN and Ti(C,N) [12, 20-24].

Carbides which are coated in PVD processes are also very suitable for strongly interrupted cutting operations, including such operations on high-strength as well as on quenched and tempered steels. One reason for this is the minimal degree to which the properties of the substrate are affected during the coating process (Fig. 9.7). The dependencies in Figures 9.7 a, b, c show that although the coating properties improve as the substrate temperature increases, the flexural strength decreases. This characteristic of coated carbides is decisive with regard to their ductility, however. In addition to the minimal influencing of the substrate, the compressive stress

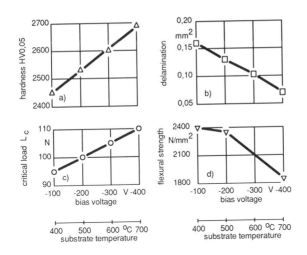

Fig. 9.7 Influence of temperature on arc PVD coating

in PVD coatings also has a positive effect on ductility.

The overall result of these effects is illustrated by the VDI strip-turning test, which has been designed primarily to test ductility requirements (Fig. 9.8). PVD-coated cutters reveal substantial advantages in the area of ductility at low cutting speeds. It is notable that the performance characteristics of a subsequently heat-treated PVD-coated carbide at low cutting speeds are similar to those for CVD-coated carbides.

In the past, the advantages of hard coatings were discussed primarily from the point of view of protection against abrasion. The applicability of coated tools at low cutting speeds in the range of built-up edges also focusses attention on the aspect of reducing adhesion. The hard layers deposited in low-temperature processes, today primarily TiN, Ti(C,N) and (Ti,Al)N, are chemically stable and non-reactive, and thus contribute to the drastic reduction in adhesion between the tool and steel materials. The area of application for carbides coated by means of these processes extends into the field of coated HSS cutting materials [25-27].

Despite all the advantages offered by PVD-coated carbides when high ductility requirements apply, it must not be forgotten that they are faster-wearing in smooth cutting operations than CVD-coated carbides, due to the comparatively low layer thickness of 3...5 μm.

Industrial application of the stated coating processes has led to a diverse range of different coatings and coating sequences. The situation

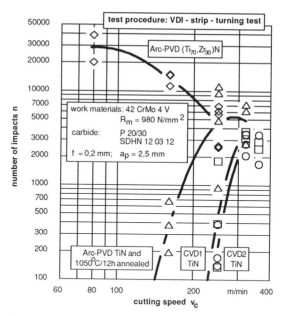

Fig. 9.8 Ductility test with coated carbides

may be confusing for the user. He must therefore understand the essential connections between coating methods and properties of the coated products, in order to be able to determine right at the definition stage for the machining task suitable cutting materials on the basis of the dominant type of stress to which the tools are to be subjected.

(c) Cermets

The term "cermet" refers to cutting materials which consist primarily of a titanium-based hard phase and a nickel-cobalt binder phase. Such cutting materials are characterized by a high level of wear resistance and a minimal tendency to adhere to steel materials. Recent years have witnessed improvements above all in the ductility of cermets, as a result of which certain types are already suitable for use in interrupted cutting operations and for milling. Cermets will acquire increasing importance in the future, on account of the high availability of their constituent materials [9, 12-14, 16-18, 28].

Cermets have formed the subject of much discussion in recent years, in publications and at conferences. Extensive tests have been carried out at the test laboratories of large-scale users, in order to determine the

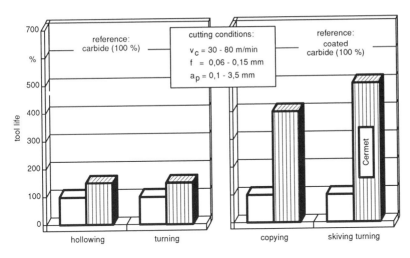

Fig. 9.9 Substitution of carbide by cermet (source: Mercedes Benz)

performance capabilities and operating limits of cermet cutting materials. Some users have replaced carbides with cermets on a substantial scale, though this substitution process is far from complete.

Apart from such cases, cermets have as yet failed to establish themselves on a broad front on our market. This has also been observed by German cermet manufacturers. On some foreign markets, through to the Far East, the demand for their products is greater than on the domestic market. The lack of awareness of the performance capabilities of these cutting materials is cited as the reason for this situation. Additionally, cermets underwent a very rapid course of development in the past, and the properties of the cutting materials which are grouped together under the term "cermet" are substantially more heterogeneous than those of the carbides.

Today, high-quality cermets are competing with uncoated and coated carbides in the area of steel machining. They can be employed for turning steel and cast materials at high cutting speeds, but are also suitable for speeds below 100 m/min (Fig. 9.9). On account of the high edge stability and wear resistance of cermets, these cutting materials continue to appear particularly suitable for finishing cuts with small chip-cross-sections. When soft steel workpieces, such as extruded parts, require to be turned instead of ground, cermet should be considered first of all as a cutting material (Fig. 9.3). The thermal shock characteristics of cermets have improved,

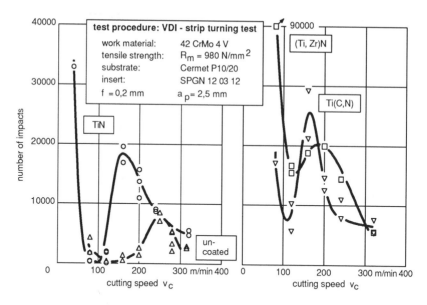

Fig. 9.10 Ductility test with coated cermets

which means that they can also be used in processes involving cooling.

Despite the high wear resistance of cermets and their low adhesive tendency, coatings also offer advantages on these cutting materials. The average service life improvement in comparison to uncoated cermets is 50%. Low-temperature processes, such as PVD and PCVD, are employed preferably for coating cermets [13, 28-36].

Cermets coated with TiN, Ti(C,N) and (Ti,Zr)N revealed marked advantages over the uncoated substrate in the VDI strip-turning test (Fig. 9.10). The coatings not only increase the tool service life, expressed here via the number of contacts with the tool (Number of impacts, n), they also result in a shift of the peak performance characteristic first of all into a lower cutting speed range (v_c =160...200 m/min).

Below this cutting speed range, the performance characteristic first of all falls abruptly, then rises again steeply at v_c=80 m/min. The reasons for this behaviour under the conditions of strongly interrupted cutting operations are as yet unknown. On the basis of the results it can nevertheless be established that the coating of cermets should be developed further, that coated cermets are not necessarily to be considered as finish-cutting materials for high cutting speeds and that some cermets at

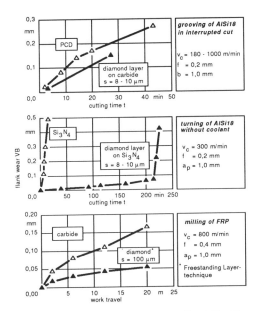

Fig. 9.11 Applications for diamond coatings (source: Sandvik, Okuzumi)

least are able to compete with carbides with regard to ductility.

(d) Diamond-coated tools

On account of its high hardness and wear resistance, diamond is an ideal cutting material for subjection to high levels of abrasive stress, provided that the contact zone temperatures remain well below 800°C and the materials to be machined have no affinity with carbon. Geometrically simple cutting tools produced in natural diamond and polycrystalline diamond cutting materials are firmly established in the machining of green compacts made of carbide and ceramic, in aluminium machining, in stone machining, in the machining of fibre-reinforced and filled plastic and in wood machining. To date, it has not been possible to produce cutting tools in diamond with complex-shaped cutting edges, for reasons of cost [37, 44]. This restriction to the deployment of diamond as a cutting material may be overcome in the future. Work has been in progress for several years now on the development of various CVD processes for depositing diamond on metallic and ceramic substrates. The first products have already been launched on the market, in the form of indexable inserts, drilling and milling tools.

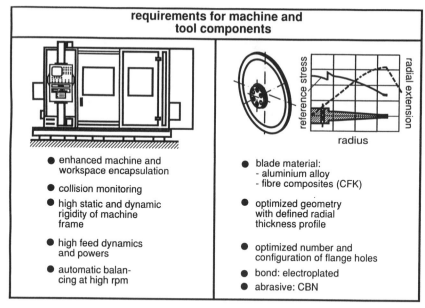

Fig. 9.12 Requirements relating to machine and tool components for high-speed grinding (source: IPT)

In addition to the direct depositing of diamond on substrates (carbides, ceramic) which have already been processed into tools, a further process which is applied is the so-called free-standing layer technique. This process involves depositing diamond layers with a thickness of some tenths of a millimetre onto a substrate, the surface of which may be flat or spherical. After coating, the thick (free-standing) diamond layer is detached from the substrate. It can then be joined to a tool blank, via soldering, for example, and finished by grinding. Some examples of applications and performance characteristics for diamond-coated tools are shown in Fig. 9.11.

(e) Tools for high-speed grinding

Developments and innovations in the area of production methods are often motivated by efforts to provide efficient processes suitable for a broad range of applications. High material-removal rates are to be combined with good workpiece quality, enabling the roughing and finishing of components to be carried out without rechucking. Various approaches apply to broadening the applications of production processes, depending on the level of development of the process concerned.

Grinding is traditionally a precision process, usually involving a low material-removal rate. Development and innovation work thus concentrates on increasing this characteristic, without permitting any significant reductions in precision. Once this is achieved, the process concerned attains a broader scope of application than previously. For the user, this means more freedom in selecting processes and configuring production sequences. Machining times, production stages, throughput times and costs can be reduced.

A key area of research aimed at combining high material-removal rates with high precision is CBN high-speed grinding. Integration of the process strategies "high-speed grinding" and "quality grinding" for a cutting speed range above 250 m/s imposes high requirements on the machine components and the grinding tools [45, 46]. The example discussed below illustrates current development trends and results obtained with high-speed grinding technology for cutting speeds up to $v_c = 350$ m/s.

In order to protect the machine operator, rigid cover of the machinery and collision monitoring of the quickly rotating grinding wheel are necessary. To guarantee safe, reliable operation of the process and reproducible results, the machine frame must be designed so as to ensure a high level of static and dynamic rigidity. High-power drives with short response times, automatic balancing systems and approach control systems are required (Fig. 9.12).

Customary designs of grinding wheels are unable to meet the requirements resulting from the high cutting speeds. At high speeds of up to 30000 min-1, centrifugal force is the dominant stress factor. Due to the peak stress which occurs in the area of the central borehole which is normally used to fix grinding wheels on the spindle, this form of fixing is not suitable for high-speed grinding wheels. Instead, the wheel is flanged onto the spindle without a central borehole, with the aim of keeping the level of stress as low as possible and avoiding excessive stress. A constant stress characteristic is attainable via geometrical adaption of the contours of the grinding wheel or optimization of the shape of the grinding wheel. In the area of the abrasive coating in particular, the adaption of the shape of the wheel body should minimize expansion. This is necessary in the interests of the service life of the wheel and in order to attain high levels of machining accuracy. A special aluminium alloy or fibre composites may be employed for the body of the wheel. The abrasive coating of such wheels consists of CBN abrasive grain electrodeposited in nickel [47].

The results of grinding operations confirm the high performance

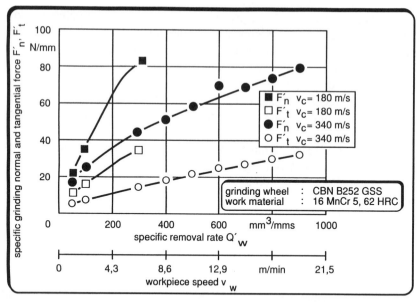

Fig. 9.13 Three fold increase in Q′$_w$ by doubling the cutting speed (source: IPT)

capabilities of high-speed grinding. The high circumferential speeds of the grinding wheel enable extremely high levels of material removal without any thermal effect on the workpiece. The produced component quality can be maintained or even improved in comparison to conventional grinding processes, while operating at substantially higher material removal rates. Increasing cutting speeds are accompanied by a corresponding reduction in grinding forces, whereby the level of reduction rises still further as the material removal rate increases. The specific material removal rate can thus be increased three-fold by doubling the cutting speed, without any increase in the normal and tangential grinding forces acting on the component (Fig. 9.13) [48].

A further important aspect of high-speed grinding is the transverse and rotational accuracy of the employed tools. Inadequate levels of transverse and rotational accuracy lead to disproportionately negative effects on the results of machining operations, particularly at high cutting speeds. The grinding wheels with electrodeposited grits which are employed primarily for high-speed grinding are not designed for profiling. Their profiles must therefore be subject to narrower tolerances than the shaping elements of the tools which they produce (Fig. 9.14). This is the essential prerequisite

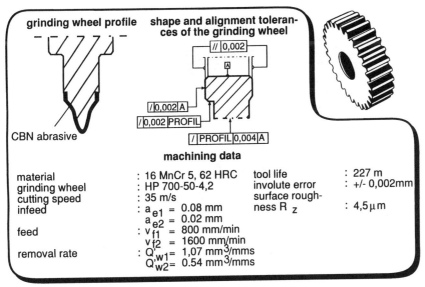

Fig. 9.14 Accuracy requirements for CBN grinding wheels (source: Winter & Sohn)

to enable sustained use of the high wear resistance of the CBN grit to produce dimensionally and geometrically accurate workpieces [49].

9.3.2 Process selection and process design

(a) Utilizing the full potential of hard coatings

Today, more coating systems are available on HSS, carbide and cermet substrates than ever before. Consequently, it is becoming increasingly more difficult for the user to select the optimal cutting materials for the specific tasks concerned or to apply the selected cutting materials in an optimal manner.

This problem area is illustrated by the example of PVD-coated HSS indexable inserts. The different thermophysical characteristics of the three carbide coatings TiN, Ti(C,N) and (Ti,Al)N in conjunction with the varying types of stress which apply in smooth and interrupted cutting operations result in varying performance characteristics (Fig. 9.15).

Due to their better thermal insulation, (Ti,Al)N coatings permit higher cutting speeds in smooth cutting operations than the two other coatings. Ti(C,N) coatings are most suitable for interrupted cutting, as their lower

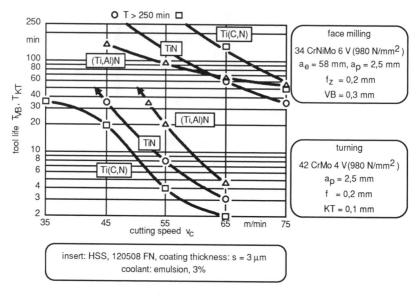

Fig. 9.15 Turning and milling using HSS cutters with different PVD coatings

level of thermal insulation influences the chip formation process in such a manner as to shift the zone with the highest level of stress away from the sensitive cutting edge and into the face of the tool [50, 51].

This knowledge of the respective properties of the carbide coatings can be utilized in two directions: for the process-oriented selection of coatings and for the coating-oriented configuration of processes.

If turning is to be carried out with Ti(C,N)-coated HSS tools, for example, high cutting rates should be attained by increasing the chip-cross-section, and not by increasing the cutting speed, in order to keep the cutting temperature at a low level. When milling with (Ti,Al)N-coated cutting tools, the highest possible feed rates are to be applied, in order to shift the stress zone away from the cutting edge and into the face of the tool.

It is to be assumed that this knowledge can also be employed for the further development of coatings on other substrates. With regard to thermal shock, which leads to comb-cracks when using coated carbides and cermets in interrupted cutting operations, coatings with good thermal insulation should be advantageous.

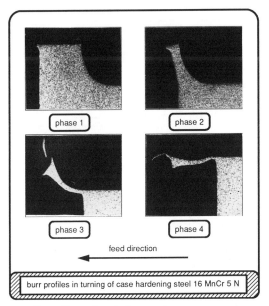

phase 1

phase 2

phase 3

phase 4

feed direction

burr profiles in turning of case hardening steel 16 MnCr 5 N

Fig. 9.16 Phases of burr formation during turning

(b) Minimization of burring via appropriate process design

Burring during the machining of metallic materials represents a serious problem in many branches of industry. Particularly in the case of components with small allowances which are subject to high requirements regarding freedom from burrs, up to 20% of production costs may be accountable to deburring operations. Apart from the costs, increasing component quality requirements and the demand for attractive workplaces are additional factors necessitating a solution to the problem of burring.

Considerable efforts have always been undertaken to remove burrs by manual or mechanical means. The option of designing tools and processes in such a way as to minimize the possibility of burrs occurring in the first place, on the other hand, has received very little attention. But this option does offer good potential for suppressing or minimizing burring. This objective can only be attained, however, when the mechanisms which lead to burring are taken into account when selecting tools and configuring processes.

Burrs occur when ductile materials are able to deform in an uncontrolled manner as a result of process forces. The necessary conditions for such deformation apply above all when tools emerge from a workpiece. The residual cross-section of the material to be removed is

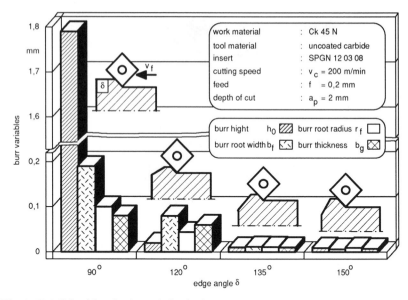

Fig. 9.17 Minimal burring in optimized edges

weakened to such an extent that chip formation does not occur, instead of which the material is plastically deformed by the tool. Fig. 9.16 shows phases of these processes by reference to the example of turning.

One method of counteracting burring is to chamfer the workpieces, as is again illustrated by reference to the example of turning (Fig. 9.17).

In the case of face milling, similar effects can be attained by selecting an appropriate cutter emergence angle and by cutting out not in a straight motion but in circular arcs. Drilling offers potential for reducing burring by optimizing the cutter emergence angle, the drill geometry and the cutting conditions. The aim of these measures must be to reduce the feed forces in the tool emergence phase. The use of cooling lubricants may also help to reduce burring, when it leads to a reduction in the temperature of the machined material in the area in which burring occurs, thereby diminishing the susceptibility of the material to plastic deformation [52].

(c) Application of hard machining
The production sequence soft machining, heat treatment, grinding originates from the fact that for a long time no cutting materials were available which were suitable for machining steels and cast iron with hardnesses of 60 HRC and more. The development of oxide ceramics, and

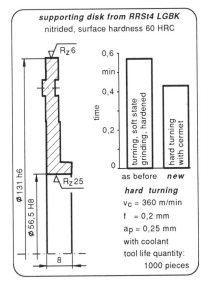 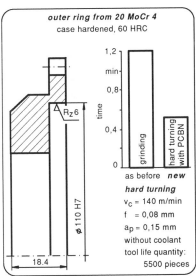

Fig. 9.18 Substitution of grinding by hard turning (source: Mercedes Benz)

above all the availability of CBN, the second-hardest cutting material after diamond, made it possible to turn, mill, drill, broach, etc. hardened iron-based materials. Today, PCBN cutting materials compete with ultra-fine-grain carbides, oxide ceramics and also cermets for such machining processes.

Hard machining must be carried out at the limit of the cutting material's stress resistance capacities. A cutting tool with a negative-acting geometry is advantageous for the purposes of chip formation. This induces compressive stress in the shearing zone, which helps to ductilize the material to be removed and enhances the formation of continuous chips. The temperature increase in the shearing zone as a result of the introduced mechanical energy supports this effect.

Due to the extreme specific stress, the cutters react extremely sensitively to any change in cutting conditions during hard machining. The range of applicable cutting parameters is consequently substantially smaller than for soft machining. Only within the optimal range are satisfactory machining results obtainable. Outside of this range, the tool life shortens drastically.

Hard machining with geometrically defined cutting edges often provides the key to new production sequences or to eliminating rough-

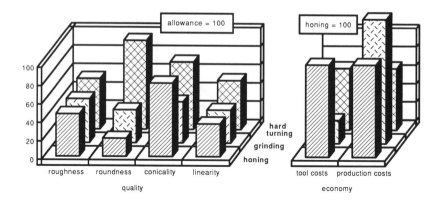

Fig. 9.19 Comparison of honing, grinding and hard turning processes (source: Mercedes Benz)

machining and preparatory operations on workpieces while still in soft state.

A pronounced trend towards the use of such technologies is to be observed at present in the automobile industry. Fig. 9.18 shows examples from the production of components for automatic gear units. The time savings attainable by switching to hard turning are considerable. The example on the left shows the results of hard turning with cermet.

A comparison of the honing, grinding and hard turning processes (Fig. 9.19) shows that hard turning is clearly the most economical option. In this example, which concerns a gearwheel made of case hardened steel 20 MoCr 4 E, hard turning can be carried out one-third more cost-effectively than honing, which serves as the basis for this comparison. Analysis of the quality characteristics in relation to the respective processes shows that attaining the required roundness in particular can be problematic with hard turning, as this characteristic virtually exhausts the available tolerance range. Experience with other applications for hard turning processes shows that in the majority of cases deviations from the required roundness and cylindricity are accountable to the chucking devices and not to the process itself.

(d) Overcoming the limitations of angular infeed grinding
In the course of grinding processes, the majority of the mechanical energy involved is transformed into heat, due to friction and separating actions

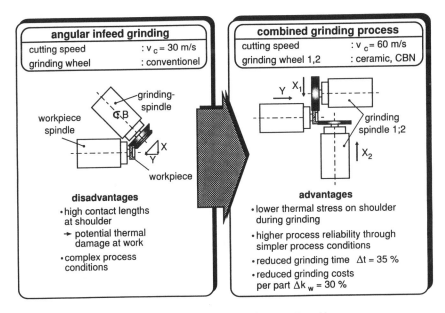

Fig. 9.20 Substitution of angular infeed grinding (source: Bosch)

during chip formation. As a result, there is a fundamental danger of an unacceptably large thermal effect on the surface layer. Apart from the dimensional, geometrical and surface quality, the reliable avoidance of thermally induced damage to the surface layer is acquiring increasing importance for the functional characteristics of ground workpieces. Weight reduction and the full utilization of the stress resistance capabilities of components are requirements whose influence extends into the areas of process configuration and process control.

In this context, all grinding processes are critical in which high contact zone temperatures, long contact times or both occur. This always imposes restrictions on improvements in the performance capabilities of grinding operations, as is illustrated particularly clearly by angular infeed grinding. In this process, external cylindrical infeed grinding and lateral cross grinding are interlinked (Fig. 9.20). The main area at which a danger of thermal damage to the surface layer of the workpiece applies is the plane shoulder. Large contact lengths apply here, and the transportation of cooling lubricant into the contact zone is subject to considerable difficulties. As a result, damage-free grinding is impossible on the plane shoulder above certain material-removal rates. As cost factors prohibit lowering the material-removal rates sufficiently to ensure that no impermissible influencing of the surface layer occurs, the ground workpieces have to be examined for surface damage. Such tests interrupt production operations, as no equipment is currently available for on-line testing and the parts therefore have to be removed from the production line.

In order to counteract these disadvantages, a supplier to the automobile industry has substituted the angular infeed grinding process with two individual processes involving the use of two grinding spindles displaced at an angle of 90 degrees (Figure 9.20). Both the cylindrical and the plane surfaces are machined by means of straight infeed grinding. The new machine further enables the use of the ultra-hard abrasive CBN. These measures have led to a marked reduction in the grinding times.

In spite of marginally higher machine costs, the reliability of the production process was increased substantially, while the workpiece costs were reduced by 30%. The result was a substantial reduction in the scope of testing required for the components. This example illustrates that process analysis and appropriate implementation of the resultant knowledge enable production sequences to be configured in such a manner as to fulfil the requirements both for increased process reliability and for lower production costs [53].

(e) Application of CBN grinding technology

CBN is hard, wear-resistant and possesses good thermal conductivity. This combination of properties offers advantages over conventional grain materials for grinding purposes. In spite of these advantages, CBN grinding wheels have yet to become established on a broad basis. This is due to the high purchase costs. It would be erroneous to consider this as the only factor when deciding on whether to deploy CBN grinding wheels. The question as to whether savings can be attained with CBN is to be clarified for each individual case by means of a careful analysis of the requirements concerned.

CBN grinding technology must not necessarily involve high cutting speeds. Multilayer CBN grinding wheels are deployed at cutting speeds of 80...140 m/s. Bonding materials for such wheels are ceramic, synthetic resin or metal. All types of grinding wheels can be profiled and are thus suitable for flexible application.

The hardness and wear-resistance of the CBN grinding material can be utilized to effect drastic increases in the volumes of material removed, provided that the grinding wheels are optimally prepared for the

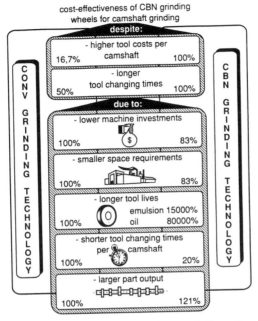

Fig. 9.21 Advantages of CBN grinding wheels for camshaft profile grinding (source: Winter & Sohn)

application concerned and rigid, high-performance machines are available. The example of camshaft profile grinding (Fig. 9.21) clearly shows that the high procurement costs for CBN grinding wheels are more than compensated by advantages such as drastically increased tool life, reduced tool-changing time per workpiece and increased output. In this example, the grinding time per workpiece is around 40% lower than for grinding with conventional grinding wheels [54].

9.3.3 More advanced erosion processes

The main areas of application for the previously presented processes and production sequences are in parts production. As has been demonstrated, high-quality and reliable tools are required for these purposes. This applies not only to cutting and grinding processes but also to casting, hot and cold forming, sheet metal forming and the production of moulded plastic parts.

Spark erosion processes (EDM) are often employed to produce dies,

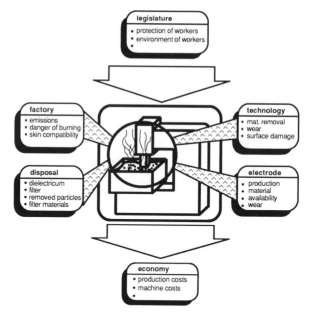

Fig. 9.22 Factors influencing the economic efficiency of spark-erosion processes

moulds, punching tools and drawing tools. These processes are not subject to any restrictions with regard to the hardness of the materials to be processed. The decisive precondition for the use of spark-erosion processes, either sinking or cutting with a wire electrode, is an electrically conductive workpiece. The further development of both processes is pursuing the same aims as those which apply to cutting production processes: increased performance, process reliability and quality in conjunction with reduced costs.

(a) EDM-sinking with an aqueous working medium

Spark-erosion machining is carried out in a working medium which exerts a considerable influence on the material-removing process and the economic efficiency of the process (Fig. 9.22). Up to now, oil-based working media have been used for spark-erosion. Such media possess a poor level of environmental compatibility and poor efficiency, however. An ecological and economical alternative to conventional oil-based dielectrics is provided by aqueous working media which consist of water and an organic, water-soluble component.

The working medium carries out highly important functions for the

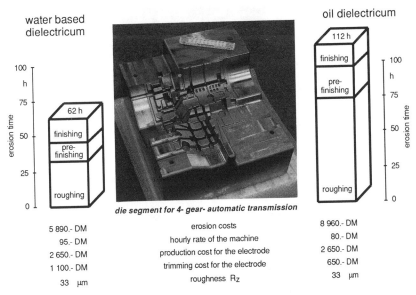

Fig. 9.23 EDM-sinking with different dielectrics (source: AEG-Elotherm)

material-removal process, as it is responsible for cleaning the working area, constricting the discharge channel and cooling. Oleaginous and aqueous working media differ substantially with regard to cooling capacity, as the water component has a strong cooling effect resulting from evaporation processes. The working area is thus cooled so intensively that small electrode geometries can be subjected to a high current. Furthermore, less gas forms in the discharge area, resulting in greater process forces. The material which is melted via the thermal effects of the spark discharge process is thus removed more efficiently from the discharge crater. In the area of rough machining, reductions in erosion times are attained which result in marked savings in the time required for rough machining and preliminary finishing in the production of dies (Fig. 9.23) [55-58].

This also has sustained effects on the economic efficiency of the process, as shown by the example of the mould segment for an automatic gear unit. Higher hourly machine costs for the aqueous working medium, resulting from higher plant costs and a more complex treatment process for the water-based medium, must be taken into account. However, the vast reduction in erosion times compensates for both the higher hourly machine costs and the higher costs for electrode reconditioning, resulting in a

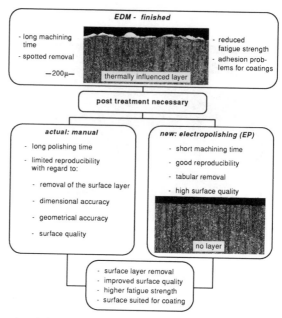

Fig. 9.24 Manual and electropolishing postprocesses

marked improvement in economic efficiency when using aqueous media.

(b) Combination of spark-erosion and electropolishing processes

The surfaces and surface zones of spark-erosion machined workpieces possess process-induced characteristics. The surfaces are characterized by crater-like structures. The surface zones are hardened, the microstructure is transformed and subject to high internal tensile stress. In some cases, the surface zones even have cracks. Erosion processes are frequently employed in the area of die and mould production. Such tools are subject to high levels of alternating mechanical and thermal stress in operation. It is an established fact that poor surface qualities, "white" surface zones and internal tensile stress are very harmful when such load applies. It is therefore necessary to smooth the surfaces, at the same time completely removing the affected surface zone. To date, it is virtually impossible to mechanize this work, due to the high hardness of the surface zone, the complex geometries of dies and moulds and the resultant poor accessibility.

The development of aqueous dielectrics for spark-erosion machining has opened up a new way of solving such problems - electropolishing (Fig.

9.24). The material-removal mechanism of this process is based on the principle of the anodic dissolving of metal with the aid of an electrolyte and a direct current. This process is carried out at temperatures below 50°C. It causes no thermal damage to the workpiece and induces no internal stress.

On technological and economical grounds it is expedient to carry out the EDM and the electropolishing (EP) process on one plant. This eliminates problems with regard to rechucking and down-times. Also, there is no need to invest in two plants. However, as non-conductive fluids (dielectrics) are required for the EDM process and conductive fluids (electrolytes) are required for electropolishing, a change in the working medium must be effected between the two machining stages. This can be carried out much more simply when using two water-based fluids than when using an oil dielectric and a water electrolyte.

The maximum material removal required in the EP post-process of eroded workpieces is 0.5 mm. Such small allowances do not result in any gap widening during electropolishing, as a result of which the dimensional and geometrical accuracy of the polished parts is high. Development of the combined EDM-EP process is far from complete. It harbours great potential for the reliable, fast and cost-effective post-process machining of the surfaces of eroded moulds and dies.

(c) Precision machining via wire EDM
Wire-EDM is employed primarily in the production of precision active elements in the field of punching and blanking tool construction. As in the case of EDM-sinking, the material-removal mechanism is based on the principle of the local melting, evaporation and ejection of material. Consequently, surfaces cut by means of spark-erosion processes also possess crater-like structures. The surface zones of the workpieces are thermally influenced and transformed as a result of the heating and cooling pWire-EDM is employed primarily in the production of precision active elements in rocesses which follow in quick succession in the course of material removal. As punching and blanking tools are subjected to high levels of static and dynamic stress, it is practically always necessary to remove these surface zones. This is a time- consuming and expensive matter [59, 60].

In order to avoid such finishing processes, wire-EDM trim cutting is employed (Fig.9.25). In this process, the main cut is followed by several trim cuts with successively reduced lateral infeed and discharge energy. This results in improved surface quality, and the thickness of the

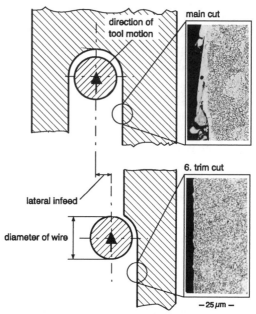

Fig. 9.25 Wire-EDM trim cutting - Principle and surface zone

transformed (white) surface zones decreases. The quality attainable via this trim cutting technique is usually sufficiently high to eliminate the need for any additional finishing measures.

The effects of the respective trim cuts on the workpiece are not identical. Up to and including the third trim cut, a new, thinner surface zone is produced by removing the former surface zone. This process not only improves the surface quality and accuracy of the workpiece, but also reduces the internal tensile stress in the surface zone (Fig. 9.26). From the fourth trim cut on, only levelling of the surface is carried out. In the course of this levelling process, no completely new surface zone is formed, as a result of which the existing surface zone becomes increasingly thinner [61]. Virtually no change occurs with regard to the internal stress, however.

The trim cutting technique not only results in improved surface quality and better dimensional and geometrical accuracy of the workpiece, it also improves the component performance characteristics decisively (Figure 9.26, right). The fatigue strength under reversed bending stress factors increases markedly as the number of trim cuts rises.

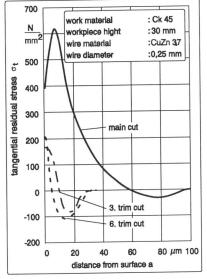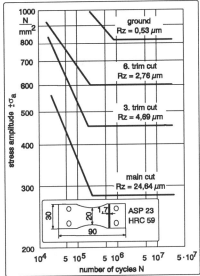

Fig. 9.26 Wire-EDM - Internal stress levels and component performance
characteristics

(d) Automatic control for spark-erosion processes

In spark-erosion machining, the gap width must be continuously adapted to
the requirements of the erosion process. Control and monitoring facilities
are therefore essential. Consequently, strategies for automatic process
control are at a more advanced level of development than in the case of
cutting production processes. This applies equally to erosion with a wire
electrode and EDM-sinking.

In the case of erosion with a wire electrode, the conditions in the
working gap are largely constant. This is due to the motion of the wire
electrode, which ensures that electrode wear cannot alter the conditions in
the working gap.

These conditions enable the machines to be equipped with facilities for
automatic process control. The input data includes the workpiece
geometry, the tool material, the machined material, the geometrical
accuracy and the required surface quality for the workpiece. In conjunction
with technological data stored in the machine, this data is processed to
create a specific NC programme for the workpiece concerned and
transmitted to the machine controller. The process then runs automatically,
without any intervention by a machine operator. The example of wire-

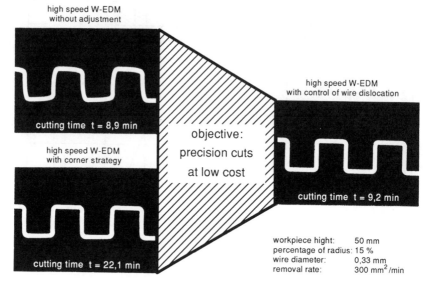

Fig. 27 Contur accuracy in wire-EDM high-speed cutting processes (source: AGIE)

EDM high-speed cutting illustrates the effects of such systems on precision and machining times (Fig. 9.27) [62].

In contrast to wire electrodes, the electrodes employed for EDM-sinking possess three-dimensionally complex shapes. When such electrodes penetrate into the workpiece, the active area of the electrode changes. This requires continuous adjustment of the process parameters to the area of the electrode which is currently actively involved in the removal of material. The creation of programmes for such machining operations requires extensive computation, in order to adapt the power input of the machine.

For optimal execution of such operations, systems are available today which adapt the process parameters to the currently effective electrode surface area automatically, after input of the workpiece/tool pairing and the penetration depth. This enables substantial reductions in the time required for programming. The risk of programming errors is practically eliminated. The optimized process sequence reduces the machining time. In addition, local excessive wear of the electrode is avoided.

9.4 Summary

Production plants are currently required to undertake
drastic reductions in manufacturing costs. The
objective is provided by safe, reliably and simple pi
production sequences. The essential potential for a(
available, but it must be exploited more purposefull
involves incorporating the know-how of employees .....
and implementing processes on an inter-departmental basis. The processes
and production sequences must be analyzed sufficiently to enable the
development of safe and reliable methods for manual or automatic process
control and process monitoring.

Examples of grinding, of hard and soft machining with geometrically
defined cutting edges, of dry machining, of the use of new dielectrics in
spark-erosion, of new cutting strategies for wire-EDM and of the
configuration of production sequences show the approaches which must be
pursued in order to attain the objective of safer, more reliable and more
cost-effective production operations in the short term.

10

SUPERIOR TOOLS AS A BASIS FOR TOMORROW'S MANUFACTURING

10.1 The future scope of requirements for production engineering

The specific performance capabilities of a product play a decisive role in determining its success or failure on the market. Key terms such as power/weight ratio, size, reliability and service life are prime factors shaping the scope of requirements for present and future products, together with economic efficiency and environmental compatibility.

While public attention often focusses on the products with their "technical innovations", it is not rare for the actual innovations behind such improvements in performance capabilities to take place out of the public eye. New materials, defined component configuration and, above all, an optimized and innovative production process, are the prerequisites for technically sophisticated products.

As an example, Fig 10.1 shows the development of the running performance of automobile gear boxes. To date, increased wear on the bearings resulting from inadequate lubrication supply has been the prime cause for the failure of gear boxes. By specifically influencing the surface during production, the performance of the lubricating system and,

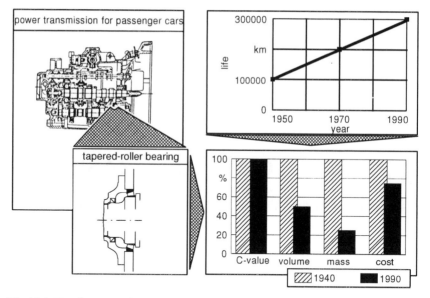

Fig 10.1 Development of the lifetime of automobile gear boxes (based on [1])

subsequently, the service life of the bearing can be decisively improved [1].

In future, the scope of surface modification will be substantially broader than the generally recognized requirement for the minimization of surface roughness. Defined "production" of the surface microstructure is required in this area, together with specific influencing of the surface layer (Fig 10.2). Whereas the prime objective of reducing surface roughness is to maintain narrow tolerances in conjunction with increased endurance strength and improved wear and corrosion characteristics, the defined modification of microstructures is aimed above all at attaining favourable tribological and fluidic properties. In pursuing these aims, production engineering will have to move away from the acceptance of self-determining surface structures and seek a way of specifically producing determined structures. As a supplementary measure to defined surface modification, specific influencing of the surface layer also serves to improve endurance strength and improve wear characteristics.

The characteristics of the surface are determined not only by the material which is employed, but also and in particular by the applied production process and production technology. Along with the machine performance characteristics and the machinin parameters, the deployed

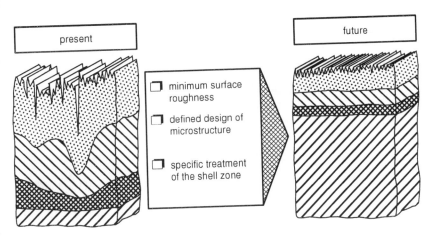

Fig 10.2 Requirements to be met in the field of production engineering in order to attain an "optimal" component structure

tool has a decisive influence on the attainable surface topography and surface quality.

The characterization of optimization strategies for conventional tools reveals parallels with the machining task and the requirements relating to the workpiece: optimized geometry and specifically selected base and cutting materials in conjunction with defined structuring of the tool surface are vital prerequisites for efficient production operations.

But beyond this - on the basis of the stated requirements regarding the performance capabilities of future production processes - similar developments are also required for the manufacture of conventional products in some areas of production engineering, whereby the need is for "visions" to be found, rather than mere additional improvements. In the area of finishing, for example, the main emphasis will no longer be on "conventional" machining. Rather, the defined modification of surfaces is required, with the aid of new technologies such as ion implantation, laser or electron beam treatment. The production of nanometric structures on the surfaces of components will also open up previously untapped potential for conventional products.

Beyond the range of conventional products, microsystems engineering products must be given prime consideration when analyzing future main areas of development. Rapid growth is forecast in this decade for the combining of microelectronic, optoelectronic and micromechanical

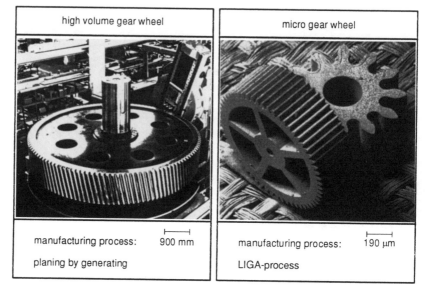

high volume gear wheel	micro gear wheel
manufacturing process: 900 mm	manufacturing process: 190 µm
planing by generating	LIGA-process

Fig. 10.3 Gearwheels for large-scale mechanical engineering and for microsystems engineering (MicroParts GmbH)

components to form one operational system [2,3,4,5,6,7,8,9,10]. The comparison of macro-gearing with a micro-gearwheel shown in Fig. 10.3 vividly illustrates the attainable dimensions and structures which will challenge production engineering in the future.

While the design and production of microelectronic and, in some areas, optoelectronic components are already well developed and such components can be produced in large series at low costs, this situation does not apply to three-dimensional micromechanical components. Due to the minimal size of such components and the substantially greater surface/volume ratio, experience and laws relating to thermal conductivity, strength, friction, etc. in the macroscopic area are not necessarily applicable to the microscopic range [11]. Neither have the possible influences of the production process on these characteristics been adequately researched.

Production engineering for the manufacture of three-dimensional micromechanical components is based either on semiconductor technology [4,8] for the manufacture of silicon-based components or on the use of the lithographic/electroplating/shaping process, which also enables other metals, plastics and ceramics to be processed [4,12]. The driving force behind the development of these processes has not been production engineering, but rather the area of chemistry and physics. Now that these processes have left the laboratory stage, more attention must be paid to aspects such as economic efficiency and quality as they are implemented in series production operations [13]. Here again, production engineering will thus be challenged to concern itself with and utilize new technologies which have not previously been introduced on a general basis.

On the basis of the above-stated situation, a general distinction can be made between three types of development for tools and processes:

In order to take due account of the increased and further increasing requirements for economic efficiency and quality, **"conventional tools"** which are employed to produce **"conventional products"** will be modified by means of new processes. The technologies employed to achieve this modification will include processes which act on the surface in particular.

In order to enable conventional products to be manufactured with new properties or high performance capabilities, it will sometimes be necessary to employ **new materials** which, in turn, will require **new processes** for economic processing.

Innovative products with completely new properties will necessitate the application of **new technologies** not included in the current repertoire of production engineering processes. Such technologies include lithographic processes to produce optoelectronic micromechanical components.

On the basis of the above-stated directions of development, the following selected examples of processes illustrate relevant aspects of the future face of production engineering. In addition to the further development of processes which are already established in industry, technologies are also presented which are presently still in the initial stages of industrial application.

10.2 The improvement of conventional tools via new processes

The suitable processes here primarily include surface modification processes, i.e. coating and structuring processes. Besides established techniques, such as laser surface treatment, modern thin-layer technologies constitute the prime area of interest.

The world market for thin-layer plants is currently estimated at DM 10 thousand million, and the products manufactured directly with these plants

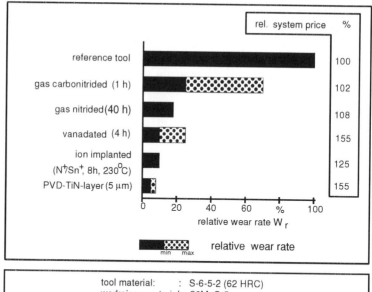

Fig. 10.4 Comparison of different wear protection processes for extrusion dies (based on [15,16]).

at approximately DM 100 thousand million [14].

Thin-layer processes are employed in such diverse fields of application as optics, electronics and mechanical engineering. Wear protection layers on tools are particularly well known. Fig. 10.4 shows the influence of various wear protection processes using the example of an extrusion die.

The diagram shows both the reduction in wear in relation to the untreated die (on a scale of 100%) and the total costs for manufacture of the die and the surface treatment.

In addition to known processes such as nitriding, unconventional processes still in the initial stages of industrial application, such as ion implantation, have been examined, along with other processes which have already undergone further development, such as physical vapour deposition processes (PVD), whereby these two last-mentioned processes provided very good results, also with regard to economic aspects.

The general aim in the area of coating technology is to deposit layers with characteristics designed specifically to meet the demands of the

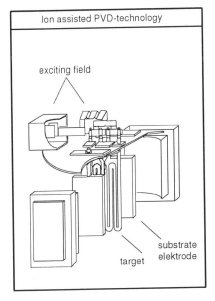

Ion assisted PVD-technology

exciting field

target   substrate elektrode

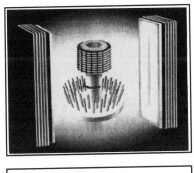

- lower thermal load of the workpiece
- homogeneous layer structure due to surface activation
- highly wear resistant
- highly corrosion resistant
- small coefficient of friction

Fig. 10.5 Wear protection via modern PVD technology (based on Leybold)

application concerned, without affecting the base material. To this end, particular efforts are being undertaken to lower the coating temperature.

Fig. 10.5 shows an example of the design of a modern PVD plant employed to deposit wear protection layers on tools.

The improved sputter ion plating process shown here involves activating the surface of the workpiece to be coated via ions from a plasma surrounding the workpiece.

Activation of the surface is necessary in order to produce an impervious and firmly bonded layer. The processes employed to date obtain the energy required for activation via the high substrate temperature. The decisive factor affecting the development of the layer's structure is the relationship between the substrate temperature, T_s, and the melting temperature of the coating material, T_m. In the case of non-particle-assisted processes, an impervious and smooth layer does not form until zone 3, whereby $T_s/T_m > 0.45$ (Fig. 10.6) [17,18]. When mechanically resistant layers are applied to substrates made of steel or carbide metal, this may change or damage the substrate [19,20].

In the presented process, activation of the surface is attained via particle bombardment. This process enables the growth of an impervious and firmly bonded layer at substantially lower temperatures (zone T,

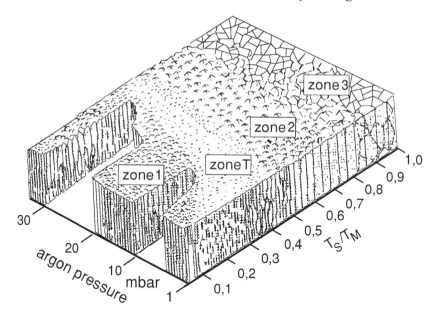

Fig. 10.6 Structure of PVD layers deposited via particle-assisted processes (based on [17,18])

$0.15 < T_s/T_m < 0.4$), depending on the partial pressure of the sputtering gas. The properties of the substrate remain essentially unaffected.

Fig. 10.6 shows the structure of the layer in relation to the temperature when particle activation of the surface is carried out. The "T zone" is accessible only when particle-assisted processes are employed.

As a result of the further development of this technology, the economical coating of large batches is now possible [21]. Additional aspects are the possibility to structure layers in accordance with the requirements of a specific problem and the high environmental compatibility of the process, as no toxic substances are employed or produced.

Recently, particular attention has been devoted to diamond layers deposited on tools or components. In addition to applications such as optoelectronics, these layers are of particular interest for the coating of forming and cutting tools. In this area too, current efforts are aimed at reducing the present coating temperature of approx. 900°C to around 550°C. This would enable high-speed steels to be coated without any loss of hardness, as a result of which expensive carbide tools could be

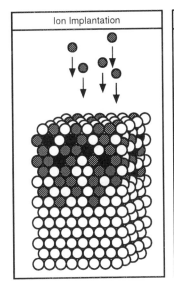
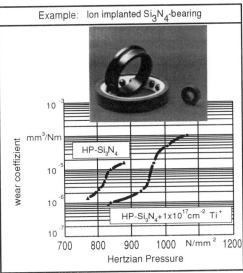

Fig. 10.7 Wear reduction on rolled ceramic specimens by ion implantation
(based on BAM)

substituted by cheaper tools produced in coated high-speed steels for certain applications. It would also be possible to produce low-wear tools with greater flexibility regarding geometry.

Ion implantation and the ion beam mixing and ion beam assisted deposition (IBAD) processes derived from ion implantation enable a further reduction in the treatment temperature in comparison to the PVD process, thereby broadening the scope of materials suitable for processing. Fig. 10.7 shows the ion implantation process in diagram form, together with the results of wear measurements on an ion-implanted rolled ceramic specimen. A corresponding practical example would be an ion-implanted ceramic bearing.

A common characteristic of the ion-assisted processes is that they are based on the use of particle radiation, which requires the process to be carried out in a high vacuum.

The ability to precisely control the process parameters results in a high level of process reliability. As the above example shows, this process is not restricted to the use of conductive substrates. It is equally suitable for metals, ceramics and plastics.

The mechanisms responsible for the reduction in wear when ion

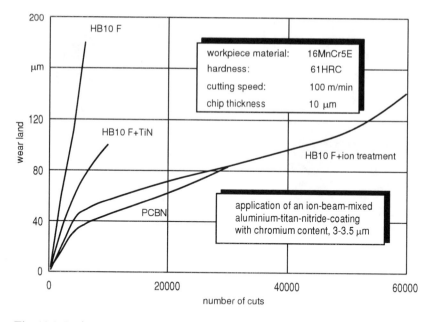

Fig. 10.8 Prolonged lifetimes for milling tools (according to WZL)

implantation is carried out are varied, and a precise knowledge is lacking in some areas. Mechanisms requiring particular mention include increased hardness of the surface layer due to lattice imperfections and the formation of compounds, the induction of internal compressive stress, improved corrosion resistance and reduction of the coefficients of friction [22,23].

Implantation of the ceramic specimen presented in Fig. 10.7 was carried out at 1×10^{17}Ti+ ions per cm² and an accelerating voltage of 200 keV. The wear tests on the rolled ceramic specimen were carried out in ungreased condition at room temperature (Si_3N_4 against Si_3N_4), and at a normal force of 290N.

The rate of wear for the ion-implanted specimen is several times slower than for the non-ion-implanted reference specimen. The causes of this behaviour in Si_3N_4 are yet to be fully understood. Aspects under discussion are the induction of internal compressive stress and the formation of lubricious oxides.

Ion implantation, which originates from the field of semiconductor technology, where it is used in series production, is attaining increasing importance in the area of mechanical engineering. The process is of particular interest for precision components and tools subject to high levels

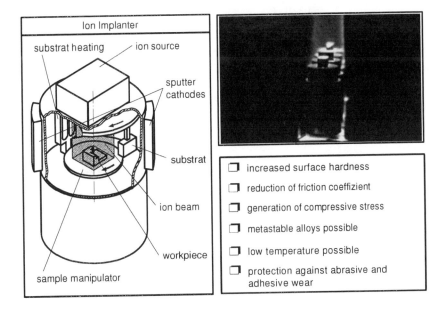

Ion Implanter

substrat heating — ion source

sputter cathodes

substrat

ion beam

workpiece

sample manipulator

☐ increased surface hardness

☐ reduction of friction coeffizient

☐ generation of compressive stress

☐ metastable alloys possible

☐ low temperature possible

☐ protection against abrasive and adhesive wear

Fig. 10.9 Wear reduction by ion beam processes (according to Leybold)

of strain, as work can be carried out at low temperatures, no dimensional changes occur and very good bonding of the modified surface layer with the substrate is guaranteed.

This characteristic is also utilized when producing layers with simultaneous bombardment by high-energy ions. The resultant coatings possess excellent bonding strength, rendering them particularly suitable for use on milling tools (Fig. 10.8).

This figure shows the results of service life trials carried out on carbide tools with various coatings. The results obtained with a tool made from polycrystalline boron nitride (PCBN) are shown for comparison. It is clear that under the treatment conditions applied here the carbide tool with the ion beam-assisted coating offers substantial advantages with regard to service life.

Fig. 10.9 shows a plant employed to produce such layers. The main components are the vacuum chamber with the appurtenant set of pumps, the ion source with acceleration section, the specimen manipulator and the gas supply. The plant further incorporates components for ion-assisted coating.

While the previously described processes are aimed at changing the

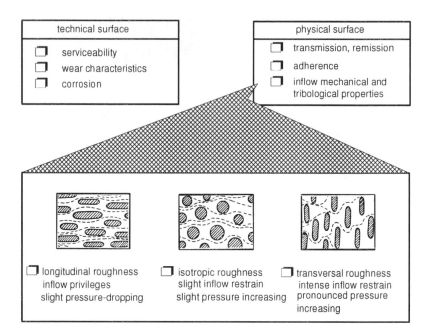

Fig. 10.10 Influence of the physical and technical surface on component
characteristics (according to [24])

chemical/metallurgical structure of the surface layer, processes for the
defined structuring of component or tool surfaces, with the aim of
influencing tribological characteristics, for example, are attaining
increasing importance. In the course of such processes, the surface is
provided with deterministic structures on the micrometre scale, enabling
the flow characteristics of lubricants to be influenced, for example
[24,25,26).

Fig. 10.10 shows the component characteristics which are influenced
via the physical or technical surface. Particular emphasis is attached to the
influencing adhesive and fluidic characteristics via specific structuring of
the surface, as illustrated by several simple structures. The hatched areas
are to be interpreted as elevations, around which a lubricant, for example,
may flow. In practical applications, determination of the form and size of
the required structure may require extensive mathematical operations,
however.

Fig. 10.11 shows an application from another area which has already
been implemented. This example concerns structuring of the surface of

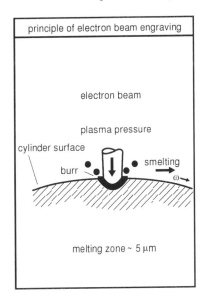

principle of electron beam engraving

electron beam

plasma pressure

cylinder surface

burr

smelting

ω

melting zone ~ 5 μm

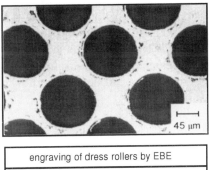

45 μm

engraving of dress rollers by EBE

❏ rollers for aluminium and steel sheets

❏ EBE 300.000 pans/sec, laser 50.000

❏ size and position of pans controllable

❏ enhanced rheological properties of the sheet

❏ enhanced adhesion of lubricants and coatings on the sheet

Fig. 10.11 Structuring of surfaces to improve flow and adhesive characteristics (according to Linotype-Hell)

temper passing rollers with the aid of an electron beam. This structuring processes improves the flow characteristics of the rolled metal and the adhesion of a subsequently applied film of paint or lubricant [27,28].

Similar structuring can also be carried out with a laser, with the additional advantage that the treatment process can be carried out in the atmosphere. Disadvantages are the reduced speed of the surface treatment process and the occurrence of oxidized burrs.

A common feature of all the processes referred to by way of example is that they involve specific modification of the surface. The processes employed to this end incorporate the use of particle radiation as a "tool", in order to ensure the required level of precision for the treatment process. The processes furthermore subject the workpiece to only minimal levels of thermal stress, rendering plastics and heat-sensitive alloys suitable for surface modification.

Optimal application of these tools requires an extensive understanding of their mechanisms of action and the stress factors which occur when the components are deployed. Primary requirements for the future, therefore, are the theoretical formulation of these variables and implementation of the resultant knowledge in specifications for surface treatment. One aim to be pursued in the course of this work is configuration of the characteristics of the modified surface on a localised basis, to enable adaption to the different types of stress which apply in different local areas [29]. This requires mathematical formulation of the stress factors, implementation in specifications for surface modification and production of the appropriate surfaces.

10.3 New machining processes for new materials

The growing problem of the environment has resulted in corresponding changes in consumer behaviour and legislation. The increasingly severe greenhouse effect, deforestation and the problem area of refuse disposal have led to more stringent maximum levels for exhaust gases and regulations on the recyclability of products.

Fig. 10.12 shows the development trend for pollution caused by motor vehicles by reference to the example of carbon monoxide (CO), and the causes behind the measured and forecast reduction [30].

Possible methods of reducing fuel consumption are increased efficiency, improved aerodynamics of vehicles and reductions in the weight of vehicle components. Initial measures in this direction are the use of components made of structural ceramic in motors [31,32,33] and the use of aluminium in structural components and panelling [34,35,36,37,38], whereby the use of aluminium simultaneously takes account of the requirement for recyclability.

In some instances, the use of such materials requires the deployment of tools with previously unutilized mechanisms of action, in order to enable a treatment process in accordance with the specific requirements of the material concerned. The area of production engineering is thus challenged to develop new concepts and solutions which will enable economical implementation of the new concepts in series production operations.

In the automobile industry, for example, there is a need to produce cooling and hydraulic circuits in plastic-coated aluminium tubing, for reasons of weight reduction and corrosion resistance. However, the material combination of aluminium and plastic render mechanical and thermal processes unsuitable for forming and joining operations.

One approach to solving this problem is provided by a process which operates without mechanical contact between the tool and the component to be formed - magnetic forming (Fig. 10.13).

This process utilizes the energy contained in an electromagnetic field

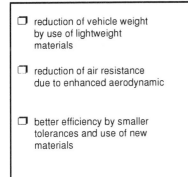

- reduction of vehicle weight by use of lightweight materials

- reduction of air resistance due to enhanced aerodynamic

- better efficiency by smaller tolerances and use of new materials

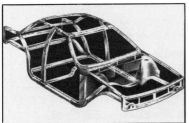

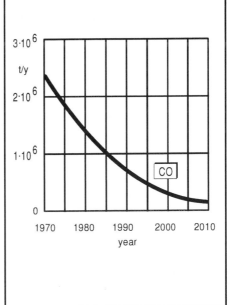

Fig. 10.12 Development trend for exhaust emissions from passenger vehicles (according to [30])

for the purposes of forming. Sudden discharge of the capacitor via the coil generates an electromagnetic field [39,40,41]. When a component made of a conductive material (e.g. an aluminium tube) is located in the coil, interaction occurs between the current induced by the electromagnetic field and the field itself. This results in a build-up of radial pressure in the wall of the workpiece, leading to deformation of the component in a few milliseconds after the yield point has been exceeded.

As there is no mechanical contact between the "tool" and the workpiece, no wear occurs to the tool. The level of process reliability is very high, as the process parameters can be controlled and monitored very precisely. No lubricants, which would necessitate subsequent cleaning of the surface, are required.

Other workpiece geometries can be formed by selecting a suitable geometry for the coil. Even non-conductive materials can be formed, using a conductive driver. This process is also suitable for solving joining problems encountered with combinations of materials.

While aluminium is already an established material in other areas of

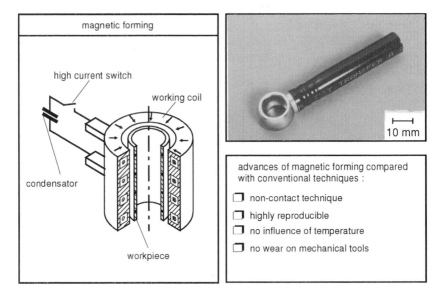

Fig. 10.13 The magnet forming process (based on Puls-Plasmatechnik GmbH)

automobile construction, ceramic still represents an exotic alternative. The original wish to produce a "fully ceramic combustion engine" has given way to a more realistic approach. Interest now focusses on the development of individual components. Work is currently in progress on developing ceramic inlet and outlet valves to a standard suitable for series production. These are lighter than metal valves and thus enable the improvements specified in Fig. 10.14. The resultant savings in fuel are small in percentage terms (0.4%), but nevertheless substantial.

The need for economical processing of such components suggests that it may be expedient to substitute the grinding process employing cooling lubricants, which is used exclusively at present, with a turning process which does not employ cooling lubricants. However, the material properties of ceramic will then render it vital to introduce additional energy into the surface layer for the purpose of material removal. This energy can be introduced in the form of laser radiation. Fig. 10.15 shows the principle of laser-assisted machining (LAM).

Polycrystalline mechanically resistant materials such as CBN or PCD are employed as cutting materials; the attainable surface quality is around $R_z < 3$ µm. The use of geometrically defined cutters also allows a greater scope of freedom with regard to the geometric design of the workpiece. In

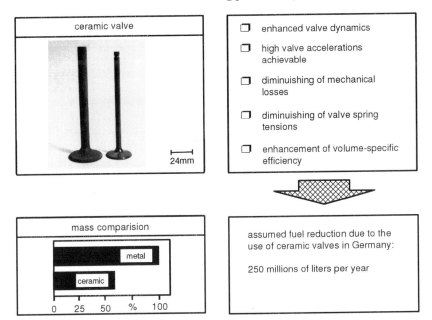

Fig. 10.14 Advantages of using ceramic valves

addition to ceramic materials, this process is also suitable for machining high-strength steels in the field of engine construction.

The high level of investment and the large area of space required by the combination of lathe and CO_2 laser represent an obstacle to the introduction of this technology. Fig. 10.16 presents a potential solution to this problem. The figure shows the present-day configuration, consisting of a lathe and a high-power laser with subsidiary units, and a further development whereby a miniaturised high-power solid-state laser [42,43,44] including subsidiary units is integrated directly into the lathe.

In addition to the minimal size of such a laser, additional advantages are its high efficiency level (approximately 6 times higher than the CO_2 laser), its higher service life (approximately 2-10 times longer than the CO_2 laser), maintenance-free operation and the substantially lower price. One watt of laser power for a CO_2-laser presently costs DM 200-400; one watt of laser power or such a diode laser would cost only around DM 100, whereby a further rapid fall in prices is to be expected according to the development in the field of micro- and optoelectronics [45].

The increasing use of new materials such as ceramics or composite materials is taking place against the background of rising requirements

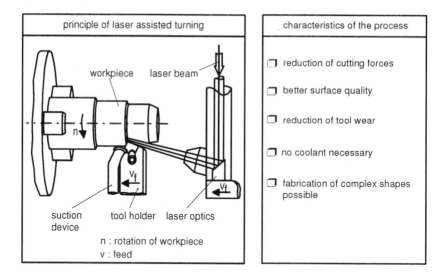

Fig. 10.15 The principle of laser-assisted machining (based on IPT)

imposed on the performance capabilities of products. In the light of this situation, production engineering is required to provide the appropriate tools to enable economical machining in accordance with the specific requirements of the materials to be machined.

The majority of the tools which are able to meet these requirements are tools which employ beams, in particular the laser, the flexibility of which renders it suitable for a broad range of applications. The laser can be employed, for example, not only as a supporting tool in material removal operations, as shown in the presented example, but can also be used directly for the purpose of surface structuring. In contrast to material removal, a further alternative application involves the specific build-up of material, via laser sintering, for example.

10.4 New technologies for the manufacture of innovative products

The increasing fall in prices in the area of electronic and, more recently, optoelectronic products enables the use of highly developed components in low-price consumer goods. As a result, established products, primarily from the field of precision mechanics, are being ousted from the market. Examples of this effect are the replacement of record players by CD

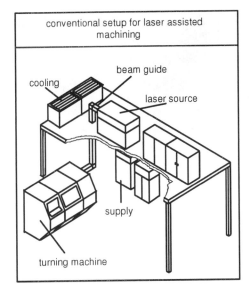
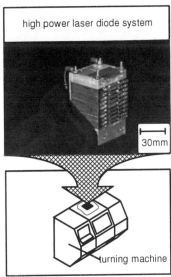

Fig. 10. 16 Laser-assisted machining using the diode laser (based on ILT, IPT)

players by CD players or of Super 8 cameras by video cameras. It is the microelectronic or optoelectronic components of these products which determine their quality and price.

But these trends are not restricted to the consumer goods industry. The main area of deployment for micro- and optoelectronic components will initially be in the field of communications engineering, for which a worldwide investment volume of approx. DM 1600 billion is forecast for the next 10 years [48]. Fig. 10.17 shows a further product based on optoelectronic components, in the form of a newly developed fibre gyroscope, which will provide a replacement for conventional precision-mechanics gyroscopes in established segments of the market as well as opening up new areas, such as robot control or the control of driverless transport systems.

The special advantage of this optoelectronic system, which is now in series production, is its robustness, which indicates that it will be suitable for deployment in the rough conditions of everyday industrial operations [46,47].

The primary components of the fibre gyroscope, the semiconductor diode, detector and evaluation unit, are microelectronic and optoelectronic units, for the manufacture of which conventional production engineering is

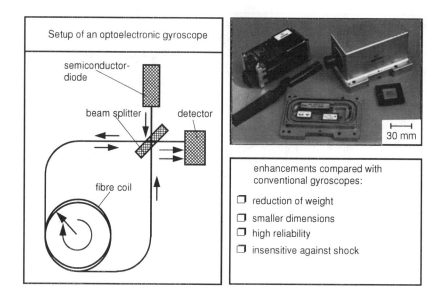

Fig. 10.17 Gyroscope (based on SEL-Alcatel)

unable to offer any adequate machining technologies.

To a certain extent, the production technologies employed today to manufacture micro- and optoelectronic components are related to those presented in section 2: These are also primarily layer technologies and tools which specifically influence the surface [49,50]. As an example, Fig. 10.18 shows a diagram of the lithographic process which is employed in semiconductor and microsystems engineering. An interferometer for longitudinal measurement in high-precision machine tools which has been developed on the basis of this technology is already available on the market [51,52].

Apart from the established and mastered technology of chemical etching to structure surfaces, the main focus of interest in the field of microelectronics is on innovative lithographic processes. The reason for this interest lies in the structure size which is currently attainable: due to the wavelength involved, conventional exposure processes based on ultraviolet light enable a minimum structure size of only approximately 300 nm. Finer tools, such as electron or ion beam treatment or x-ray lithography, are essential, in order to attain smaller structure sizes for higher levels of component density. Structural widths in the range below

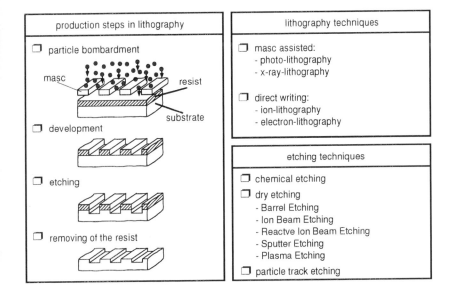

production steps in lithography	lithography techniques

Fig. 10.18 Lithographic and etching processes in production operations

100 nm appear to be attainable here.

Very similar processes are employed by the area of microsystems engineering, which is presently in its initial stages of development [3,4]. As a supplementary technology to precision mechanics, microsystems engineering will initially be applied primarily in the areas of sensor technology, actuator technology and medical technology.

The processes can be differentiated according to the following manufacturing variants:

- the production of planar structures (e.g. sensors)
- the production of three-dimensional structures (e.g. motors)
- the production of three-dimensional structures in metal, plastic and ceramic (e.g. gearwheels, motors)

The first two variants are based on processes from the field of semiconductor technology or the further development of such processes. The range of machinable materials is restricted to silicon, however.

The last-stated variant generally employs LIGA technology, which enables greater freedom with regard to material selection and geometric design. Production of the master mould does require a substantial scope of production technology, however, as the x-ray radiation required for the

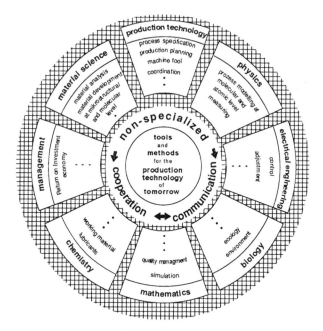

Fig 10.19 Interdisciplinary cooperation in the field of production engineering

lithographic process has to be generated in an electron synchroton.

The advantage of the above-described processes is the large number of workpieces which can be machined simultaneously in one machining operation. At present, approx. 16000 sensors can be produced in a single operation on a silicon disc of four inches in diameter [4]. Correspondingly large production volumes are possible at low unit costs, on an appropriately designed plant.

Although the technology for producing microstructures in silicon for sensors has been developed to a standard suitable for series production, intensive research work is still required in order to attain technical optimization of the processes. The specific areas requiring research are component layout, the applicability of macroscopic laws to the microscopic area, quality assurance and the economic efficiency of the processes. Industrial implementation of the processes demonstrated at laboratory level is necessary, in order to produce reliable and low-cost microsystems.

10.5 Summary

Increasing the performance capabilities of products is a decisive factor in determining their success on the market. Properties such as corrosion resistance or resistance to abrasive wear are acquiring importance in efforts to enhance the specific performance characteristics of a product, including its service life. As these characteristics are determined by the surface of the component, processes and tools which enable specific and controlled modification of the surface will acquire increasing importance in the production engineering of the future. The objectives in this area go beyond the simple minimization of surface roughness and are aimed rather at attaining defined surface modification in accordance with the specific requirements of the application concerned, while at the same time specifically influencing the surface layer.

The future design and production of surfaces will be based to an increasing extent on mathematical models, rather than empirical specifications. However, the formulation of such mathematical models requires a sound understanding of the processes which occur in the material during production and deployment. On the basis of this knowledge, the objective of production engineering will be to select and develop new processes which enable the defined structuring of surfaces.

Due to the small dimensions of the required structures and layers and the need to avoid changing the properties of the substrate, particle-assisted processes are attaining increasing importance in the area of surface modification. Such processes enable activation of the surface and specific, controlled machining into the submicrometer range.

The machining of materials which are rendered necessary by the increasing requirements relating to the performance capabilities of products often requires processes which incorporate new, previously unutilized mechanisms of action. The majority of the tools suitable for these areas of applications are tools which employ radiation, in particular the laser. This versatile tool is suitable for a broad range of applications, such as assisting the conventional machining of materials with poor machining characteristics.

With regard to the manufacture of completely innovative products - examples being optoelectronic sensors and mechanical components in the micrometer range - technologies will be applied which are currently employed in the field of semiconductor engineering. Beyond conventional exposure and etching processes, electron and ion beam machining and x-ray lithography in particular should be mentioned in this connection.

Due to the complex interrelationships and the various mechanisms of action involved in these processes, close interdisciplinary cooperation will be necessary, as illustrated in Fig. 10.19. Theoretical assessment of the processes which occur during manufacture and deployment of a product goes beyond the scope of a purely empirical approach, to enable optimization of the production process and/or the product itself. However, appropriate knowledge from the most diverse specialised areas is required for the purposes of this theoretical assessment. A primary task for production engineering in developing future products and processes will be to collate this knowledge and to render it usable for a specific problem.

11

INTEGRATED QUALITY INSPECTION

11.1 Introduction

In Europe, the German industrial companies in particular have developed a virtually perfect system for detecting quality defects on products, whereby this system involves high costs for equipment and personnel. After completion of the product, a decision is required above all as to whether the quality characteristics of the produced workpiece (e.g. a turned part) are within or outside of the specified tolerances, and whether a part which is not in compliance with the specified limits can be saved by reworking.

At the same time, the requirements relating to the quality of the produced workpieces have also risen. Whereas the AQL for sub-contracted parts was still measured in percentage points in the last decade, today failure rates are accepted only in the ppm-range - particularly for parts to be supplied to leading companies. Neither perfect testing and inspection processes at the end of the manufacturer's production line nor an exaggerated scope of inspection work at the customer's incoming goods department represent appropriate measures to fulfil these requirements.

A change in the scope of tasks for production-integrated metrology is essential: It is no longer sufficient to integrate more or less complex manually or computer-controlled measuring instruments into production and assembly lines and to file the obtained measurements without inputting them into the production system, as is common practice today. Rather, the measured data and signals from machine-linked or machine-integrated measuring instruments must be fed back into the production process in such a manner as to enable them to render a substantial contribution towards stabilizing and improving the capabilities of the machining processes.

11.2 Production metrology today

The method of fault detection at the end of the production line which is still practised today ties up vast capacities and ultimately causes enormous costs in the form of wasted production capacity. To enable all the workpieces to be delivered without any defects, either the quantity of parts machined at the beginning of the production process must be increased by the expected failure rate, or additional machine capacity has to be provided, in order to rework the defective workpieces or to produce completely new workpieces in their place.

Valuable time is also lost. By the time the finished workpiece has passed through the queue in front of the integrated measuring instrument, numerous other parts have been produced with the same machine settings - and the same deviations from the specified requirements - without the need for intervention having been recognised. Should the workpiece require tempering in a climatic chamber prior to undergoing precision measurement, the entire batch may be defective, as a result of the delay in measuring.

To facilitate faster reactions to geometrical deviations, the metrological devices can be increased to the extent that several measuring machines are used for the 100% inspection of the manufactured workpieces. Capable processes only require a reduction in the workpieces to be measured by introducing a quality control system employing statistical methods. However, this reduces the basis of knowledge for decisions which the inspector's acceptance or rejection is dependent upon.

11.2.1 Quality improvement via process-integrated control

In order to reduce the costs which are caused by both reworking and the production of replacement parts and by more stringent measuring cycles, quality assessment must not be restricted to the finished product alone. The production process must be controlled during machining of the workpiece

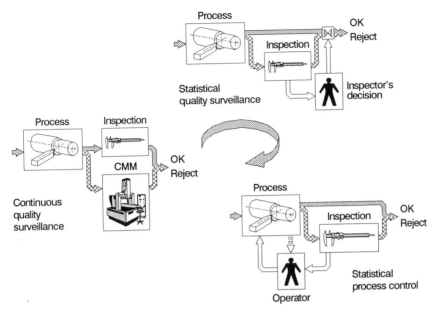

Fig. 11.1 Surveillance and control of production processes

in such a manner as to minimize the number of defective units which are produced. For this purpose, the quality capability of all the components involved in the process requires to be stabilized.

An essential requirement for any strategy for preventing defective products is the feedback of measurement results to the machining facilities. The aim is to stabilize the quality of the relevant process parameters within permissible tolerance limits by means of correcting or controlling mechanisms derived from the deviations established in monitored product and process characteristics. As in the case of statistical process control, the function of the final controlling element can be carried out by a trained operator, who corrects the machine parameters on the basis of measurement results and acoustic and visual signals (Fig. 11.1) [1].

The capability of the production process also influences the selection of testing and inspection equipment which is suitable for post-process surveillance of the finished product. In principle, it is possible to minimize the proportion of workpieces whose defects are not detected in the final inspection and testing process via an extreme increase in the capabilities of the testing and inspection equipment. This will ensure that workpieces whose dimensions are at the limits of the tolerance range can be assessed

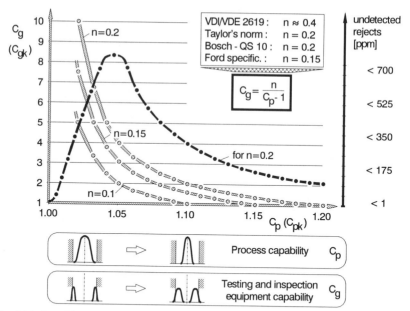

Fig. 11.2 Possible methods of reducing undetected rejects

as acceptable or rejected with a high degree of precision. However, such an approach will also result in avoidable costs, both for the defective parts which are produced and for procurement and maintenance of the measuring equipment.

A much more sustained reduction in the scope of testing and inspection operations - and a subsequent reduction in costs - can be attained by increasing the process capability. An advantageous secondary effect of this option is that the tolerance range for the production process is no longer fully exhausted (Fig. 11.2). As an increase in process capability also extends the permissible dispersion range for the testing and inspection equipment when a fixed tolerance range applies, simpler and consequently more cost-effective measuring instruments can be deployed [2,3].

A more reliable and objective method, however, involves attaining the required control over the process by continuously monitoring the process characteristics with the aid of sensors integrated into the process machinery and by controlling the process parameters. The necessary manipulated variables are adapted to the prevailing machining conditions, as a result of which any deviations from the required state are corrected immediately.

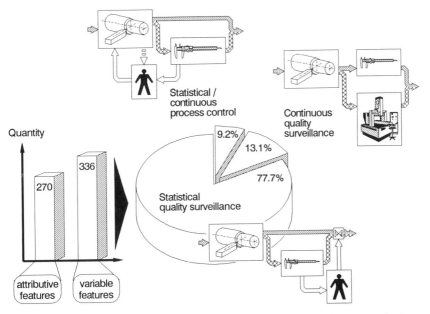

Fig. 11.3 Respective levels of deployment of the testing and control processes in the area of production [1]

As the results of production operations will involve only a minimal scope of dispersion when the critical process parameters are set correctly, the scope of additional measuring operations at the end of machining operations can be reduced. Final inspection and testing cannot be completely omitted, however, as a company must always be able to furnish evidence of conformity with customers' specifications and statutory requirements.

11.2.2 Process control in practice

A look at industrial practice reveals that the surveillance processes clearly predominate over controlling processes. A study carried out at a company which manufactures generators for automobile engines revealed that a total of 606 assessed inspection and testing sequences contained only 336 variable tests. The statistical surveillance processes accounted for the major proportion (77.7%) of these tests, with the continuous surveillance processes representing 13.1% of the total. Statistically or continuously

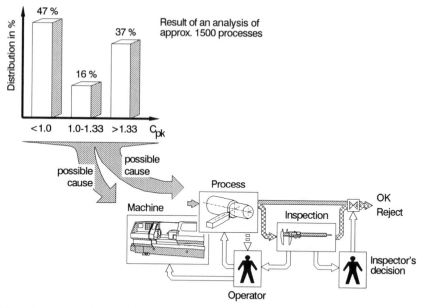

Fig. 11.4 Factors for consideration with regard to the improvement of process capability

controlled production processes accounted for a mere 9.2% of the inspection and testing sequences (Fig. 11.3) [1].

The results of these studies clearly show that there is an enormous need for measures to introduce control processes which intervene to correct the production process. First of all, however, a basis requires to be established which will enable effective deployment of a control strategy. The prime prerequisite for any type of quality control system is to ensure that there is a clear understanding of the connection between the configuration of a quality characteristic on the product - e.g. the dimensional, geometrical or positional stability - and the process parameters which correlate with these characteristics. Only then will it be possible to implement control mechanisms on the basis of this knowledge which activate correlating process control variables to reduce the level of deviation, when deviations occur in product characteristics of relevance to the quality of the product.

A current study of a large automobile group shows that the capabilities of the production process have to be optimized. Of more than 1500 processes analysed in the machining area of mechanical engineering, almost half (47%) were unable to fulfil the requirements regarding

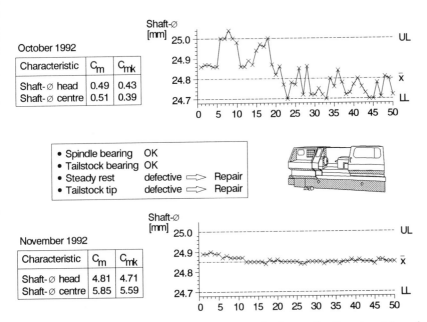

Fig. 11.5 Influence of machine components on production quality (based on Mercedes Benz AG)

compliance with specified tolerances. 16% of the processes had C_{pk} values varying from 1.0 to 1.33, while only 37% of the production processes continuously exceeded the required C_{pk} limit of 1.33 (Fig. 11.4).

One of the possible causes may be malfunctions in the deployed machine tool. During analysis of the production processes, for example, a lathe for machining rear axle shafts was identified whose machine capability, C_m, had fallen to extremely low values. Short-term analysis revealed that strong dimensional fluctuations occurred in particular with regard to the longitudinal turning of shafts. As an initial measure, an inspection of the machine was carried out, revealing severe defects on the steady rest and the tip of the tail stock. The causes of the fluctuations in production quality were eliminated and the machine capability of the turning station substantially improved by means of simple repair measures (Fig. 11.5).

As the identification of quality defects on the product after the completion of machining via precise but cost-intensive post-process measuring equipment alone is not sufficient to guarantee a constant

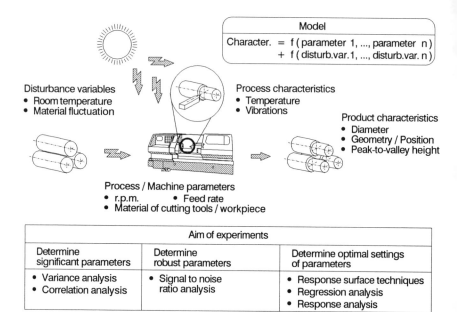

Disturbance variables
• Room temperature
• Material fluctuation

Model

Character. = f (parameter 1, ..., parameter n)
          + f (disturb.var.1, ..., disturb.var. n)

Process characteristics
• Temperature
• Vibrations

Product characteristics
• Diameter
• Geometry / Position
• Peak-to-valley height

Process / Machine parameters
• r.p.m.   • Feed rate
• Material of cutting tools / workpiece

Aim of experiments		
Determine significant parameters	Determine robust parameters	Determine optimal settings of parameters
• Variance analysis • Correlation analysis	• Signal to noise ratio analysis	• Response surface techniques • Regression analysis • Response analysis

Fig. 11.6 Objectives of statistical design of experiments

standard of production quality, corrective intervention in the process is required at an earlier stage. The aim must be to continuously detect the current forces and temperatures in the course of the machining process and to employ this information to stabilize the production process via control mechanisms.

11.3 Controlling processes by monitoring states

Bringing the testing and inspection time forward from the product characteristic to the correlating process parameters results in more than just the earlier identification of rejects: Continuous process surveillance employing sensors enables trends to be identified immediately and compensated by manual intervention on the part of the operator or - the objective of many research projects - by appropriate reactions of the machine control system. To prevent over-reaction of the process, however, intervention is to be carried out only when specific limits for the process or machinery are exceeded.

A further reason for the increased use of process-monitoring sensors

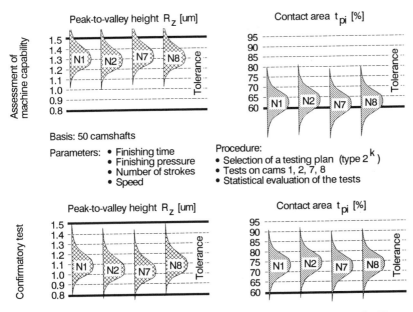

Fig. 11.7 Process optimization via statistical test planning (based on Mercedes Benz
AG)

results directly from technological and design improvements in the field of
tool, drive and plant configuration. Today, high cutting speeds are attained
primarily on fully encapsulated machines. As a result, the machine operator
is becoming increasingly less able to monitor the process and to react
sufficiently quickly and adequately to sudden changes in the process
resulting from malfunctions.

In order to ensure constant production quality, therefore, control
circuits must be installed which intervene to improve the process when
statically or dynamically defined action limits are exceeded. But a quality
characteristic often depends not on only one, but on several parameters
simultaneously. In such cases it is often impossible to specify fixed action
limits for the individual parameters, as the permissible range is always
affected by the levels of the other parameters.

Furthermore, the connection between the process variables and the
quality characteristics of the product is often unknown. This leads either to
important process parameters being ignored or an unnecessarily large
number of parameters being monitored. As a result, irregularities fail to be
reported and necessary warnings fail to be emitted - or are triggered by

parameters which are of no relevance to quality.

Prior to introducing statistically or continuously operating control strategies, the stage at which a measuring process is necessary and the information to be obtained from the measurement result are to be examined for each process. Essential requirements for any control process are control elements by means of which a process at the warning limits can be corrected.

While it is possible to respond directly to overload signals from sensors, by reducing the feed rate or increasing the cutting speed, for example, process assessment or a suitable testing method is necessary in order to determine the relationship between a dimensional fault on the workpiece and the generating process variables. The object of the statistical testing method is to investigate the primary connections via well-selected tests in such a manner as to enable a mathematical model to be evolved which represents the relationships between target variables and influencing factors (Fig. 11.6) [4].

A practical example illustrates the difficulties which require to be surmounted in analysing the connections between the process variables and the result of production operations. In the course of assessment of the machine capability of a camshaft grinding machine it was established that although the variation of peak-to-valley height and the percentage bearing area was in accordance with expectations, the mean value of the measured values obtained via statistical test planning deviated substantially from the required value.

In order to carry out process optimization with the methods of statistical test planning, it was first of all necessary to specify the influencing variables which are primarily responsible for the result of production operations: finishing time, finishing pressure, number of strokes and speed. On the basis of this information it was possible to draw up an appropriate testing plan, which indicated an optimal combination of settings for the variables affecting the result of production operations after statistical evaluation of the implemented tests. A confirmatory test showed that the measures which had been undertaken were suitable in order to reduce the peak-to-valley height of the ground cams and simultaneously provide a decisive improvement with regard to the percentage bearing area (Fig. 11.7).

After the relationships between the target variables and the influencing factors of a production process have been determined using the methods of statistical test planning, for example, and rendered accessible in the form of a mathematical model, it is possible to evolve and implement continuously controlled production processes. These processes utilize and process measured data both from manually and CNC-controlled coordinate and multipoint measuring devices and from process-integrated sensors which are installed in the machining area of the machine.

11.4 Sensors for process surveillance

A large number of sensors are available for detecting process-related variables, whereby a substantial proportion of these sensors are of sufficiently robust design to enable reliable operation under extreme ambient conditions in the machining area of machine tools. In addition to state sensors operating on a binary basis, which account for the majority of detectors employed, sensors which enable the measurement of parameters such as the power consumption of axle drives, machining temperatures, process forces, process moments and vibrations are being employed to an increasing extent [5].

11.4.1 Force-measuring sensors in machine tools

Many known process monitoring systems are based on the measurement of the machining force components or the effects of these components on parts of the machine tool. A widespread variant are piezoelectric three-component force measuring elements, which may be fitted under the cutting tool, for example, for turning operations, or under the workpiece in the case of milling and boring. These elements are inexpensive, robust, available in many designs, and provide an excellent basis for the comparative assessment of other sensors due to their linearity.

Strain gauges, which utilize the change which occurs in the resistance of a metallic or semi-conductive material during extension, are also inexpensive and available in many practical configurations. However, precise analysis of extension is often necessary in order to locate suitable measuring positions, in the course of which feedover of different force components is often unavoidable. In view of the special knowledge which is required with regard to the fitting of strain gauges and their mechanical vulnerability, it is recommendable to deploy strain gauges in the form of preconfigured components (e.g. torsion measuring shafts) in rough practical operation.

Problems arise when employing force-sensitive monitoring systems,

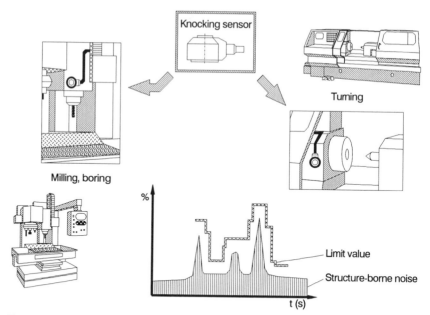

Fig. 11.8 Structure-borne noise sensors in machine tools

due to the increasing requirements regarding the surface quality and dimensional and geometrical stability of the workpieces. Consequently, in the case of turning and milling in particular, the permissible level of tool wear is relatively low, as a result of which the wear-induced increase in the machining force components often remains within the background noise range of the evaluation circuit.

A trend towards increasingly smaller allowances is also to be observed. In the case of rough parts produced via forming processes in particular, dimensions which virtually correspond to the final geometry (near net shape geometry) are required, as a result of which the machining cross-sectional area of the cut and, subsequently, the cutting forces are reduced. In the case of drilling, too - particularly with regard to small holes - the wear-induced increases in feed force and torque are very minimal or undetectable.

A more suitable alternative is provided by force-sensitive detectors for fracture monitoring. Sudden, pulse-shaped changes in force occur, which are simple to analyse and provide a reliable indication of the failure of a tool.

11.4.2 Structure-borne noise sensors in machine tools

Noise emissions occur when the structure-borne noise resulting from machining operations is released into the environment. An increasing level of tool wear generally causes a rising signal level. In the case of turning and milling, this is due to the changing temperature of the contact zone between workpiece and tool and the change in the shape of the chips, while in the case of boring the primary cause is to be found in the increasing dynamics of the process.

The knocking sensor, which was developed for the purpose of identifying knocking combustion in petrol engines, has become established as the most effective sensor for structure-borne noise (Fig. 11.8). A favourable aspect is the fact that the fitting force has no identifiable effect on the frequency sensitivity. The sensor also remains unaffected by drift, as only dynamic signal components are evaluated. Due to their robust design, knocking sensors are largely resistant to chips and cooling lubricants, which means that they can be installed in the vicinity of the machining process.

A problematic factor with regard to the evaluation of noise signals, however, is sources of noise which also emit high-frequency signals. These may be electronic components, rolling bearings or pulses emitted during rapid traverse movements of machines. Incorrect signals must be prevented via suitable signal filtering and plausibility checks.

11.4.3 Optical sensors in machine tools

In principle, optical sensors are suitable for identifying dimensional and geometrical deviations and macroscopic surface damage on tools and on finished workpieces. In principle, this information can be further applied to quality-related process parameters, such as tool wear. However, unfavourable ambient conditions in the machining zone - such as oil mist and chips - and the inadequate scope of image analysis facilities which are available today prevent the broad-scale use of such sensors.

Optical systems based on CCD arrays have already been developed which are able to determine the current wear state of tool cutters after depositing the tool in the magazine [7,8]. In addition to the current two-dimensional sensors, sensors with three-dimensional vision will acquire increasing importance in the future, which evaluate the images of two fixed-position cameras via triangulation or via structured illumination of the workpiece and subsequent photographing by means of the split-beam method or a grid-type aperture.

11.4.4 Technical realization

Technical implementation of such a continuous process monitoring system requires the employed machines, controllers and analytical computers to be equipped with the necessary measuring transducers, interfaces and programmes. Such a complex configuration, which requires solutions to be found on a step-by-step basis via iterative procedures and extensive trials, cannot be carried out by the operating company which is dependent on the trouble-free operation of its production facilities. Rather, it is the component manufacturers who are required to provide the performance capabilities required by the process.

Effective configuration of the process-integrated sensors requires the special knowledge of the machine designer. This applies in particular in the case of force-sensitive transducers which are integrated into the force flux of the machining process and thus influence the rigidity of the machine. Many other sensors are simple to adapt, but must be positioned correctly, if they are to function correctly. All sensors must be designed to cope with the extreme ambient conditions in the working area of machine tools and should be simple to remove for the purpose of easy replacement.

The special knowledge of the producer of the controller is also indispensable. He must either wire up the necessary interfaces for recording and processing the sensor signals in fixed configuration or offer a modular concept which enables any sensors to be connected via interface cards. An advanced stage of development has been attained with regard to the interface for the higher-level master computer. The scope of language may require to be adapted here, to enable transmission of the recorded sensor information.

It is to be expected that generation of the programmes for the higher-level analytical computer will involve the greatest scope of development work. In the course of process analysis work, the described correlations between the production parameters and the result of the process require to be identified and implemented in a programme designed in accordance with the specific machine tools involved in such a manner as to enable the device control system to respond to deviations of the sensor signals from the required values by altering the "correct" controller outputs. This work can be carried out by university research institutes, engineering companies

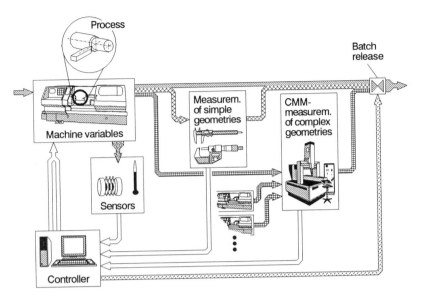

Fig. 11.9 Continuous process control

or the software departments of companies, whereby prime importance should be attached to integrating the specialist knowledge which the machine operator possesses - often in intuitive form.

11.5 Post-process measuring equipment monitors capable processes

Measuring operations after the completion of machining always remains necessary, irrespective of whether a machine tool is equipped with process-integrated sensors which enable the compensation of errors in the production process. However, the increased level of process capability does lower the required capability for the measuring equipment (see Fig. 11.2), enabling the use of simpler measuring instruments.

The long-term aim must be to develop continuously controlled processes which obtain their information from all areas of production operations: sensors provide process data, hand-held measuring instruments supply data on deviations in the dimensions and shape of the finished geometry shortly before the end of the production stage concerned, while CNC coordinate measuring devices and automated multi-point measuring devices record dimensional, geometrical and positional deviations of the

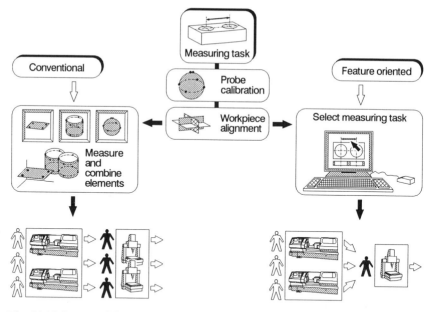

Fig. 11.10 Increased flexibility via feature oriented programming (based on: Leitz
Meßtechnik GmbH)

workpiece geometry (Fig. 11.9).

Both hand-held measuring devices and multi-point measuring devices
adapted to the machining application concerned are employed for the
purpose of geometrical testing after machining of the workpiece has been
completed. Batch size, the scope of testing per workpiece and the testing
time for a geometrical feature vary considerably, as a result of which the
economic efficiency of the testing equipment must be assessed on an
individual basis. The complexity of the measuring task also requires to be
taken into account. Complex measuring instruments which operate in a
closed coordinate system are essential, whenever a three-dimensional
relationship requires to be established between individual geometries
[9,10].

Universal coordinate measuring devices are suitable for virtually every
geometrical measuring application, but require a high level of investment in
terms of both equipment and personnel training. Considerable scope is
available to the expert with regard to the configuration of a measurement
procedure for a specific part, but for those lacking the necessary
knowledge there is substantial potential for errors. As a result, expensive

measuring programmes have to be purchased for complex workpieces in particular. In many cases, a complete set of similar programmes has been developed for related parts, such programmes becoming unusable when only minimal changes occurred to the design of the workpiece to be measured.

Not only have the costs of programme development been reduced to a minimum since "feature oriented programming" has rendered it possible to operate coordinate measuring devices in such a manner that the desired measurement procedure can be selected with the assistance of a graphic display. The simplified mode of operation also enables a marked increase in the workpiece throughput rate. Flexible access to all stored measuring programmes means that the time-consuming procedure of changing the programme for each new workpiece is no longer necessary, as a result of which trained staff are able to carry out any measuring operations on various parts (Fig. 11.10) [11,12].

11.6 Monitoring of quality control equipment

By means of continuous surveillance of process states with the aid of sensors mounted close to the manufacturing process, production centres are able to produce workpieces with small geometrical errors, whose quality is supervised with post-process metrology devices. To use these control mechanisms in the long term, the reliability and also the quality capability of all the machine tools and sensors involved in the process and of the hand-held measuring instruments and automatic measuring devices which are employed must also be checked in the course of routine monitoring procedures.

Technical disturbances generally impair the capability of testing equipment. Ambient influences such as deviations from the reference temperature of 20°C and ground or air vibration in particular cause stochastic measuring errors which are difficult to correct. Long-term subjection to ambient influences often leads to systematic errors in the testing equipment, as material measures are maladjusted or the calibration capacity of the testing equipment is impaired by material fatigue.

Impaired capability in testing equipment may also be caused by the user, however. Because it is assumed, due to ignorance or negligence, that testing equipment is
- calibrated and
- reliable,

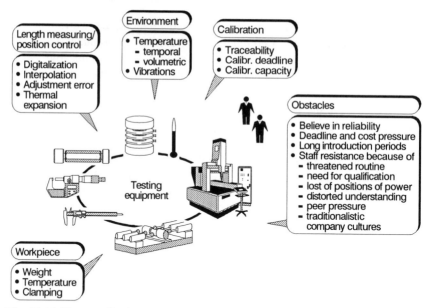

Fig. 11.11 Factors affecting the capability of testing equipment

- not subject to any ambient influences and
- impervious to operating errors,

it is often wrongly deployed, incorrectly stored or the date for recalibration is overstepped (Fig. 11.11).

This has serious consequences with regard to the informative value of the measurements carried out with the non-calibrated measuring equipment. Under certain circumstances, a good workpiece may be rejected or - even more seriously - a dimension which is actually outside of the tolerance limits may be declared to be within the tolerance limits. Special effort is therefore necessary to promote the consistent application of monitoring processes which monitor the current measurement uncertainty of the testing equipment and to provide a certificate confirming the general capability of testing equipment.

11.6.1 Monitoring processes

Depending on the degree of automation of the components, it may be necessary to carry out monitoring of the operational and test equipment involved in the production process in off-line mode outside of the

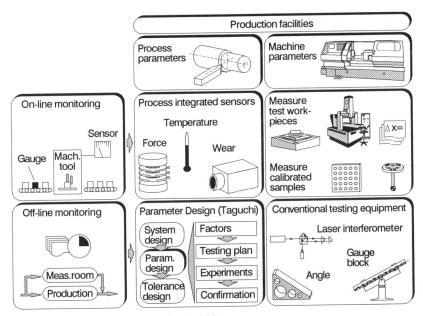

Fig. 11.12 Monitoring of production facilities

production line, or it may be possible to integrate monitoring operations into the production process, e.g. via the on-line measurement of suitable calibrated standards. The most diverse processes are suitable for carrying out these monitoring operations, though to date only a small number of processes have become established in practice.

No general distinctions apply with regard to the reliability or informational value of off-line and on-line monitoring. Great differences do apply, however, concerning the time required by the two processes. While off-line monitoring requires a large number of calibrated testing instruments which may take several days to assess a machine, on-line monitoring by means of a calibrated test block for general or specific measuring tasks rarely require more than half a day. The scope of measuring operations involved in the testing process is based on the size of the test block and is therefore generally smaller than for off-line monitoring.

These circumstances cause the employment of the off-line and on-line monitoring method. No industrial company will be able to renounce a complete off-line analysis of all manufacturing and control devices in the course of a yearly inspection. In order to perform a quick on-line inspection of the machine capability, it is sufficient to manufacture a probing workpiece once a week or to integrate a calibrated gauge in the material flow of a coordinate measuring machine once a day. However, an on-line supervision cannot replace the complete off-line inspection because of the resultant reduced validity.

11.6.2 Monitoring of production facilities

Monitoring of the production facilities has prime priority, as the process capability must be ensured, if the objective of preventing errors is to be attained. The process parameters can be recorded with the aid of measured variable transducers, whereby the mutual influencing of the target values can be analysed via appropriate testing methodology. In addition to simple state-detecting devices which operate on a binary basis, machine tools also incorporate numerous other sensors in various designs for measuring forces, temperatures and wear. These sensors record the process parameters during the machining process and, in turn, require to be monitored as testing equipment.

In off-line configurations, the kinematic motional sequences of the machine tools are recorded by conventional testing devices, which record the deviations and store them as a correction matrix. In on-line mode, this matrix can be obtained with the aid of a test piece which is moved into the machining area on a machine pallet and measured with a sensor which is installed in place of the tool. However, it is also possible to establish the deviation matrix for correction of the machine motions by producing a specimen workpiece with the machine tool which is to be examined and then - in off-line mode - measuring and analysing this specimen workpiece on a coordinate measuring device (Fig. 11.12).

11.6.3 Monitoring of testing equipment

A distinction between on-line and off-line procedures also applies with regard to monitoring the process capability of testing equipment. The object of such monitoring processes is to monitor the reliability of the sensors which are employed to monitor the process capability of machine tools. For this purpose, the sensors must either be removed from the production equipment and tested off-line under defined conditions or - more practically - replaced by sensors which have already been tested

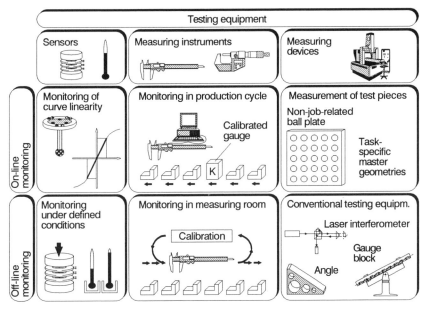

Fig. 11.13 Monitoring of testing equipment

(Fig. 11.13).

Hand-held measuring instruments can be tested both off-line, in the measuring room, and on-line, via measurement on calibrated gauge blocks. The purpose of monitoring in the measuring room, particularly with regard to hand-held measuring instruments, is to carry out general calibration, while monitoring measurement on the production line calibrates the testing equipment primarily for use as comparators, i.e. the measuring instrument is calibrated for one special measuring task only, as opposed to all measuring tasks.

Essentially, the monitoring of universal coordinate measuring instruments is subject to the same procedure as applies to the monitoring of machine tools. Off-line monitoring of the overall scope of equipment to be measured is carried out with conventional testing equipment, while calibrated ball plates in various sizes have become established for the purposes of on-line monitoring. In the face of ever more complex testing tasks on the free-form surfaces of, for example, toothing and compressors, more recent job-related test pieces have been developed which provide a much more comprehensive picture of the capabilities of coordinate measuring devices [13].

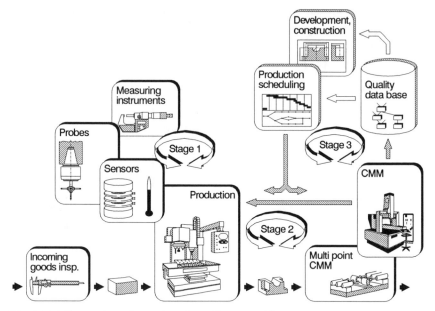

Fig. 11.14 Integration of metrology in quality control loops

11.7 Summary and outlook

Against the background of ever more complex production processes and an increasing level of cost-consciousness in all areas of industrial companies, the objective of measuring processes is shifting from mere fault identification after the completion of machining operations - which requires very expensive equipment and processes and is still unable to detect all faults - towards process-integrated sensor technology deployed in a preventative mode. Continuous process monitoring enables trends to be identified immediately and corrective intervention to be applied as soon as action limits are exceeded.

However, the connections between the process variables and the quality characteristics of the product are often unknown. As a result, irregularities fail to be detected and necessary warnings are not emitted - or are triggered by parameters which are of no relevance to quality. In order to obtain an efficient process control system, it is therefore essential to determine the most important influencing variables via process assessment or a suitable testing process.

The integration of sensors into the process variables and the resultant ability to increase the level of process capability results in a change in the tasks to be carried out by the post-process measuring processes, which remain necessary. As there is a reduction both in the scope of testing to be carried out and in the requirements relating to the quality of the deployed measuring equipment, increased requirements can be placed on flexibility, in order to enable existing measuring equipment to be employed for several parallel tasks. In this way, the investment costs for post-process measuring equipment can be reduced substantially.

In addition to ensuring the process capability of machine tools, the reliability of the deployed measuring equipment must also be monitored on a regular basis. In this connection it is essential to ensure that the capability of the testing equipment can be verified at all times via reference to national standards (traceability).

Long-term research must concentrate on the implementation of large quality control circuits involving the collection of an extensive scope of measured values in data bases and subsequent influencing of the indirect production areas of "Development and construction" and "Production scheduling" (Fig. 11.14). A substantial scope of development work is required in order to analyse the principles behind the actions of the various influencing variables and to evolve computer-controlled countermeasures for such complex interrelated actions. Although several individual systems have been developed, an all-purpose solution for these quality control circuits, to meet the requirements of an open range of workpieces to be produced by means of various machining processes, is still in a state of development.

Part Four

PRODUCTION PLANTS: MEETING ECOLOGICAL AND ECONOMICAL REQUIREMENTS

12 Fast - precise - clean: the machine tool in the context of conflicting interests of economy and ecology
13 The open controller - key element of high-performance production facilities
14 Robotics in production systems

12

FAST - PRECISE - CLEAN: THE MACHINE TOOL IN THE CONTEXT OF CONFLICTING INTERESTS OF ECONOMY AND ECOLOGY

12.1 Introduction

The technical development of machine tools is currently in a phase marked equally by rising technical requirements (e.g. increased reliability, increased precision, increased material-removal rates and improved environmental compatibility) and by the very bad economic situation which presently prevails. The development trend which is currently to be observed for all types of machines thus arises from the attempt to attain what is technically feasible at an acceptable level of costs.

The collapsing markets in eastern Europe, recessionary trends in Germany and other industrial countries and the somewhat restrained signals of growth in the US economy have led to drastic falls in orders in the German machine tool manufacturing sector. While all manufacturers are affected equally by the global economic situation, the situation for the German machine tool construction sector is additionally aggravated by specific national factors. An analysis of the reasons for the tense economic situation reveals the four core areas shown in Fig. 12.1 [1, 2]:

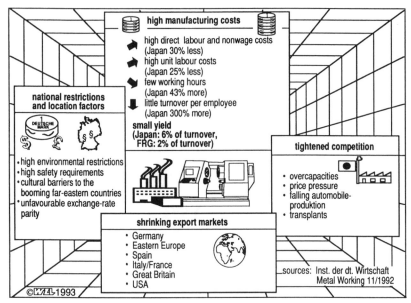

Fig. 12.1 Reasons for the tense situation in the machine tool manufacturing sector
(Source: Institut der deutschen Wirtschaft, Metal Working 11/92)

1.   The unit labour costs are higher than those of rival companies in Eastern Asia. They result from high wages in combination with lower annual working hours, make industrial production more expensive and enable only minimal profit margins on turnover.

2.   Due to the national debt resulting from reunification, the level of interest rates is high in comparison with important rival countries. In addition to the counter-cyclical effect of these interest rates, this situation also results in a strong mark, which further diminishes the export chances of German manufacturers. In addition to these direct costs, the high national environmental and safety standards also worsen the competitive situation. Finally, excess capacities were built up in the boom years which now have to be dismantled in a painful process. Any alternative markets are restricted to the south-east Asian countries, to which German companies have difficulty in gaining access, due to both cultural barriers and their medium-sized structures.

3.   In the medium to long term, the demand for automobiles in Western Europe will stagnate or even decline, which means that the

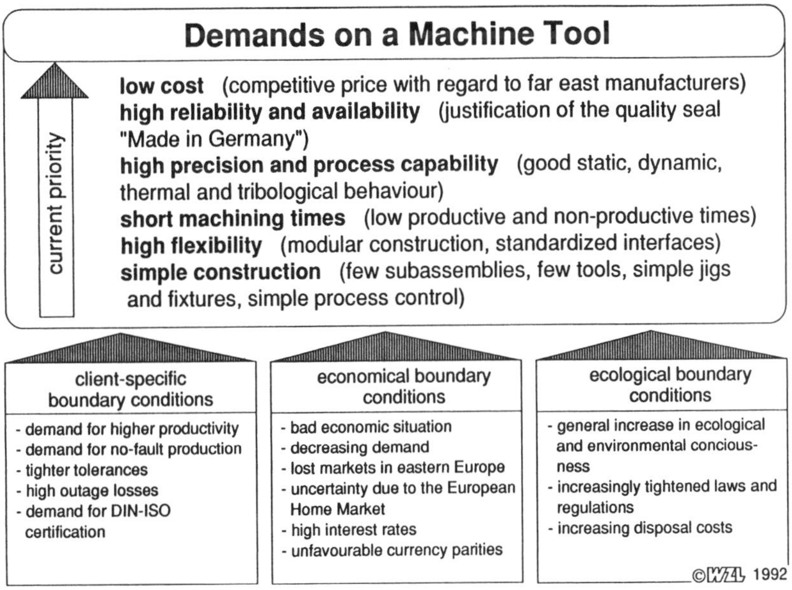

Demands on a Machine Tool

current priority ↑

low cost (competitive price with regard to far east manufacturers)
high reliability and availability (justification of the quality seal
"Made in Germany")
high precision and process capability (good static, dynamic,
thermal and tribological behaviour)
short machining times (low productive and non-productive times)
high flexibility (modular construction, standardized interfaces)
simple construction (few subassemblies, few tools, simple jigs
and fixtures, simple process control)

client-specific boundary conditions	economical boundary conditions	ecological boundary conditions
- demand for higher productivity - demand for no-fault production - tighter tolerances - high outage losses - demand for DIN-ISO certification	- bad economic situation - decreasing demand - lost markets in eastern Europe - uncertainty due to the European Home Market - high interest rates - unfavourable currency parities	- general increase in ecological and environmental concious-ness - increasingly tightened laws and regulations - increasing disposal costs

©IWL 1992

Fig. 12.2 Conflicting demands on machine tools

investment activities of the most important customer for machine tools and its ancillary suppliers are likely to be restrained. In the area of the automobile components industry in particular, the widely propagated low vertical range of manufacture has led to a network of small and medium-sized companies which, in view of these prognoses, are unwilling to tie up capital by investing in modern production plants. The price pressure resulting from excess capacities in the machine tool manufacturing sector cannot be absorbed by protectionist measures undertaken by the EC for the domestic industry. The German machine tool manufacturing industry is strongly dependent on exports and thus profits primarily from free world trade, in addition to which the domestic market is now also being supplied with Japanese transplants from Great Britain.

4. The collapse of the Soviet Union and the entire planned economy of the Eastern bloc has resulted in the almost entire loss of a market which in the past often behaved countercyclically to trends in the Western economic area and thus helped to cushion economic downturns. And, due to the prevailing economic situation, there has

©IWZL 1993	cutting machines with geometrically defined edge *turning, milling, drilling, sawing and broaching machines*	cutting machines with geometrically undefined edge *grinding, honing and lapping machines*	erosion machines *EDM and laser machines*	forming and cutting machines *presses, hammers, rolling, bending, drawing, punching and nibbling machines*
development aims	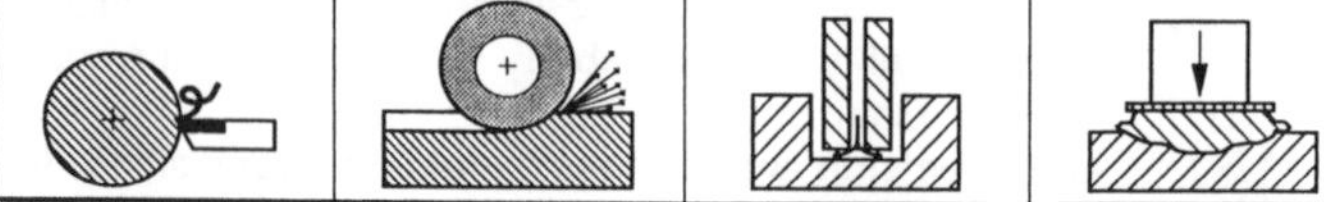			
costs	standardization and modularization of the machine concepts		Soft-Tooling (creation of contours using NC-axes without forming tools)	high wear-resistant worktool materials
	reduction of the tool variety	reduction of the grinding wheel change time		process integration in sheet metal machining centres (several bending operations, punching and nibbling)
	hard machining			
reliability	more complex sensors (process monitoring)	more complex sensors (scratching, sparking)	measurement of tool wear	process force monitoring
precision	g e n e r a l l y : h i g h e r a c c u r a c y d e m a n d s			manufacture of burr-free parts close to final contour (Near-Net-Shape technology)
	statistische Abnahme	eccentric grinding with highly dynamic x-axis	CNC-4/5-axis-erosion	
machining speed	g e n e r a l l y : h i g h e r m a c h i n i n g s p e e d s			
	faster workpiece and worktool change	CBN-grinding	higher cutting rates due to gap regulation	High Speed Blanking
ecological friendliness	generally: abandoning environmentally damaging and unhealthy additives in the media			
	ecological chip treatment, dry cutting			"white lubricants"

Fig. 12.3 Development trends for metal-working machine tools

also been a severe decline in demand in the Western world.

12.2 Development trends in the machine tool manufacturing sector

The above-described economic conditions have led to a shift in priorities with regard to the requirements placed on a machine tool (Fig. 12.2). Despite numerous other factors, from the customer's point of view a competitive purchase price in comparison with rival products from the Far East must be regarded as the dominant criterion determining purchasing decisions, in relation to which the other requirements are of only secondary importance. In contrast, the manufacturers traditionally pursue longer-term technical objectives, such as reliability, precision and high machining speeds, which may increase the overall profitability of a machine in the long term, but which also result in a higher purchase price.

Development in the field of machine tools is subsequently progressing at a moderate rather than a rapid pace. The expectations and actual capabilities of HSC technology (HSC stands for high-speed cutting) inrecent years, for example, have shown thatall areas ofpotential have their

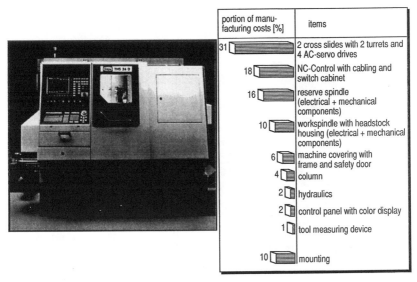

portion of manu-facturing costs [%]	items
31	2 cross slides with 2 turrets and 4 AC-servo drives
18	NC-Control with cabling and switch cabinet
16	reserve spindle (electrical + mechanical components)
10	workspindle with headstock housing (electrical + mechanical components)
6	machine covering with frame and safety door
4	column
2	hydraulics
2	control panel with color display
1	tool measuring device
10	mounting

Fig. 12.4 Cost structure for a double-slide double-spindle CNC turning machine (based on: Traub)

limits. Although great interest remains in exploiting the speed limits which are attainable with the available tools, the restrictions which apply in practice, such as undesired thermal effects, dynamic forces, high contouring errors of positional control systems and the danger of inertial forces to operator and machine constitute physical hurdles which cannot be circumvented, but can only be surmounted via the rigorous implementation of a large number of small and sometimes cost-intensive design stages.

The isolated, purely technical assessment of a design is also a thing of the past. Medium- and long-term competitive capacity is attainable only via overall consideration of the product, from production through operation to removal from service (recycling of old parts, scrapping and disposal). In this context, environmental obligations, which the machine tool manufacturing sector is unable to evade in either economical or ecological terms, has become a focal point of interest.

Fig. 12.3 provides an overview of the current development trends for metal-working machine tools. These trends are a reflection of customer requirements and often involve conflicting aims. Today's customer requires low investment and operating costs together with high precision and machining speeds, and takes reliability and environmental compatibility for

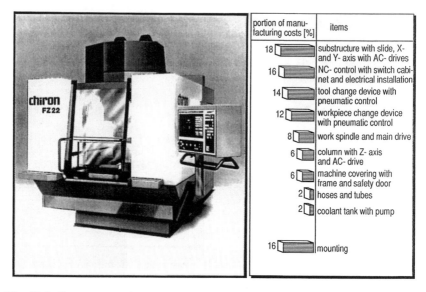

portion of manu-facturing costs [%]	items
18	substructure with slide, X- and Y- axis with AC- drives
16	NC- control with switch cabi-net and electrical installation
14	tool change device with pneumatic control
12	workpiece change device with pneumatic control
8	work spindle and main drive
6	column with Z- axis and AC- drive
6	machine covering with frame and safety door
2	hoses and tubes
2	coolant tank with pump
16	mounting

Fig. 12.5 Cost structure for a CNC machining centre (based on: Chiron)

granted. As the manufacturing company has no precise knowledge of the subsequent operating conditions for the machine, it is compelled to ensure that the design takes due account of all possible extreme stress situations, in order to guarantee the required availability for the machine. This involves higher costs, however. Large manufacturers from the Far East avoid this conflict by meeting the customers' requirements for ever faster and more accurate machines only to a limited extent, whereas the small and medium-sized European manufacturers are unable to risk such a market policy determined by strategic considerations.

The following sections refer to case studies to illustrate various approaches, each of which is being adopted in practice to improve one of the aspects of costs, reliability, precision, machining speed and environmental compatibility.

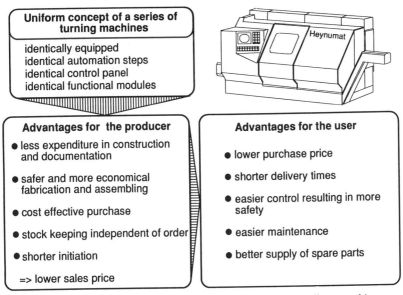

Uniform concept of a series of turning machines

identically equipped
identical automation steps
identical control panel
identical functional modules

Heynumat

Advantages for the producer

● less expenditure in construction and documentation

● safer and more economical fabrication and assembling

● cost effective purchase

● stock keeping independent of order

● shorter initiation

=> lower sales price

Advantages for the user

● lower purchase price

● shorter delivery times

● easier control resulting in more safety

● easier maintenance

● better supply of spare parts

Fig. 12.6 Advantages of a uniform design concept (based on: Heyligenstaedt)

12.3 Increasing the economic efficiency of a machine tool

Analysis of the cost structure for a machine tool shows that the overwhelming majority of material, design and production costs are accountable to the feed slides, tool turrets, tools and workpiece spindles (Figs. 12.4 and 12.5 for a turning machine and a machining centre respectively). The assembly costs also play an important role. In the interests of cost reduction it is always expedient, therefore, to create pre-assembled groups of parts or to use externally purchased parts in which several functions are integrated. By comparison, the costs relating to the machine base, panelling for the working area and auxiliary facilities such as hydraulics and chip conveyors are of secondary importance.

The following areas of potential are available for reducing the manufacturing costs of a machine tool:
- fully modular configuration of assemblies and machines,
- reduction in the number of assemblies and parts,
- the integration of several machining processes in accordance with the requirements of the application concerned and
- the utilization of new machine elements which are lower-priced or

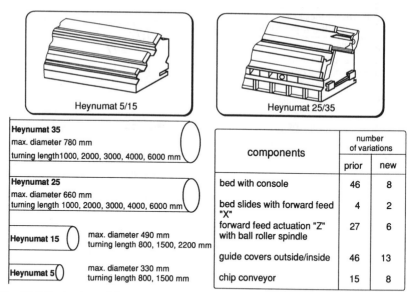

Fig. 12.7 Reduction of the number of variants of a turning machine (based on: Heyligenstaedt)

- simplify production and assembly operations.

Figs. 12.6 and 12.7 show an example of the implementation of a completely modular concept for a series of machines [3]. The aim pursued here was a drastic reduction in the number of variants by means of standardized levels of performance and automation, a uniform user interface and identical functional modules. The basic element of all the machines belonging to the series of four models is a four-rail base, which is offered in only two cross-sectional sizes and up to five base lengths. The additional costs incurred due to the resultant oversizing of the small models are more than balanced out by the savings in the areas of design, documentation, purchasing, production, assembly and storage. The main effect for the user is a lower purchase price. Users who possess several machines from the series further benefit from the facilitated operation and maintenance of the machines and the improved supply of replacement parts resulting from the uniform concept applied for the operating and programming interface and for the mechanical and electrical components.

For some machining applications, the unit production costs can be reduced by integrating several complementary machining processes in one

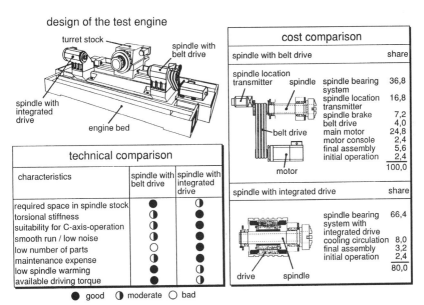

design of the test engine

technical comparison		
characteristics	spindle with belt drive	spindle with integrated drive
required space in spindle stock	●	◑
torsional stiffness	◑	●
suitability for C-axis-operation	◑	●
smooth run / low noise	◑	●
low number of parts	○	●
maintenance expense	◑	●
low spindle warming	●	◑
available driving torque	●	◑

● good ◑ moderate ○ bad

cost comparison

spindle with belt drive	share
spindle bearing system	36,8
spindle location transmitter	16,8
spindle brake	7,2
belt drive	4,0
main motor	24,8
motor console	2,4
final assembly	5,6
initial operation	2,4
	100,0

spindle with integrated drive	share
spindle bearing system with integrated drive	66,4
cooling circulation	8,0
final assembly	3,2
initial operation	2,4
	80,0

Fig. 12.10 Different spindle concepts for a CNC machine tool (based on: Heyligenstaedt)

the decisive factors here. With regard to technical aspects, however, the required installation space for the spindle and the lower driving torque of conventional motor spindles at the same radial and axial rigidity must be assessed as negative factors.

Remarkable progress has been achieved in recent years in the field of linear guideways, where changing technical and economical requirements have led to modified design concepts. Whereas the hydrodynamic guideway was previously the dominant variant on account of its good rigidity and damping, roller guides are now being used to an increasing extent in machines with fast traversing axes. These roller guides result in a substantially lower level of friction, which reduces the temperature increase in the frame components and also enables more precise traversing of the smallest distances. Simplified production and assembly operations are a further reason for their increasing deployment (Fig. 12.11). These guides enable a reduction in overall costs of approx. 20% in comparison with slideways.

Increasingly, measuring systems are being integrated into assemblies in order to increase the robustness of the overall system, to reduce assembly

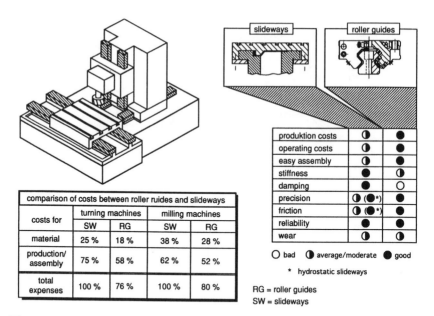

comparison of costs between roller ruides and slideways				
costs for	turning machines		milling machines	
	SW	RG	SW	RG
material	25 %	18 %	38 %	28 %
production/ assembly	75 %	58 %	62 %	52 %
total expenses	100 %	76 %	100 %	80 %

produktion costs	◐	●
operating costs	◐	●
easy assembly	◐	●
stiffness	●	◐
damping	●	○
precision	◐ (●*)	●
friction	◐ (●*)	●
reliability	●	●
wear	◐	◐

○ bad   ◐ average/moderate   ● good

* hydrostatic slideways

RG = roller guides
SW = slideways

Fig. 12.11 Comparison of the characteristics and costs of different guide principles (based on: Traub, Chiron)

costs and to reduce the possibility of errors in the assembly process. Examples are ball bearings with integrated force sensors and rotation-angle sensors or clutches with integrated torque-measuring devices. A technical comparison of commonly employed principles for measuring displacement in machine tool axes (Fig. 12.12) shows that a new measuring system integrated into the guideway, which operates by scanning a magnetic scale, possesses similar characteristics to the glass linear scale [5]. With these systems, the dimensional information is applied to a magnetic ruler which is fixed to one of the two guide rails (Fig. 12.13). In comparison with the concept involving a separate linear rule, the costs for designing, producing and assembling the elements attached to the rule are saved here.

A further example of reduced costs via a reduction in the number of parts and the integration of several functions into one pre-assembled module is provided by the pneumatic cylinder for the tool-changing device of a machining system, shown in Fig. 12.14. This tool-changer incorporates a separate gripper with lifting cylinder for each tool, in order to minimize the time between machining operations. The previous concept (on the left of the figure) involved conveying the compressed air from the

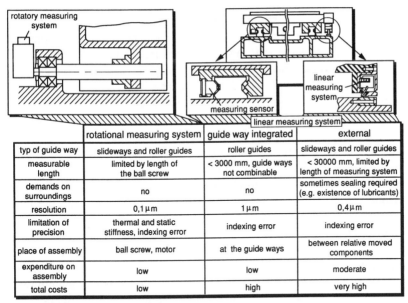

	rotational measuring system	linear measuring system guide way integrated	external
typ of guide way	slideways and roller guides	roller guides	slideways and roller guides
measurable length	limited by length of the ball screw	< 3000 mm, guide ways not combinable	< 30000 mm, limited by length of measuring system
demands on surroundings	no	no	sometimes sealing required (e.g. existence of lubricants)
resolution	0,1 µm	1 µm	0,4 µm
limitation of precision	thermal and static stiffness, indexing error	indexing error	indexing error
place of assembly	ball screw, motor	at the guide ways	between relative moved components
expenditure on assembly	low	low	moderate
total costs	low	high	very high

Fig. 12.12 Measurement of linear movements on machine tools: Requirements, principles, accuracy

central supply point to a valve block. This valve block distributes the compressed air to the two sides of the piston for the forward and return stroke via two connections per cylinder. The position of the piston is monitored by limit switches which are attached to the cylinder with clamping screws. The limit switches require a separate distribution box and two electric terminals for each cylinder. The disadvantages of his onfiguration are the fact that the scope of assembly and adjustment work for the limit switches increases with the number of tools, the tendency of the switches to be shaken loose in operation and the extensive electric and pneumatic cabling and tubing.

The new configuration (shown on the left of the figure) consists of a preassembled cylinder unit in which the valves and limit switches are already integrated. Each cylinder now has only one terminal for pneumatic and electrical connection. The preassembled module reduces the times required for final assembly of the machine and subsequently lowers the job throughput times and the costs of tied-up capital. Also, parts management and maintenance are simplified for a small number of more complex component units.

A further measure for optimizing costs in the area of production engineering is the substitution of processes. In addition to classical approaches involving, for example, the replacement of cutting by precision casting in order to save raw materials, hard machining has also acquired increased importance here. Hard turning (Fig. 12.15) as a substitute for external and internal cylindrical grinding requires practically no special machine equipment and offers several process-related advantages [6]. For example, no lubricant is required - from a technological point of view, at least. Rough- and finish-machining can be carried out on one machine. Furthermore, turning requires less energy than grinding. A disadvantage which has yet to be overcome, however, is the substantial production of heat, which leads to thermo-elastic deformation of the workpiece and machine when no lubricants are employed. As the diagram on the left of the figure shows, an increase in the temperature of the workpiece and the tool of almost 10°C occurs in a turning process of approx. 80 seconds' duration under the specified machining conditions. The resultant thermoelastic deformation in the diameter leads to a dimensional error of over 10 µm, which renders the process unsuitable for fine machining in its present form. Additional measures are thus required here to enable cooling and/or compensation of the heat-induced faults.

The laser has opened up new areas of applications both as a supporting facility for processes (e.g. when turning ceramics with carbide tools) and in the substitution of processes (hardening, cutting and welding as supplementary functions in conventional machine tools) [7]. The standard laser today is the CO_2 laser with a beam power of up to 10kW. The prevailing trend is towards higher beam powers (levels up to 50kW are currently under discussion) and, in particular, more compact designs (Fig. 12.16), which will open up totally new possibilities with regard to integration into machine tools and handling devices. The newest development in this area is the diode laser, which measures only a few decimetres, and which is expected to be more favourably priced than current devices, once it goes into series production. This laser consists of a large number of diodes attached to a heat sink, the light from which is focussed into a beam. The present capacity for positioning the diodes relative to one another permits only an unsatisfactory concentration to a light spot of 2 - 3 mm, however, whereas a size of 1 µm is attainable with a CO_2 laser. Intensive work is currently in progress in this area.

guiding track

guiding shoe

hard magnetic ruler

sensor head

Fig. 12.13 Linear guideway with integrated linear measuring system (based on: Schneeberger, Heidenhain)

12.4 Improving the reliability of a machine tool

The superior reliability and availability of German machine tools in comparison with products from rival markets is one of the prime areas of potential for securing the market positions of German companies. In addition to producing assemblies which are susceptible to wear in durable design, the use of sensor technology to monitor processes has acquired particular importance in recent years (Fig. 12.17). In addition to diagnosis of the operating characteristics of rolling bearings and gearwheels, further development work has been carried out in the area of the monitoring of process forces and process noise. The sensor technology integrated into the turning part of a milling spindle which is shown in the diagram enables the monitoring of structure-borne noise, axial and radial forces and the torque of the cutting process [8]. Installation of the sensors in the direct vicinity of the cutting process means that the connection between measured signal and tool state can be analyzed effectively and in reproducible manner. The breakage of a cutting edge, for example (see right-hand side of figure) results in a serious change of the torque spectrum.

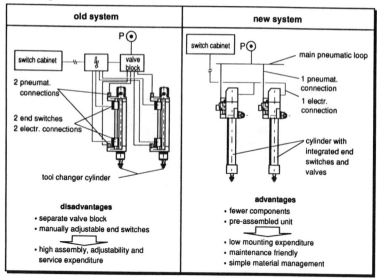

Fig. 12.14 Reduction in the number of components as illustrated by the example of a cylinder of a tool-changing device (based on: Chiron)

In the case of high speed spindles, bearing failure resulting from brief drops in the lubricant supply level - as a result of incorrect operation or defects in the lubricating system, for example - is a serious problem. The use of ceramic balls in so-called oil-off tests (Fig. 12.18) has demonstrated that the safety reserve can be increased substantially via the combination of different materials and the resultant reduction in the affinity of the materials in rolling contact [9]. In the described test, a conventional steel bearing and a bearing of identical design with ceramic balls were run for 48 hours at 10000 1/min, with oil-air lubrication. The supply unit was then switched off, to simulate an error-induced interruption in the lubricant supply. Both bearings ran for 100 hours without failing. The speed was therefore increased to 14000 1/min. The steel bearing then failed immediately, while the hybrid bearing ran for a further 58 hours. Wear analysis (see scanning electron microscope photographs on the right of the figure) revealed that the raceway surfaces of the steel bearing were completely destroyed, while those of the hybrid bearing were virtually unaffected. The strong increase in temperature in the hybrid bearing had evidently been caused by cage friction.

A further problem area which often receives inadequate attention in the design phase concerns the sealing of spindle-bearing systems.

The primary objective pursued in the design of a spindle seal is to

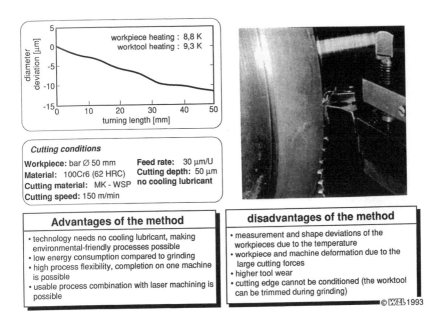

Cutting conditions

Workpiece: bar ∅ 50 mm	**Feed rate:** 30 µm/U
Material: 100Cr6 (62 HRC)	**Cutting depth:** 50 µm
Cutting material: MK - WSP	**no cooling lubricant**
Cutting speed: 150 m/min	

Advantages of the method	disadvantages of the method
• technology needs no cooling lubricant, making environmental-friendly processes possible • low energy consumption compared to grinding • high process flexibility, completion on one machine is possible • usable process combination with laser machining is possible	• measurement and shape deviations of the workpieces due to the temperature • workpiece and machine deformation due to the large cutting forces • higher tool wear • cutting edge cannot be conditioned (the worktool can be trimmed during grinding)

© WZL 1993

Fig. 12.15 Turning of hardened workpieces

protect the bearing from dust and cooling lubricants and to prevent the leakage of lubricant. The space available for these purposes is always highly restricted, as any increase in the projection of the spindle is accompanied by a decrease in rigidity, particularly on the working side of a spindle-bearing system. Standard configurations involving contact or non-contact sealing elements are available. The field of applications for contact seals is subject to fundamental restrictions relating to friction and the dependency of the sealing effect on the speed. At lower speeds in particular, standard non-contact sealing systems are unable to provide adequate sealing against foreign matter. By systematically applying the five action principles of repelling, throwing off, recirculating, collecting and discharging, labyrinth seals of optimized design can be developed which maintain a tight seal both when the spindle is at a standstill and at maximum speeds (Fig. 12.19) [10].

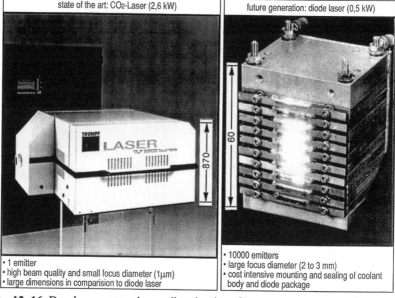

state of the art: CO$_2$-Laser (2,6 kW)	future generation: diode laser (0,5 kW)
870	60
• 1 emitter • high beam quality and small focus diameter (1μm) • large dimensions in comparision to diode laser	• 10000 emitters • large focus diameter (2 to 3 mm) • cost intensive mounting and sealing of coolant body and diode package

Fig. 12. 16 Development trend regarding the size of lasers (source: ILT Aachen, Trumpf)

12.5 Increasing the precision of a machine tool

Apart from the continual rise in cutting speeds and feed rates, the other outstanding trend with regard to the machine performance characteristics is an increase in the required level of precision. However, there is often no direct connection between the accuracies specified by customers and actual requirements. Rather, a large proportion of these precision requirements are accountable to restrictions in tolerances resulting from the recent introduction of statistical machine acceptance testing throughout the automobile industry. On the one hand, the specified tolerances for the parts to be manufactured are indeed becoming increasingly smaller, while on the other hand a machine capability test imposes higher requirements on the stability of the produced workpiece dimensions [11-13]. Fig. 12.20 shows the example of the restricted tolerance for a shaft with a tolerance of 1/10 mm. The user of the machine tool may, for example, require a critical machine capability index of $C_{mk} \leq 2.0$ to be attained in acceptance testing. For this purpose, the machine manufacturer is to produce a sample of 50 parts. The machine capability index is determined as follows:

milling spindle withstructure noise, force and torque measurement devices

course of the torque during machining with a 6-edge tool before and after tool breakage

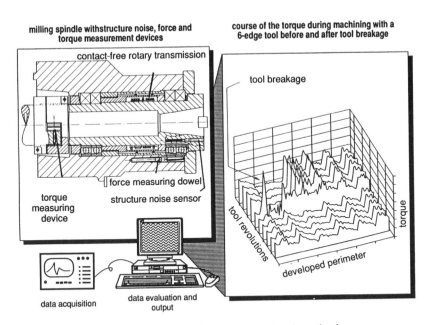

Fig. 12.17 Machine integrated sensors for process and tool monitoring

Machine capability index

$$C_m = \frac{T}{6s},$$

critical machine capability index

$$C_{mk} = \frac{\Delta_{crit.}}{3s}$$

whereby T: tolerance zone

s: standard deviation of produced parts

$\Delta_{crit.}$: smallest difference between mean value for produced parts and tolerance limits

A machine capability index of $C_m \leq 2.0$ means that the range of six-fold standard deviation may cover only half the tolerance zone. For this example, this means that the maximum permissible standard deviation is 1/120 mm. For a sample of 50 parts, the statistical mean of the difference between the largest produced dimension and the smallest (range) constitutes 75% of six-fold standard deviation. The range of the produced ample must therefore be no greater than 38 µm. When the mean value for

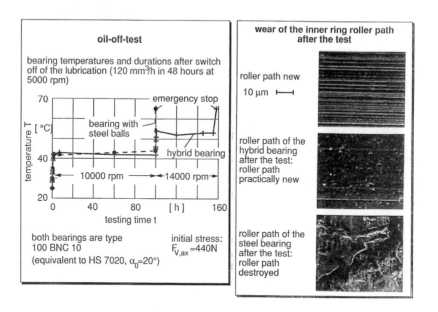

Fig. 12.18 Increased reliability of a bearing via the use of ceramic balls

the produced parts moves away from the tolerance centre, the critical capability index, C_{mk}, comes into effect. If the mean moves away from the tolerance centre by 10%, that is, by only 1/100 mm, this will result in a further 20% reduction in the allowed tolerance. The permitted range will then be only 30 μm. Statistically determined values fluctuate randomly around a mean, irrespective of the sample size. If the manufacturer wishes to be at least 90% certain of passing an acceptance test, he must design the process in such a manner that the dimensions utilize only 85% of the maximum theoretically permissible range. Ultimately, the process and the machine must be designed so as to ensure that the difference between the largest and smallest produced dimension is 26 μm. Critical specified tolerances are usually within a range of a few hundredths of a millimetre. When a tolerance zone of 1/100 mm and the same assumptions as stated above apply, the range must be less than 3 μm. This is often within the range of inaccuracy of the employed measuring equipment.

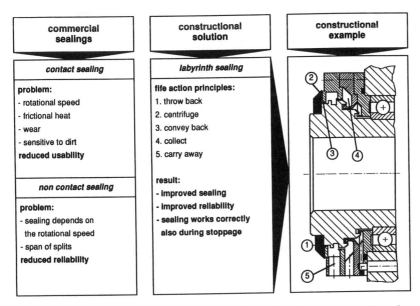

commercial sealings	constructional solution	constructional example

contact sealing	*labyrinth sealing*
problem: - rotational speed - frictional heat - wear - sensitive to dirt **reduced usability**	**fife action principles:** 1. throw back 2. centrifuge 3. convey back 4. collect 5. carry away

non contact sealing	**result:** - improved sealing

problem:
 - sealing depends on
   the rotational speed
 - span of splits
 reduced reliability
	- improved reliability
	 - sealing works correctly
	   also during stoppage

Fig. 12.19 Protection of the spindle bearing from dirt and cooling lubricants (based on: IMA Stuttgart)

A higher level of precision is often attainable via more rigid and/or more precisely running linear guideways and spindle-bearing systems.

In the area of linear guideways, various guiding principles are available which fulfil technical, economic and ecological requirements to varying degrees. The previously mentioned roller-guided rails are becoming increasingly widespread in small and medium-sized machine tools, on account of their low friction, their subsequently enhanced positioning capabilities and their simple production and installation [14]. Fig. 12.21 compares the rigidity and damping characteristics of the available systems.

Roller-based guideways differ primarily with regard to the type of rolling elements employed. Both balls and rollers are used in X and O configuration. The point contact which occurs with balls leads to relatively high levels of surface pressure in the contact zone. The rigidity of the ball/socket contact is consequently relatively low and is determined to a decisive extent by the ball diameter and the osculation of the running groove. By comparison, the surface pressure is lower when rollers are employed, due to the linear contact. The rigidity is correspondingly higher,

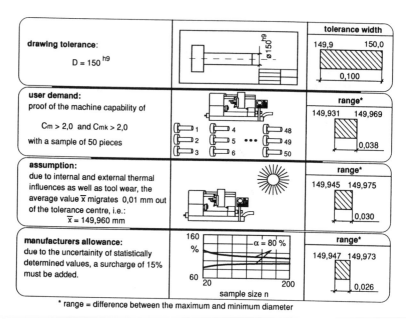

Fig. 12.20 Restriction of tolerance by 3 quality levels as a result of statistical acceptance testings

and depends on the roller diameter and the effective bearing length of the rollers. As the results of tests under lateral load shown in the figure indicate, the rigidity levels of roller-guided rails may even exceed those of a slideway of identical size, whereas their damping capacity is well below that of a slideway. The situation is improved when a supplementary damping carriage is employed. In this connection it should be noted, however, that a damping carriage is effective only at resonant frequencies at which relative displacements occur in the guideways.

With regard to the straightness of linear motion, the roller-based rail is superior to a hydrodynamic slideway, whereby the ball-guided rails represent a more favourable option than the roller-guided rails, as balls can be produced with a higher level of geometrical accuracy than rollers (Fig. 12.22).

Under laboratory conditions, a total deviation of 2.1 μm was measured for the roller guide, compared to 1.6 μm fr the ball guide. High precision requirements, such as apply to the cutting of laser reflectors, for example, can be fulfilled with aerostatic slide systems (total deviation 0.5 μm). Fault

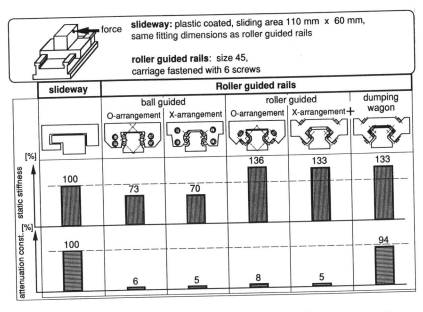

Fig. 12.21 Comparison of the rigidity and damping characteristics of different linear
guideways

analysis reveals, however, that the 3rd and 4th degree fault, which is
characteristic of the guide system (it occurs, for example, as a result of the
frequency at which the rolling element enters the load plane and the
qualities of the relatively displaced surfaces) is not dominant. Rather, it is
the faults caused by production and assembly errors, in the form of 1st
degree faults (e.g. non-parallel installation of guideways) or 2nd degree
faults (e.g. incorrect fitting of the guideways resulting in waviness and
screw clearance) which frequently predominate [15].

An undesirable effect of increasing machining speeds is the increase in
the temperature of a machine tool, as the resultant thermoelastic
deformation leads to lasting impairment of the accuracy of machine tools.
Whereas users were previously prepared to accept a heating-up phase for
the machine prior to beginning a precision machining operation, the level
of acceptance for such a phase is now on the decline. Furthermore, as a
result of the decreasing batch sizes, a machine often has no opportunity to
attain a thermally stable state. In addition to the minimization of
thermoelastic displacement via appropriate design measures to eliminate or

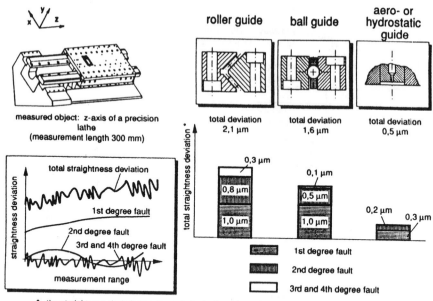

Fig. 12.22 Accuracy levels attainable with precision guideways (based on: IPT Aachen)

isolate heat sources, compensation via control measures is now acquiring increasing importance (Fig. 12.23) [16, 17]. Such measures involve recording the temperature-induced displacement characteristics of machine in a learning phase. A personal computer then calculates which temperature measuring points possess a good correlation to the measured displacements and specifies a compensating equation on the basis of these measuring points. This compensating solution can then be tested under various operating conditions, by continuing to record the relevant temperatures with the PC, determining the compensation values and transmitting these to the NC controller. When a suitable compensating solution has been found, this can be directly integrated into the machine controller. By way of qualification it should be pointed out that this procedure is suitable only for compensating linear displacements between the workpiece holder and the tool holder in the directions of the feed axes (in the case under discussion: x and z direction). The inclination of a spindle is thus difficult to compensate and usually requires a 5-axis machine.

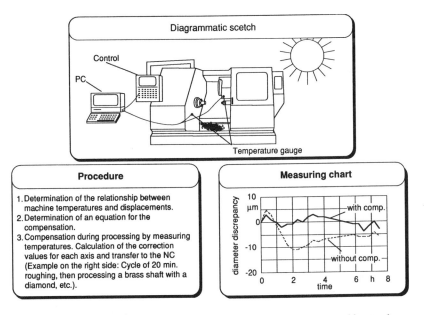

Fig. 12.23 Indirect compensation of thermoelastic displacements on machine tools

Due to the long free-standing extended lengths of the stands involved, large boring machines and portal milling machines, for example, are sensitive to fluctuations in the ambient temperature. The front and rear sides of stands possess different time responses, due to the differing wall thicknesses. As a result, the stand in the presented example inclines forward in the morning, while in the evening the front side of the stand is warmer, as a result of which the stand inclines backwards. These inclinations are particularly unpleasant as, in conjunction with the large lever arms, they lead to large deviations, particularly when boring beads. In the example shown in Fig. 12.24 it was attempted to minimize thermal deformation and to balance out the varying thermal capacities of components with different wall thicknesses via appropriate insulation of the stand components.

To this end, the temperature characteristics of cuboids of varying base areas were measured with and without insulation. Even a relatively thin insulation layer of 25 mm resulted in a substantial decline in the temperature fluctuations of the measured cuboid. After implementing this knowledge in the form of an insulation system for the stand of a portal

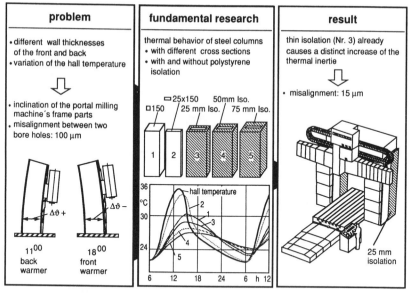

Fig. 12.24 Thermal insulation of stand components against fluctuations in hall temperature (based on: Waldrich Siegen)

milling machine, the misalignment between two holes drilled by means of the bead-drilling process was reduced from 0.1 mm to below 0.015 mm [18].

Spark erosion has long been established in the area of tool and mould manufacturing. This technology is particularly suitable for forming materials which are difficult to cut. However, the result of such machining processes can be influenced substantially by varying the large number of process parameters which are available. The special knowledge of experienced operators is required here. On account of the long machining times it is desirable to carry out the actual machining processes at a constantly high level of material removal and accuracy during low-manned shifts. The accuracy of the produced workpiece depends primarily on the level of tool wear. A statistical gap control system developed at WZL is employed to enable good machining results during low-manned shifts (Fig. 12.25). Research tests have shown that the ignition delay provides a good means of assessing and controlling the process. The ignition delay is consequently measured over a large number of ignitions and statistically analyzed, in order to regulate the machine's feed axes. At a roughly

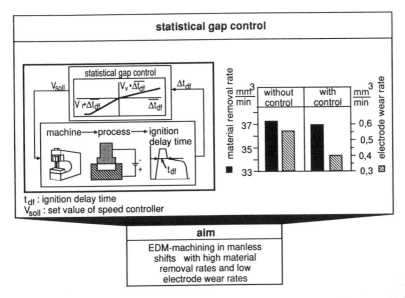

Fig. 12.25 Gap control for spark-erosion machining

constant material-removal rate, the tool wear can be reduced by over 30% [19].

12.6 Increasing the machining speed of a machine tool

The trend towards higher spindle speeds and feed rates reflects the wishes of manufacturers to tap, for their own ranges of parts, the available performance capabilities of modern cutting materials by means of high-speed cutting (HSC), the advantages of which have been demonstrated for several applications involving high machining times. However, the technical measures necessary to this end inevitably lead to a loss of reliability and a substantial increase in investment costs.

As the bed, stand, slides and headstock are subject to comparable stress in an HSC machine and in a conventional machine, the customary design principles apply with regard to dimensioning. As a result of the high rapid traverse rates, the inertial forces of the components during acceleration and braking acquire increased mportance. Materials which

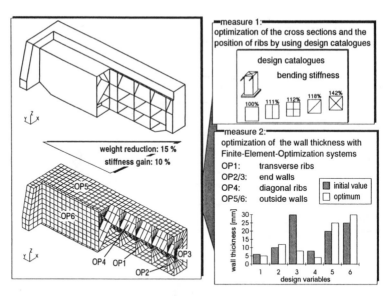

Fig. 12.26 Optimization of the transversal bar of a portal milling machine

are lighter than steel but possess similar stability (e.g. Al-Ti alloys or fibre composites) are suitable here. Also, the required component rigidity can be attained with a minimal use of material, via optimized shaping of the components [20]. This approach was adopted in dimensioning a portal finish-milling machine for machining deep-drawing and forging dies. To this end, the transversal bar of the machine shown in Fig. 12.26 was first of all modified on the basis of general design principles, in order to improve the ribbing configuration. Wall thickness optimization was then carried out with the aid of the finite element method. Starting from initial wall thicknesses based on previous knowledge, it was established in the course of the calculation process that optimized material distribution applies when the outer walls possess approximately five times the wall thickness of the transverse and diagonal ribs and three times the wall thickness of the end walls. This resulted in a weight reduction of 15% in conjunction with an increase in rigidity of around 10%.

For some years now, rolling bearings have been in standard use for machine tool spindles. HSC machining requires an increase in the upper speed limit, to enable cutting to be carried out at the economically interesting cutting speeds which are possible with modern cutting

type of bearing / properties	7020C	7020C WZL	71922C	HS7020C	7020C hybrid	HS7020C hybrid
rotational speed parameter n*dm [10^6 mm/min] with grease lubrication	1	1	1	1,2	1,2	1,4
rotational speed parameter n*dm [10^6 mm/min] with oil-air-supply	1,5	1,9	1,5	1,8	1,9	2,3
dynamic load rating C [kN]	75	75	60	41,5	2)	24
radial stiffness of a bearing pair at low pre-load UL [N/µm]	670	670	650	528	2)	2)
axial stiffness of a bearing pair at low pre-load UL [N/µm]	115	115	110	104	2)	2)
moment of friction at n = 10^4 rpm [Nm][1] grease lubrication (32 cSt) Frad = 0 Fax = UL	0,152	0,152	0,158	0,112	0,139	0,105
relative costs (March 1993)	100%	110%	100%	125%	250%	315%

©WZL1993   1) calculation by WZL   2) no manufacturer data available

Fig. 12.27 Comparison of rolling bearings for fast-rotating machine tools

materials. The primary requirements here are improvement of the tribological characteristics of the bearing and the reduction of friction and wear [21, 22]. As only the most minimal quantities of lubricant are required to separate the surfaces which are moved relative to one another when a suitable lubricant is selected, and in view of the fact that the friction moment of a bearing rises drastically as the quantity of lubricant increases, all improvement measures are based on optimization of the lubricant and reduction of the required quantity of lubricant. This gives rise above all to the problem of correct lubricant supply, as a reduction in the quantity of lubricant usually also means reduced operational safety. A problematic aspect here is how to convey minimal quantities of oil to the contact point through the air curtain, which is circulating at high speed. The question of the correct method of filling greased bearings, on the other hand, has been largely clarified: The entire quantity of grease should be inserted in the bearing during assembly and displaced by the bearing itself in the course of so-called grease distribution runs. Separately filled grease supply chambers have proven inexpedient.

One method of reducing wear and improving anti-seizure performance is to vary the material pairing in Hertzian contact. Ceramic balls can be

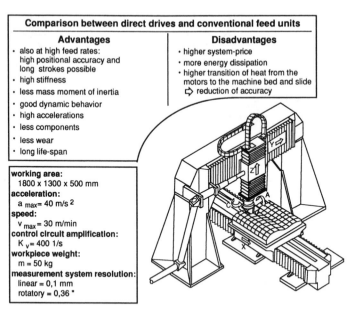

Comparison between direct drives and conventional feed units

Advantages	Disadvantages
• also at high feed rates: high positional accuracy and long strokes possible	• higher system-price
• high stiffness	• more energy dissipation
• less mass moment of inertia	• higher transition of heat from the motors to the machine bed and slide
• good dynamic behavior	⇨ reduction of accuracy
• high accelerations	
• less components	
• less wear	
• long life-span	

working area:
 1800 x 1300 x 500 mm
acceleration:
 $a_{max} = 40$ m/s^2
speed:
 $v_{max} = 30$ m/min
control circuit amplification:
 $K_v = 400$ 1/s
workpiece weight:
 m = 50 kg
measurement system resolution:
 linear = 0,1 mm
 rotatory = 0,36 "

Fig. 12.28 Direct linear drives for machine tools (based on: ZFS)

used, for example, or the raceways can be coated with carbides. Fig. 12.27 compares the characteristics of the rolling bearings available for machine tool spindles. Particularly suitable for very high speeds are the HS bearing (High-Speed) and the hybrid bearing (bearing with ceramic balls), which possess advantages over the conventional options due to the reduced centrifugal force resulting from smaller balls and the more favourable combination of materials. Today, hybrid bearings represent a very interesting alternative, due to their improved anti-seizure performance and the fall in the price of ceramic balls.

Conventional drive elements for the linear traversing axes of machine tools, such as ball-and-screw spindle drives, belt drives and rack-and-pinion drives, permit only limited traversing speeds at specified accuracy levels. The energy required to accelerate a slide or the energy which is released during braking or in the event of a crash depends primarily on the mass moment of inertia of the rotating components (electric motor, ball-and-screw spindles, clutches, belt pulleys). Direct linear drives are not subject to such restrictions. They further possess the additional advantage that they require less components, the service lives of which are longer, as

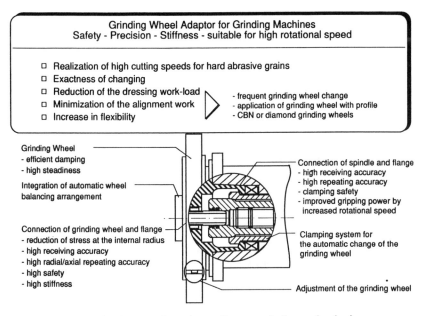

Grinding Wheel Adaptor for Grinding Machines
Safety - Precision - Stiffness - suitable for high rotational speed

- □ Realization of high cutting speeds for hard abrasive grains
- □ Exactness of changing
- □ Reduction of the dressing work-load
- □ Minimization of the alignment work
- □ Increase in flexibility

- frequent grinding wheel change
- application of grinding wheel with profile
- CBN or diamond grinding wheels

Grinding Wheel
- efficient damping
- high steadiness

Integration of automatic wheel balancing arrangement

Connection of grinding wheel and flange
- reduction of stress at the internal radius
- high receiving accuracy
- high radial/axial repeating accuracy
- high safety
- high stiffness

Connection of spindle and flange
- high receiving accuracy
- high repeating accuracy
- clamping safety
- improved gripping power by increased rotational speed

Clamping system for the automatic change of the grinding wheel

Adjustment of the grinding wheel

Fig. 12.29 Standardized connecting adaptor for new grinding technologies

they operate on a contact-free basis and are therefore free of wear (Fig. 12.28).

However, high coil currents are required, in order to attain a high level of standard rigidity in the feed direction. As a result of these high currents, the magnetic forces of attraction acting on the guideways are five times higher than the attainable feed forces. The coil output also has a detrimental effect on the thermal behaviour of the machine, as the coil introduces heat into the bed and the slide, thereby deforming them. Maximum acceleration rates of 40 m/s² and speeds of up to 30 m/s were attained on a laser portal developed at the Zentrum für Fertigungstechnik Stuttgart (ZFS) [23].

The mechanical connections in particular are also required to maintain the required positional accuracy under the influence of high centrifugal forces. The capacities of the standard adaptor designed for automatic tool-changing, the steep taper, are virtually exhausted with regard to these requirements [24]. Under the influence of centrifugal forces, a spindle taper opens out more at the large diameter than at the small diameter. This geometrical change results in a poorly fitting taper with supporting areas at

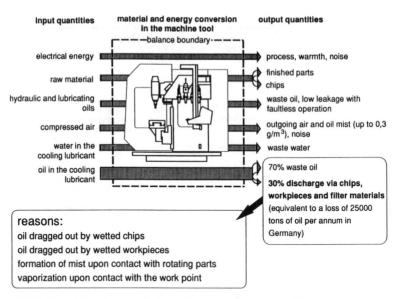

Input quantities **material and energy conversion in the machine tool** **output quantities**

r – – –balance boundary· – – – ¬

electrical energy → process, warmth, noise

raw material → finished parts
→ chips

hydraulic and lubricating oils → waste oil, low leakage with faultless operation

compressed air → outgoing air and oil mist (up to 0,3 g/m³), noise

water in the cooling lubricant → waste water

oil in the cooling lubricant →

70% waste oil

30% discharge via chips, workpieces and filter materials (equivalent to a loss of 25000 tons of oil per annum in Germany)

reasons:
oil dragged out by wetted chips
oil dragged out by wetted workpieces
formation of mist upon contact with rotating parts
vaporization upon contact with the work point

Fig. 12.30 Material and energy balance for a metal-cutting machine tool

the end of the taper and leads to axial displacement of the tool shaft into the spindle. When the machine is at a standstill, significantly higher ejection forces are then required, as the tool is jammed in the spindle. Alternatively, this widening of the taper can also result in the tool shaft being forced out of its central position by the machining forces, making operation at high speeds impossible. In extreme cases, the tool may represent a safety risk, as the tool may fly off because the clamping mechanism is no longer able to maintain the connection. Consequently, a hollow-shaft adaptor has been developed in the course of a joint project between the university sector and tool manufacturers which is able to meet the high requirements which apply here. When carrying out high-speed grinding with diamond and CBN, frequent grinding wheel changes and profile grinding require a high level of axial and radial positioning and repeat accuracy for the grinding wheel, in order to minimize dressing. Here again, conventional fixing devices for grinding wheels are no longer fully suitable for high-speed grinding, for reasons of safety and rigidity. A research project which is to commence shortly at WZL will thus be concerned with the development of a new, suitable adaptor. The starting point for this work will be the hollow-shaft adaptor which has already proven effective for milling tools (Fig. 12.29).

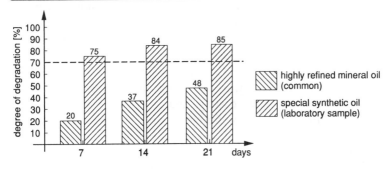

Classification of the degrading behaviour acc. to CEC-L-33-T-82

Degree of degradation of 7.5 mg test substance in 150 ml mineral nutritive solution, mixed with 1ml Inoculum (bacteria culture from sewage plants) after 21 days:

degradation   > 70 %   biologically quickly degradable
               20 % - 70 %   biologically potentially degradable
               < 20%   biologically non-degradable

highly refined mineral oil (common)

special synthetic oil (laboratory sample)

Fig. 12.31 Quickly biodegradable lubricants for spindle bearings (source: Klüber Lubrication)

12.7 Improving the environmental compatibility of a machine tool

Comprehensive assessment of the environmental compatibility of a machine tool requires the overall consideration of all aspects, from manufacture, through operation to retirement from service (recycling or disposal of the employed materials). The growing level of environmental awareness among the population in recent years, increasing disposal costs and the pressure of increasingly more stringent legislation are compelling more and more industrial companies to assess the environmental compatibility of their production operations. The design objectives for today's metal-working machine tools cannot be considered in isolation from one another, which often results in paradoxical requirements. In the case of cutting machine tools, high material-removal rates still necessitate large quantities of cooling lubricant to discharge the chips and cool the cutting site and the machine frame. Yet cooling lubricant represents the prime pollution factor in a cutting process [25, 26].

Assessment of the material and energy conversion in a metal-cutting process (Fig. 12.30) reveals the following:

The supplied electrical energy is completely dissipated. In absolute terms, however, the energy level is minimal and of no economical relevance, at an average value of less than 5% of the hourly machine costs. Furthermore, no significant savings are to be expected as a result of improvement measures, e.g. by increasing the electrical or mechanical efficiency of the machine elements. Provision of the high connected loads which are required for acceleration processes constitute a substantial cost factor, however.

In accordance with the primary function of the machine, the infed raw material is processed into finished parts, in the course of which material is always removed via cutting, forming, erosion, chemical processes, or with laser beams or water jets. Methods of reducing the level of this lost material are already being adopted today, wherever they are possible and economically viable (e.g. near net shape technology). Forging, cold massive forming, precision casting and sintering have acquired new importance as processes for preforming workpieces in such a way as to minimize the volume of material which requires to be removed in order to attain the final contours.

Most of the employed lubricating and hydraulic oil is recycled in the form of used oil and thus constitutes no primary source of pollution, at least. Although the lubricating oil used in machine tools represents only a small proportion of the total of over 1 million tonnes of oil used each year in Germany and, for tribological reasons, is on the decline, methods of environment-friendly disposal are nevertheless under consideration here. The term "biological degradability", as defined in CEC-L-33-T-82, is often employed as a basis for assessing these methods. This standard defines substances which are degraded to a level of 70% into H_2O, CO_2 and harmless inorganic residual compounds by bacterial cultures within 21 days as "quickly biodegradable". Substances which are degraded to a level between 20% and 70% are referred to as "potentially biodegradable", while substances with degradation levels below 20% are considered "non-biodegradable". As Fig. 12.31 shows, suitable, synthetic lubricants which qualify as quickly biodegradable are already available today for lubricating fast-rotating spindle bearings [27]. High-quality mineral oils, which may be quite suitable lubricants from a technical point of view, possess considerably less favourable degradability characteristics.

Cooling lubricants (oil or emulsion) are fed into the machining process in order to reduce the friction between tool and workpiece or chip, to discharge heat from the tool, workpiece and chip and to remove chips from

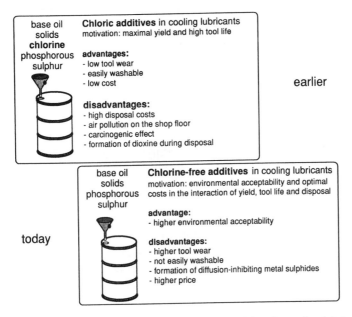

Fig. 12.32 Advantages and disadvantages of chloric additives in cooling lubricants

the machining site. The proportion of oil used here amounts to an annual total of approximately 85000 tons in Germany, around 30% of which - representing the considerable quantity of 25000 tonnes - is dragged out with chips and finished parts, atomized or evaporated.

The recycling of machine tools is no problem, as they consist primarily of cast iron or steel, which has been effectively recycled for a considerable time. The reaction resin concrete which is sometimes used for the beds of machine tools is also highly suitable for recycling, as it absorbs only the most minimal quantities of oil. In the context of the present situation, therefore, the aspects relating to the environmental compatibility of a machine tool can be reduced to the problem of the cooling lubricant which is discharged and how to dispose of these cooling lubricants.

Measures to combat the cause of this problem, such as dry machining or the use of cooling lubricants without oil content, are as yet unfeasible for many production processes, on technological grounds. In the case of dry machining, the previously described thermal effects on the tool and workpiece, diminished accuracy due to thermoelastic deformation and chip discharge problems represent additional technical obstacles.

Another approach involves measures to combat the effects, such as fully enclosing the machine, rinsing chips and workpieces, centrifuging and compressing the chips and drawing off and filtering all the oil mist and oil vapour. An important step is the substitution of aggressive additives in cooling lubricants, which recent studies furthermore reveal to be often unnecessary. Despite the undisputed technological disadvantages the Association of German Engineers (VDI) now promotes the switch-over to chlorine-free processes, resulting because of the carcinogenic effect of substances containing chlorine when inhaled or in contact with the skin and because of the formation of polychlorinated dibenzodioxins (PCDD) during disposal (Fig. 12.32). The use of quickly biodegradable vegetable oils in lubricants, such as rape oil, does not produce any significant advantages. No significant differences between cooling lubricant emulsions with vegetable or mineral oil additives are indentifiable with regard to human- and ecotoxicological properties or biodegradability. This is because the decisive aspects of both assessment criteria are not the properties of the oil in new condition, but rather the additives which are required for both oils and the reaction products which result from impurities and contact with the hot machining site. In the case of emulsions, therefore, correct monitoring and maintenance of the cooling lubricant, which have a decisive influence on the service life and, subsequently, on the length of the changing cycles, are the most important prerequisites for environment-friendly production operations. More detailed information on this subject will be provided in Chapter 5 of this volume.

Systems for cleaning cooling lubricants must also be seen in this context. For many years, filter systems with fibrous-web filters were standard fittings in turning and milling centres. Today, efforts are in progress to find new alternatives which do not require the paper web, which is classified as pollutant waste. The same filtering fineness can be attained using drum filters incorporating screens, without any additional filtering elements (Fig. 12.33).

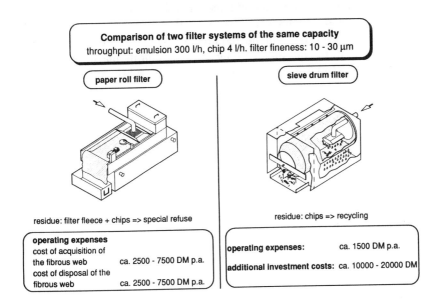

Fig. 12.33 Physical filters for separating chips and cooling lubricant (based on: Chiron)

12.8 Summary

A broad scope of requirements and conditions require to be observed when developing a machine tool. In addition to the long-term trend towards increasingly more reliable, more precise and faster machines, the growing level of environmental awareness in the population is having increasingly more tangible effects. But in the face of a generally poor economic situation, the implementation of design measures must, above all, be carried out in a cost-effective manner.

In spite of the tense economic situation, which also prevents cost-intensive research and development activities, interesting innovations are still to be found. Design concepts which render the machine tool not only more technically efficient but also more cost-effective have been made possible above all via the integration of functions and processes, the growth of sensor technology and new developments in the field of machine elements.

But the area of environmental protection in particular will acquire increasing importance in the future. Existing laws will be further tightened, resulting in more stringent requirements and high disposal costs. For both economical and ecological reasons, therefore, the machine tool manufacturing industry will be required to devote more attention in the coming years to the question of how a manufacturing machine tool can be rendered more environmentally compatible and subsequently more cost-effective.

13

THE OPEN CONTROLLER - KEY ELEMENT OF HIGH-PERFORMANCE PRODUCTION FACILITIES

13.1 The background situation

Machine tools occupy a position of key importance within the industrial sector as a whole. In turn, the functionality, attractiveness and acceptance of complex machine tools and production facilities are determined to a decisive extent by the efficiency of the deployed control systems. Modern controllers enable complex, automated machining operations, high levels of speed and accuracy, user-friendly programming and operation of the machinery and much more besides, whereby economic efficiency must always be the top priority for the user (Fig. 13.1). Depending on the number of movements to be controlled and the degree of automation of a machine, it is quite possible for electrical and control components to account for up to 50% of overall costs.

On the basis of the typical control tasks which machine tools are required to carry out, a traditional division into numerical controller,

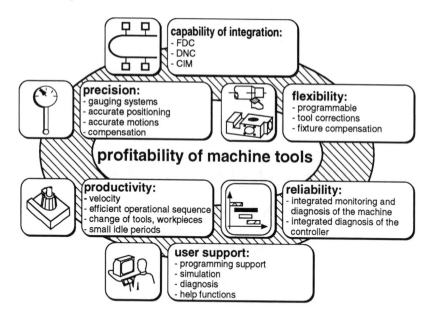

Fig. 13.1 The importance of the controller for a machine tool

programmable controller, drives and, if employed, cell or master computer has become established (Fig. 13.2) [1]. In the coming years, however, this division into hardware units is expected to give way to a more function-oriented approach, as the clear boundaries between these sub-systems which have applied up to now will disappear. The design and planning of control equipment for machine tools will develop increasingly into an area of work requiring an overall, integral approach [2].

But it is in the area of numerical control systems in particular that European control systems manufacturers are in danger of being left behind by Japanese companies. A look at the world market for numerical controllers (Fig. 13.3) shows that Fanuc alone holds around 50% of the market as the leading Japanese supplier. Whereas the products of the more than 60 European numerical control system manufacturers are used primarily on the European market, the products of the small number of Japanese NC manufacturers are well established throughout the world.

A comparison of Japanese and European numerical controllers reveals that the strengths of Japanese products lie less in their technical capabilities and rather in the very low prices which normally apply, the high degree of reliability, the high quality and the short product

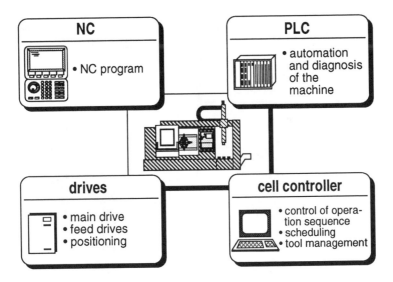

Fig. 13.2 Control technology in machine tools (overview)

development times, which result in a fast pace of innovation. From the point of view of the control systems manufacturers, the great diversity of technologies involved in the area of machine tools is a highly problematic aspect: In contrast to Japan, there are only a small number of large-scale customers for controllers, while there are many special applications which usually require only small quantities of controllers. This makes the cost-effective production of large quantities more difficult and increases the expenditure incurred by the manufacturers of control systems in developing special concepts for specific technologies (e.g. for laser machining or the machining of freeform surfaces).

In view of this market situation, which is gradually threatening to become a serious danger to the European industry, this paper is to focus primarily on the area of NC controllers. It is not appropriate, however, to consider NC controllers in complete isolation, as due account must also be taken of the other control components (programmable controllers, drives, etc.).

In the light of the Japanese dominance of the market, discussion as to how to combat this situation has been in progress in Europe for several years [3, 4, 5, 6]. Apart from numerous individual measures to optimize prices, performance, functionality, user-friendliness, susceptibility to faults

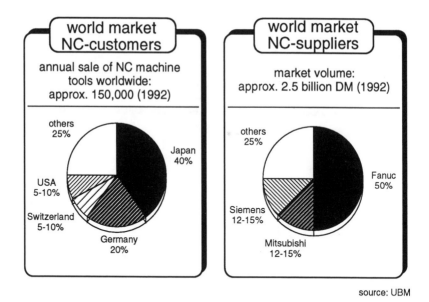

source: UBM

Fig. 13.3 Summary of the world market for NC controllers

and unit volume, a key area of effort by the Europeans will be concerned with the creation of "open" control systems (Fig. 13.4). The primary objectives pursued with open control systems are to provide flexible control functions with a broad scope of freedom for the machine manufacturer to adapt the functions to the specific requirements of the application concerned and to increase the use of standardized components. Additionally, integration into the user's existing production engineering environment is to be simplified substantially via the use of clearly defined, function-oriented interfaces.

Open controllers are currently a highly topical subject, on which numerous and sometimes conflicting opinions and information are available. In this paper, it is intended to discuss the extent to which the creation of such open control systems will be able to secure the competitive capacity of European manufacturers of control equipment and machine tools. In particular, the primary objectives pursued in the development of open controllers are to be presented and the benefits to be expected from open systems are to be assessed, in order to enable a more accurate appraisal of the lively current discussions on open control systems. In addition to presenting the current situation with regard to

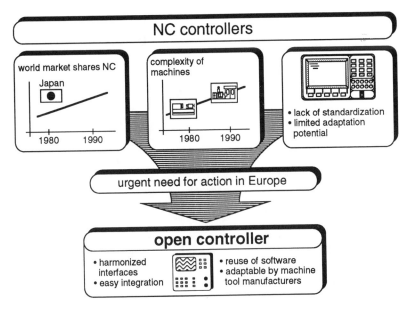

Fig. 13.4 Motivation for the development of open control systems

numerical controllers and the various philosophies advocated in public discussion, this paper is therefore also intended to help objectify the discussion on open controllers.

13.2 The scope of functions and deficiencies of modern numerical controllers

As illustrated in Fig. 13.5, today's numerical control systems incorporate a highly extensive scope of performance capabilities and functions [1], even though the scope of functions shown in the diagram is not covered by one single system, specific combinations of the options shown being implemented in accordance with the requirements of the market segment concerned. Current new developments are to be observed, for example, in the integration of PCs (e.g. for management functions or user interfaces), increasing levels of adaptability for the machine manufacturer and new interpolation processes (splines, nurbs).

But the decisive problem areas relating to numerical controllers are not so much the inadequate technical capabilities of the numerical controllers,

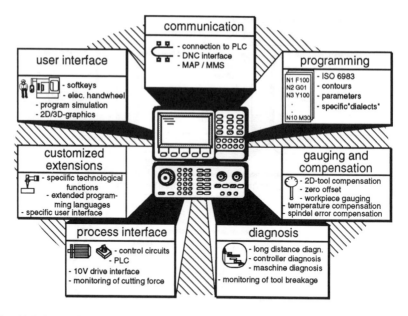

Fig. 13.5 State of the art in the field of numerical control systems

but rather the excessively high development and adaption costs and the inadequate degree of harmonization between the various manufacturers. Fig. 13.6 provides an overview of the most important current problem areas relating to numerical controllers, presenting the different points of view of the main partners involved in the value-creation process - the control systems manufacturers, the machine tool manufacturers and the end users -, as their respective interests often differ considerably.

The main problems for the control systems manufacturer are the high development costs for new controllers, above all with regard to software development: Only seldom can existing programmes from a manufacturer's previous generation of controllers be reused. Additionally, numerous machine manufacturers require time-consuming and expensive adaptions to specific technological applications.

The main areas of complaint for the machine tool manufacturer are the inadequate facilities for adapting numerical controllers to specific applications and the requirements of ultimate users. In addition to individual configuration of the user interface, the implementation of process-integrated control functions for special applications is often necessary, and to date the machine tool manufacturer is often unable to carry out this work

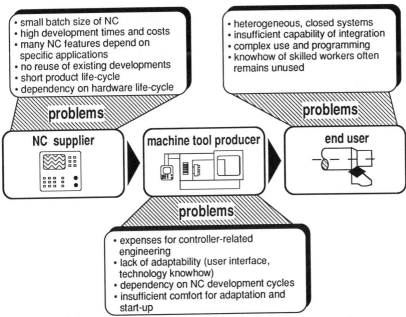

Fig. 13.6 Breakdown of the current problem areas relating to NC control systems

by himself, requiring the cooperation of the control systems manufacturer.

The greatest problems from the end user's point of view are presented by the inadequate standardization of the NC controllers and the appurtenant interfaces available from various manufacturers. This is of importance both for the user guidance functions and the integrative capacities of the controllers. Although standards do exist in many areas, the degree of practical application of these standards remains inadequate.

Considering all these factors as a whole, it becomes clear that a prime objective to be pursued in future efforts must be reduced costs, to enable standardized controllers in larger quantities produced to be applied to an increasing extent to special applications. The concept of an open controller produced to standard specifications by all manufacturers, which is considered in detail below, can help to remedy this situation by enabling the simple and cost-effective configuration of numerical controllers and therefore permitting flexible adaption to varying machining tasks. Machine manufacturers in particular often wish for a possibility of integrating their own know-how.

Prerequisite for a modular, flexible control architecture is the increased standardization of functional control modules and the interfaces of a numerical controller [7]. Fast and simple integration of the controller into the existing production environment is also required. This can be attained via the increased harmonization of NC programming and high-capacity interfaces to the user, to the process and for the purposes of inter-system communications.

The following sections discuss in more detail how these objectives can be attained with the aid of the concept of the open controller.

13.3 Essential functions of open controllers

The essential functions of an open controller are
- to enable companies other than the NC manufacturer to integrate their own functions with the smallest possible scope of work and expenditure,
- to provide standard interfaces, in order to enable or enhance the interchangeability of functional NC modules from various manufacturers,
- to provide standard interfaces for the connection of external systems, in order to enable integration of the NC into the information-based environment of the production process.

In order to gain an understanding of the discussion on "open" controllers, however, a more precise definition of the term is first of all necessary.

13.3.1 Definition of the term "open controller"

The term "openness" can be divided into two basic areas of "internal" and "external" openness (Fig. 13.7). Internal openness refers to an openness concerning the internal control functions, while external openness covers all external interfaces of the controller (e.g. interfaces connecting the controller to other controllers or to the drives). On the basis of this differentiation it can be directly concluded that internal openness is of importance primarily to the manufacturers of control systems and the machine tool manufacturing sector, while external openness is of prime importance to the end user.

Current discussion is concerned primarily with internal openness (Fig. 13.8, left). The comments of those involved in this public discussion vary greatly with regard to interpretation of the term; it can be established as a

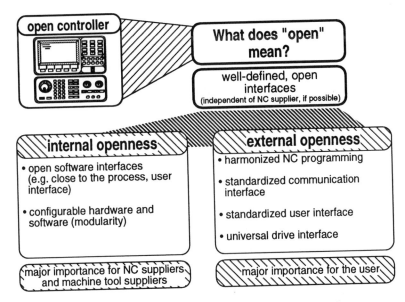

Fig. 13.7 Definition of the term "open controller"

point of consensus that a characteristic feature of open systems is that they possess a component element which is accessible not only to the manufacturer. Access to this open part of the controller is facilitated by specific aids, such as interface specifications and development tools.

Considerable differences of opinion exist, however, with regard to the questions as to

- which parts of a control system should be produced in open configuration,
- who should have access to these parts,
- whether the interface definitions valid in these areas should be standardized and vendor-neutral,
- the degree of detail to which access to internal functions via standardized interfaces should be stipulated and
- how the resultant technical, organizational and financial conditions are to be assessed.

These questions can be answered by reference to the target criteria which open controllers are required to fulfil. These target criteria can be derived directly from the above-described problems of existing systems (cf. Fig. 13.6). the aim of developing open systems in the area of internal

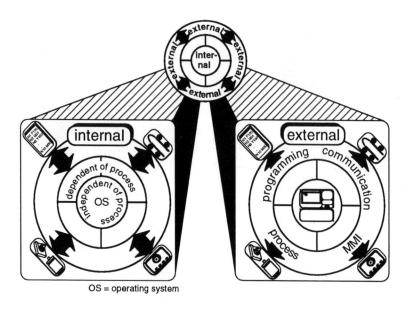

OS = operating system

Fig. 13.8 Internal and external openness

openness should be
- to enable machine tool manufacturers to integrate their own functions into various control systems in a simple and, as far as possible, standardized manner
- to enable the manufacturers of control systems to develop software as independently as possible of hardware, thereby avoiding the extremely short hardware life cycles,
- to enable controllers to be adapted in a fast, simple and inexpensive manner to the individual requirements of specific applications and
- to enable the integration of existing or newly developed functions, including such products from external suppliers, such as software companies.

Once these stated aims are accepted, several answers to the above questions can be established. Only standardized, vendor-neutral interfaces enable machine tool manufacturers to integrate their own functions into various systems at a viable cost. To enable the integration of functions offered by software companies which are independent of any controller manufacturers, a sufficient market volume is required for such software products. This is made possible by manufacturer-neutral interfaces. The functions developed by machine tool manufacturers are normally technology-oriented and generally require intervention in the control system which goes far beyond the scope of the user interface. The machine tool manufacturer must therefore be able to modify internal control functions and/or have access to internal interfaces. The functions which have been developed by the machine tool manufacturer or purchased from external companies and which require to be integrated only in the rarest instances do affect the core of the controller. Neither is user access to most of the blocks of the NC core, such as the interpolation block, an expedient measure. If necessary, such a block would be completely replaced by another block. Further sub-division of the stated level is therefore inexpedient.

In the course of the current discussion on open systems, one aspect is usually only touched upon in brief. This is the capability of the controller to be integrated in the simplest possible manner into an existing information technology environment - referred to here as external openness (Fig. 13.8, right). This capability is a prime requirement for the end user, i.e. the customer of the machine tool manufacturer and, subsequently, of the control systems manufacturer.

This area of external openness can, in turn, be sub-divided into the four main areas of

- user interface or MMI (man-machine interface)
- programming interface
- process interface and
- communications.

In all the above-stated areas, the end users wish to be able to integrate control systems into their existing production engineering and information technology environment with the minimal scope of adaption or, ideally, without requiring any adaption whatsoever.

This situation calls for two courses of action. Firstly, an increased level of standardized interfaces requires to be created in the stated areas in the future. Secondly, controllers must be designed in such a manner that their external interfaces can be parameterized simply, quickly and in the broadest possible scope. This is where the terms internal and external openness meet up once again: on the basis of systems which possess internal openness, the functions which are required for external openness can be provided and/or the necessary adjustment operations can be carried out by parties other than the manufacturer of the control system.

13.3.2 Existing standards and current developments with regard to efficient external interfaces

It is essential to incorporate existing and drafted standards and quasi-standards into the development of open systems and the interface specifications upon which such systems are based. This is the only way of ensuring the required degree of acceptance by the user. The area of hardware is consciously avoided in the discussion of open control systems. Firstly, the stipulation of standards should be avoided in this area, to prevent the blockage of new developments. And secondly, an important aspect of the development of open systems is that the software should be created with the greatest possible degree of independence from hardware, to help achieve the change-over to new, higher-capacity processors at a feasible level of expenditure.

It is noticeable that neither established nor de facto standards exist for internal openness. The software standards which apply in the area of general data-processing are unsuitable or only partly suitable for application to control systems. Initial bases for applicable standards are to be provided, for example, by the POSIX specification, which is concerned with the standardization of a UNIX operating system with real-time capability. The question as to the extent to which this specification will fulfil the requirements for real-time capability and openness is the subject of intense discussion at present.

In the past, most control systems manufacturers have attempted to develop optimal systems to meet their requirements, independently of machine tool manufacturers, end users and competitors. This does not mean that the systems manufacturers failed to take due consideration of the needs of their customers, that is, the machine tool manufacturers and end users, in designing new controllers. But they considered these needs as requirements relating to a closed system. Facilities for external intervention in the control process received little or no consideration in the specification for configuration of the overall control system. The resultant problems for both the control systems manufacturers and their customers have already been presented in preceding sections.

In the area of the user interface in particular, a number of standards have become established which are of relevance to external openness, and which thus require due consideration (Fig. 13.9). Some of these standards originate from the area of general data-processing, which is particularly interesting, as the next generation of operators have already become acquainted with the use of general data-processing equipment, i.e. in most

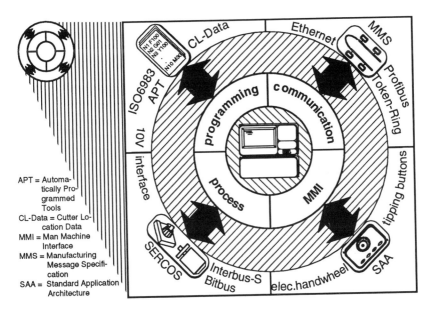

APT = Automatically Programmed Tools
CL-Data = Cutter Location Data
MMI = Man Machine Interface
MMS = Manufacturing Message Specification
SAA = Standard Application Architecture

Fig. 13.9 Incorporation of existing standards

cases PCs, as an integral part of their theoretical and vocational training.

(a) NC programming

In the area of NC programming which, as Fig. 13.9 shows, belongs to the area of external openness, various, sometimes complementary concepts apply as a result of the various paths and sources of information (Fig. 13.10). A common feature of all these concepts is the possibility of applying DIN 66025 as the input format for the core NC. This standard is comparable to an assembler programme, onto which the various higher "programming languages" are mapped.

Today there are a vast number of NC programmes in accordance with DIN 66025 or ISO 6983. These programmes constitute an important production factor for the manufacturing companies. The costs incurred in developing these programmes and the majority of the technological know-how required to this end are tied up in DIN programmes. It is therefore of vital importance for these companies to be able to continue to use these programmes in the future. However, developers and users are unanimously agreed that the interface in accordance with DIN 66025 is already unable to meet all the requirements of modern production systems. Consequently,

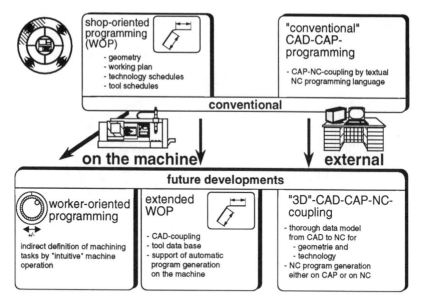

Fig. 13.10 Trends in NC programming

diverse efforts are currently in progress to devise new, more efficient concepts for NC programming and the CAM/NC system.

WZL is presently involved in the development of new, efficient concepts for NC programming and the CAM/NC system on the basis of open, modular systems, in connection with the MATRAS European research project. The main area of this development work concentrates on the definition of several interface levels, with the aid of which both new, high-capacity controllers with a high level of intrinsic intelligence and existing systems can be integrated. An important aspect is the handling of technological information. The machine can be provided with effective support as an independent production unit by transmitting the technological information defined at the level concerned to the controller together with the geometrical information. In this way, current structural changes in the area of production through to smaller, decentralized and autonomous units can be incorporated and promoted. This new concept for the area of the CAM/NC system will, of course, integrate the existing programmes created on the basis of previous standards, i.e. it will be (downward)-compatible.

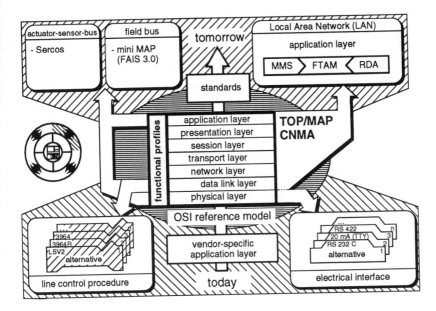

Fig. 13.11 External NC communications interfaces

(b) Open communications

The second area in which intensive research and development work is also in progress at WZL is that of communications (Fig. 13.11). Company-wide communications based on stable standards are essential for new concepts described by terms such as Lean Production or the Fractal Factory. An example of the work being carried out in this area is provided by the efforts to develop a system of production-oriented communications via MMS (Manufacturing Message Specification) and its Companion Standards [8, 9].

In the area of MAP developments (Manufacturing Automation Protocol) there were a number of problems in the past, resulting from the exclusive use of one bus system, non-downward-compatible new versions and the inadequate consideration given to production engineering aspects. The MMI specification, which is based on the MAP specification, has undergone substantial development in recent years as a result of intensive standardization work. The scope of this specification includes the Companion Standards, which serve to supplement the general functions covered by MMI with the expanded functions required for special applications. Several Companion Standards have been completed of late,

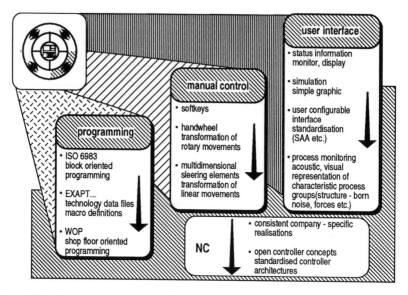

Fig. 13.12 Development trends in the area of the "man-machine interface"

including the NC Companion Standard (ISO TC184/SC1/WG3), which constitutes the authoritative standard for NC controllers.

The second shortcoming mentioned with regard to the MAP specification, the exclusive use of one bus, has recently been abandoned in favour of the de facto standard, Ethernet. All the obstacles to the establishment of MAP/MMI applications on a broader basis thus appear to have been removed.

(c) Standardized man machine interface

There are several concepts of particular importance relating to open systems in the area of the user interface. As shown in Fig. 13.12, the man machine interface can be sub-divided into the areas of

- user interface
- manual control elements and
- programming interface.

The programming interface has already been discussed with regard to aspects of data systems engineering. One of the most important requirements on the part of the end user is the wish for uniform operation of the various controllers, or a system of operation which is based on the same basic rules.

There are a number of parallels to general data-processing in the area of the user interface. With regard to open control systems, extensive simplification in comparison with today's systems would be possible via the rigorous application of existing standards alone. The SAA standard (Standard Application Architecture) serves as an example here. This standard describes the design of a graphical user interface and is in widespread use in the area of PC- or workstation-based interfaces. In particular, it facilitates the operation of changing systems substantially, as certain basic rules are maintained for all systems. Application of this standard would greatly assist the users of controllers when switching not only between products from different manufacturers but also between various products from the same manufacturer, as frequently occurs in practice.

Important work in the area of the man machine interface, concentrating on manual control elements, is in progress at WZL as part of the joint project CeA ('Computer-unterstützte erfahrungsgeleitete Arbeit' - Computer-assisted experience-guided work), sponsored by the Federal German Ministry for Research and Technology (BMFT) [10]. For the purposes of observing the process, information from the machining process is made accessible to the user on a visual, acoustic and tactile basis. For the purposes of process control, i.e. controlling the machine, the operator is provided with new control elements, some of which are multi-dimensional. An essential feature of these control elements is that they are provided with feedback mechanisms, which allow the user direct access to the process (e.g. force feedback). An important objective in this connection is to control the process via increased utilization of the operator's experience.

(d) Process interfaces

A large number of different concepts apply in the area of the process interface [11, 12]. The so-called field buses which are employed here can be roughly divided into two groups. On the one hand there are buses which are designed for specific tasks. Examples here are Sercos (drives) and Interbus-S (binary actuators and sensors). On the other hand, there are buses which are suitable for general use, such as Profibus or P-Net. In the case of all these standards, a specification may be fulfilled to varying degrees, while the products are nevertheless considered to conform to the standard concerned. A well-known example in this connection is the Profibus, the specification for which consists of a basic hardware part and a protocol part configured on this basis.

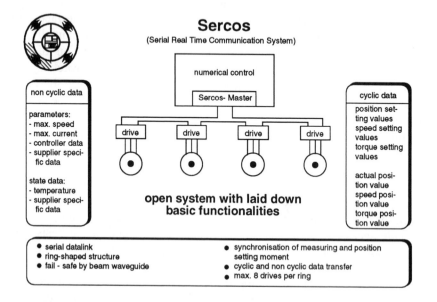

Fig. 13.13 Sercos as an example of an open drive interface

The drive interface is of particular importance in both technical and financial terms with regard to the area of numerical controllers. This interface is required to enable each machine axis to function, and several such interfaces are thus installed on every numerically controlled machine tool. All newly developed interfaces in this area are geared towards replacing the traditional analogue 10V interface with digital interfaces.

An example of an open digital interface between controller and drives is the SERCOS interface [13]. The first products based on this interface specification, which is summarized in Fig. 13.13, are already available on the market.

The main advantage of this digital drive interface is that it enables the combined operation of any drives and controllers from different manufacturers. Beyond this, this specification also supports totally new functions, such as the feedback of drive data to the NC. Interestingly enough, there are similar concepts and systems in Japan, but the configurations differ from one manufacturer to another. The question as to whether the SERCOS interface, with its gateway in the form of a currently available ASIC with a transmission rate of 4 Mbaud, possesses sufficient capacity for future applications is the subject of widespread discussion. Experience certainly needs to be acquired with various SERCOS-based control concepts, before this matter can be judged. The SERCOS initiative recently suffered a setback, however, in the form of Siemens' decision to abandon SERCOS in favour of a drive interface based on a new, extended Profibus specification.

To summarize these considerations, it may be stated that new control concepts will have a chance only if it is possible to integrate these existing standards or standards currently under development. An open concept provides a better basis for this integration process than the existing closed concepts.

13.4 The implementation of open control systems

The essential feature of most of the existing numerical control systems which are designated open is a certain degree of internal openness configured by the individual manufacturer. Control systems which possess external openness in the form of a capacity to process several programming formats (DIN 66025, DIN 66215, VDA-FS, etc.), for example, are rare exceptions [14].

The manner in which open systems should be implemented in terms of internal openness is currently the subject of much discussion. Some manufacturers tend towards the opinion that the use of PC components alone is sufficient to provide the desired openness [15]. They regard the PC as the de facto standard for the controller of the future. It is true that standard PC components are produced and offered in very large quantities and at highly favourable unit prices. But this must be seen in relative terms when considering the area of NC control systems. A control concept developed completely on the basis of PC components, with the specific requirements it imposes on the employed hardware and software, does not always represent a more favourable alternative than the systems of similar capacities which are currently available, in either technical or economical terms.

The problem area of internal control system interfaces has been recognized in recent years and examined by several initiatives, including non-European projects [16-19]. In the USA, an initiative which has become renowned under the name of "Next Generation Controller" (NGC) was started several years ago. A particularly important objective of current activities is the "Specification for an Open System Architecture Standard" (SOSAS). The work being carried out in this area concentrates on the development of specifications and interface definitions (NML - Neutral Manufacturing Language) for numerical controllers, robot controllers and measuring machine controllers. The relevance of these activities to the European area must be questioned, however, particularly in view of the fact that the proportion of American machine tool manufacturers and, in their wake, of control systems manufacturers who have been able to withstand the onslaught of their Japanese rivals has so far been minimal.

In connection with the OSACA project within the ESPRIT programme, which is discussed later in more detail, the European control systems manufacturers who still possess substantial competitive strength were recently induced to enter into cooperation with the aim of evolving a specification for an open control system architecture and, on the basis of this specification, developing a prototype system incorporating such an architecture. This decision to enter into cooperation must surely be seen in the light of the current market situation. Nevertheless, these activities provide vivid confirmation of the importance of open specifications and interfaces.

13.4.1 Open control systems configured individually by specific manufacturers

A number of control systems already exist which claim, with varying degrees of justification, to be open. The degree to which the machine tool manufacturer is able to intervene in internal functions varies among these systems. A common aspect affecting all these systems is that no neutral interface specification currently exists, which means that every concept implemented on these open systems must be completely reconfigured when transferred to a system from another manufacturer. The resultant scope of development work imposes extreme demands on the machine tool manufacturer's personnel capacities and results in high costs, despite the open configuration of the internal interfaces.

The open systems which are currently available or under development differ in particular with regard to the extent to which intervention is

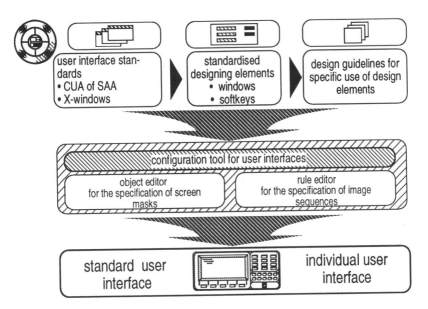

Fig. 13.14 User-configurability of the user-interface (based on Bosch)

possible by the machine tool manufacturer. Some manufacturers take the view that the possibility for the user of configuring the user interface or of integrating correction values for path calculation operations is sufficient to render these controllers open [20-22]. By contrast, some newer developments further extend the scope of intervention to include internal functions [23, 24]. Two examples help to illustrate the flexibility which is attainable above all for the machine tool manufacturer with the aid of the available systems.

The first example illustrates the configuration of individual user interfaces (Fig. 13.14). The configuration of interfaces by the user, to which all users attach great importance, is effectively supported in the presented system by the provision of a tool box. From the point of view of the user, i.e. the machine tool manufacturer, this tool box consists of an object editor and a rule editor. With the aid of these editors, users are able to create user interfaces which are optimally adapted to their requirements and which emphasize their individual configurations in a relatively simple and reliable manner. In principle, it would be possible, using these tools, to emulate the user-interfaces of any control systems, for example. Furthermore, with the aid of standardized design elements and design

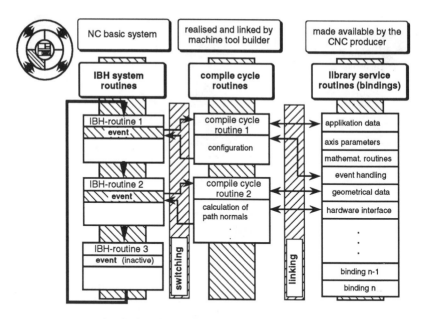

Fig. 13.15 Example of a function implemented with open internal interfaces (sensor-guided path correction based on IBH)

standards, the tools provided can be used on the one hand to ensure that design work is carried out in accordance with the user's requirements, while on the other hand supporting compliance with existing standards. Equally, the machine tool manufacturer may, of course, employ the standard user-interface as supplied by the control systems manufacturer, thereby saving the time and costs required for individual modifications.

Fig. 13.15 illustrates the incorporation of technological functions into a numerical controller, which is particularly important for machine tool manufacturers in the area of adapted concepts. Support for the machine tool manufacturer by the control systems manufacturer is of special significance here. The machine manufacturer uses the interfaces provided by the control systems manufacturer within the internal NC software in the form of so-called events. From these events, machine tool manufacturers can branch off from the control system into their own specific functions. In turn, the basic functions provided by the control systems manufacturer in the form of libraries can be employed within these routines programmed by the machine tool manufacturer. This procedure is comparable to the customary procedure for today's compilers, whereby the programmer is

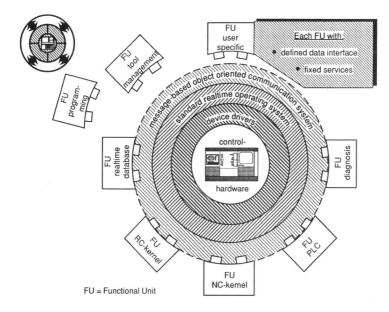

Fig. 13.16 Open, function-oriented control architecture (OSACA)

provided with a number of basic functions in the form of libraries, which the linker integrates into the operable programme.

But however great their flexibility and user-friendliness, the concepts presented here are nevertheless restricted to specific control systems configured by individual manufacturers. Open systems in accordance with IEEE must possess certain characteristics, however. They must be
– vendor-neutral,
– consensus-driven,
– standards-based,
– freely available.
The above described concepts do not fulfil these req uirements.

13.4.2 Standardized open control systems

As already mentioned, an open, functionally oriented control architecture in accordance with the requirements of internal openness is currently under development in the OSACA project. The main focus of the work being carried out in this area is on the definition of a hardware-independent

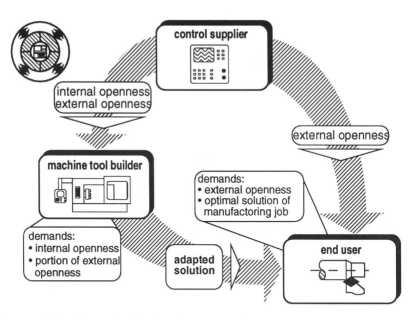

Fig. 13.17 Provision and utilization of open control systems

reference architecture. This will be based on functional units (Fig. 13.16). On the basis of a breakdown of the tasks concerned, which in addition to the NC functions also include programmable controller and cell computer functions, a hierarchic architecture model is derived involving minimal data exchange between assignable functions. In this way, it is intended to facilitate or enable the configuration of a control system with any required and optional functional units. As an important result of this work, it is intended to create the possibility of extending the functionality of control systems on a vendor-neutral basis.

Fig. 13.17 first of all provides a general illustration of how the internal and external openness of a control system interact. The user's requirement for external openness and the optimal execution of production operations can be fulfilled via adapted configurations achieved on the basis of internal openness.

Fig. 13.18 shows a possible configuration for a controller optimally adapted to the requirements of the machine tool manufacturer and the end user on the basis of open systems. In ideal cases, modules of the control systems manufacturer, the machine tool manufacturer and an independent software supplier can be integrated on the basis of a clear interface

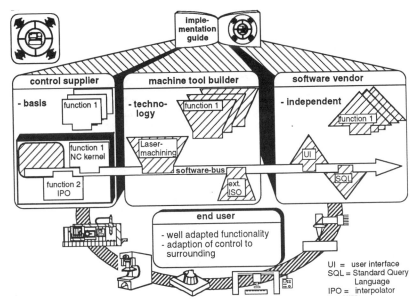

Fig. 13.18 Configuration of an NC control software concept on the basis of open systems

specification (Implementation Guide).

Fig. 13.18 also shows the possible assignment of modules to the specified partners. A genuinely open concept also requires new structures with regard to the sharing of task areas between the control systems manufacturer and the machine tools manufacturer. Provision of the hardware and a core of basic functions will remain the exclusive responsibility of the control systems manufacturer, who will further be required to provide the infrastructure for integrating external functions. This may take the form of a "software bus", for example (Fig. 13.18). Such a "software bus" would be comparable to an inter-computer bus system and would have the task of ensuring the transmission of information or the calling of information between various software packages, in the capacity of an internal communications interface within the control system. This process requires the mapping of subsets of the OSI reference model and the provision of uniform mechanisms for the exchange of information between the individual functional modules of the control system [25]. On this basis, the machine tool manufacturers can integrate their own functions. The functions of external suppliers can also be integrated in this

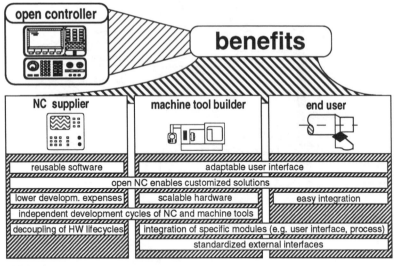

Fig. 13.19 Advantages of an open control architecture

way. A decisive aspect here is that machine tool manufacturers in particular are free to chose whether they wish to procure functions which go beyond the scope of the control base from the control systems manufacturer or from external suppliers or, alternatively, whether they wish to develop and implement such functions themselves. Examples of such functions are database applications, user interfaces or completely new elements for the purposes of user guidance and process control.

13.4.3 Utilization and assessment of open control systems

Open control systems as the response to closed systems (in particular those of the Japanese companies) will be well received wherever they reveal decisive advantages over such closed systems. In Fig. 13.19 the main advantages of open control systems are summarized and sub-divided according to their importance for control systems manufacturers, machine tool manufacturers and end users. The primary advantage for the control systems manufacturer is a reduction in manufacturing costs. The machine tool manufacturers profit above all from the possibility of integrating their

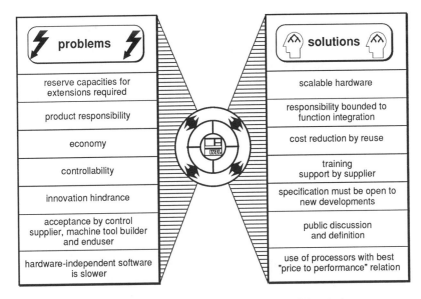

Fig. 13.20 Problem areas for open control systems and possible solutions

own technological know-how and adapting the user interfaces, thereby enabling more flexible adaption of the control system to the user's special requirements. From the user's point of view, this adaptability and the integrative capacity of an open control system are of importance.

Open systems will, in all probability, be able to display and utilize their greatest strengths in those areas in which the potential of standard controllers is virtually or completely exhausted. In such cases, the possibility of simple integration of special functions will doubtless induce potential buyers to opt for an open concept.

These new control structures do, of course, require adequate structures in the organizational, legal and technical environment. As Fig. 13.20 shows, such a fundamentally new approach to the development of a control system naturally harbours a certain degree of risk, which must be identified in good time and countered by suitable measures.

For example, the question of the liability for functions which are not integrated by the control systems manufacturer must be clearly defined. The form of support which the control systems manufacturer provides for the machine tool manufacturer will shift in the direction of intensive training. The development capacities in the control area at the machine tool manufacturing company must be expanded, to enable effective utilization of the possibilities which are provided. In return, the machine tool manufacturers will be able to integrate their individual know-how to a greater extent, in order to improve their competitive positions.

But the two most important points are without doubt the question of economic efficiency and the closely related matter of acceptance by control systems manufacturers, machine tool manufacturers and end users. In order to gain the acceptance of the respective partners, the question of economic efficiency must be considered, in realistic terms, not on the basis of an overall economic assessment, but rather by means of a positive assessment for each of the partners involved in the value-creation chain. Each of these partners will be prepared to accept and implement this new concept only when they are able to identify resultant benefits for themselves.

In addition to the question of economic efficiency, the acceptance of the new concepts by manufacturers and users is also of fundamental importance. This must be promoted via effective information management by all partners with an involvement and/or interest in the development of such concepts. Experience shows that the effective dissemination of information on available technologies decisively promotes their further development.

13.5 Summary and outlook

The current trends in the field of numerical control systems indicate that the current dominance of the Japanese control systems manufacturers threatens to develop into a monopolistic position. As explained in detail in this paper, this course of development must be countered by joint efforts on the part of European manufacturers of control systems and machine tools, and also by the end users (Fig. 13.21).

An important basis for such efforts is provided by the concept of open control systems. This concept is particularly suitable for the market for adapted configurations, which occupies a position of particular importance for European machine tool manufacturers. The open controller concept is able to reduce engineering costs substantially and thus balance out possible price advantages of Japanese controllers.

Of particular importance in this connection is increased cooperation at international level, in order to promote the development and widespread application of standards. The concept of an open numerical control system

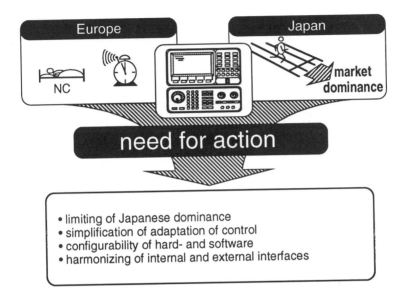

Fig. 13.21 Key aspects of the paper on "The open controller"

and the initiatives aimed at implementing such a system (e.g. the previously mentioned ESPRIT project, OSACA) are thus to be developed and promoted by suitable direct or supporting measures at all stages of the value-creation chain. This will provide an important, if not decisive contribution towards the survival of the European control systems and machine tools industries.

14

Robotics in production systems

14.1 Introduction

Since the first industrial robots were put into operation in Germany at the beginning of the 1970s [1], the users in industry and the companies manufacturing such robots have acquired a good 20 years of experience in their deployment. As a result of the strong growth in worldwide demand which has taken place throughout this period and the subsequent extensive development work, robot systems have developed from their initial form, which essentially lacked the flexibility associated with the word "robot", into standard manipulation systems which can justifiably be considered flexible with regard to the scope of motions and control functions which they are able to integrate.

However, the fact that standard robots are now available in a large number of different designs and sizes does not mean that these robots are suitable for standard deployment in any applications. On the contrary, a realistic assessment of the applications implemented to date reveals that not only the configuration of a production plant with robots and the commissioning of these robots, but also resetting for a different range of components are considerably time-consuming and cost-intensive matters. Despite the flexibility of robots in terms of their mobility and programmability, there are a number of additional factors, such as the work and expense involved in reprogramming, modifying the periphery, linking periphery and robots, etc., which stand in the way of the fast configuration or modification of overall plants. The robot must not be considered in isolation, but must always be regarded in connection with peripheral equipment, sensors, tools, programmes. Apart from a large number of limitations which still have to be accepted, even on modern robots, with regard to dynamic performance, rigidity, motional possibilities, etc., a large proportion of the problems relating to the industrial deployment of robots results from the complexity of integrating robotics into the production environment. Recognition of this problem area has led to an increasing unwillingness among users to carry out the necessary integration measures themselves. Instead, robot manufacturers are developing to an increasing extent into system suppliers who supply complete turnkey systems, including the required controllers, which are designed specifically to satisfy the individual user's requirements, on the basis of their know-how in the respective fields of application. Consequently, each system increasingly acquires the character of a specialized configuration, resulting both in higher costs and diminished expandability, with a subsequent loss of flexibility.

As already indicated, the robot itself and/or its control system are, of course, also subject to certain restrictions which limit or exclude the possibility of deployment for certain applications, or at least constitute problems in certain environments.

A brief summary of the current areas of application for robots in the field of production is followed by a critical assessment of the current situation, on the basis of the experience acquired over 20 years of robot deployment and development. The individual areas of mechanics, control systems, programming, sensors/actuators, man-machine interface and periphery are then examined in detail, and deficiencies are outlined with regard to the applications concerned. A number of concepts which enable improved or more economical production with robots for specific applications are also presented for the respective areas examined in detail. In view of the tight earnings situation which prevails throughout industry, the aspect of economic efficiency has attained a particularly high level of importance in recent years. This fact is given special consideration in the following assessments by referring to concrete fields of application in evaluating new concepts, so as to enable estimation of the costs/benefit factor. In the context of the now widespread philosophy of lean production, it is inexpedient to consider a proposed improvement measure in isolation from the application concerned.

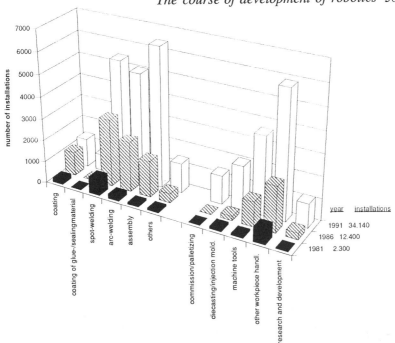
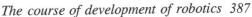

Fig. 14.1 Deployment figures for industrial robots from 1981 to 1991

14.2 The course of development of robotics and current fields of application

Today, robots are deployed in production operations for a large number of different applications, although it is evident that the scope of deployment remains largely restricted to classical fields of applications and has barely changed in the past 10 years. Fig. 14.1 shows the trends in the deployment of robots in Germany over this period.

While the number of robots deployed rose from 2300 to over 34000 between 1981 and 1991, the main areas of application remain spot and path welding, assembly, machine tool linkage and general workpiece handling. The areas of spot welding and general workpiece handling in particular accounted for the majority of the robots deployed in 1981 and may be regarded as <u>the</u> classical fields of application.

An interesting aspect is the fact that no increase worthy of mention occurred in the area of coating from 1986 to 1991. This is accountable to a saturation of the market with robots in the fields of painting and coating.

Due to the evident concentration of robot deployment on a small number of fields of application and in view of the constancy of the respective market shares for these fields, a distinction is made below between historical and more recent fields of application. In this context, historical applications are understood to be applications for which experience with the deployment of industrial robots has been available for a considerable period of time. These are the areas of welding / cutting, coating and general workpiece handling.

14.2.1 Historical fields of application

The first applications for which robots were employed were relatively simple machining operations. In view of the limitations which applied with regard to accuracy, flexibility and programming, robots were initially restricted to tasks which did not involve any high requirements in these respects and at the same time guaranteed high production quantities. A further requirement was for the technology involved in the processes concerned to be fully developed, thereby enabling relatively swift adaption of the processes to the general requirements of flexible automation. In this way, it was possible to avoid the need for large-scale modifications, new developments or additional, complex process control systems.

After initial deployment on specially selected simple tasks, continual efforts were undertaken to improve quality, speed, flexibility and economic efficiency. As a result of this development process, the robot, control and peripheral components required for the historical applications have now attained a highly advanced state of development and are practically available on the market as standard products. Current developments are aimed primarily at further improving quality and economic efficiency. Nevertheless, the intensive development work carried out in this field has as yet failed to enable complete systems to be offered in the form of standard configurations. Due to the varying requirements of the users and the continuing unsatisfactory degree of modularity for the individual components, systems designed to customer specifications, which involve a high level of work and costs for development and commissioning, are still customary.

(a) Spot welding
As already mentioned, spot welding was one of the first areas of applications in which robots were deployed, and it still represents the most common application for industrial robots. The early automation of spot

welding is accountable to three main factors. Firstly, the process imposes very minimal requirements with regard to the robot's motional performance capabilities, as neither a particularly high level of precision (when welding laps) or defined path operations of the tool tips are necessary. As a result, simple point-to-point controllers can be employed, without recourse to the interpolation process, which requires extensive computing time. Secondly, the spot-welding process itself is comparatively easy to control, and can be carried out independently of the robot after positioning the welding tongs. No complex linkage is required between robot and welding tool, therefore, and the experience acquired with manual welding tools can be referred to regarding welding parameters. Finally, the batch sizes which apply in the automobile industry - the main user of robots in the area of spot welding - are suitable for ensuring that the scope of development work and investment costs required for robot systems pay off.

Although virtually all spot-welding operations in the automobile industry are now automated, further development work is nevertheless required in this area. As in the case of practically all applications, current efforts are aimed primarily at reducing the cycle time, i.e. increasing the working speed of the robot, improving the accuracy, reducing the programming time (collision avoidance and optimization in particular are critical here) and improving the quality of the weld spots themselves via the use of spot-welding control systems. Apart from enabling documentation of the production process, more precise control and monitoring of the welding process also result in high-quality welding results which, in turn, allow a reduction in the number of weld spots and a subsequent reduction in production costs.

(b) Path welding

Path welding imposes considerably higher requirements on the automation equipment than spot welding, as precisely defined movements are required. The programmed path and path speed must be maintained with a very high level of precision, in order to obtain acceptable welding results. Path welding was nevertheless automated relatively early, no doubt due to the widespread use of the process in mechanical engineering and the numerous applications which are available. Here again, the deployment of robots was initially restricted to large-scale production operations, in order to attain economic efficiency in spite of the high costs of automation and, in particular, programming and optimization of the welding result.

In contrast to the initial applications restricted to very simple, easily accessible seams involving low tolerance requirements, the transition to

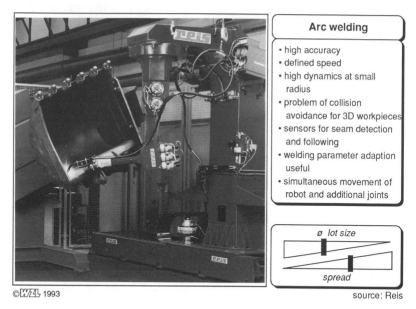

Fig. 14.2 Complete robot welding plant for dredging shovels

automation of the welding of complex 3D components, which involves an increased risk of collision and more complicated geometries, is currently in progress. The process-related tolerances resulting from deformation and preliminary working (tacking of the components by hand) mean that sensors are required, in order to obtain good welding results. A distinction can be made here between sensors for locating the beginning of the weld seam and sensors for tracking the seam during the welding process. A common procedure for locating the beginning of seams involves scanning component surfaces with a contact-tube sensor. The energized burner tip is moved towards the earthed component until mechanical contact is established, which is registered via the breakdown of the voltage. The position of the surface can thus be derived from the position of the burner tip at this time. When the beginning of the seam is in a suitable position, several search operations will enable its coordinates to be determined. Tracking of the welding joint, which is of vital importance to the results of welding operations, is generally carried out using arc sensors. Via defined oscillation of the burner at right angles to the joint, the length of the arc and, subsequently, the position of the centre of the joint can be determined by analyzing the current or voltage characteristic at the output of the

Flame cutting

- typical CNC application in the field of 2D
- robots in 3D area / for workpieces with curved contours
- sensors for edgetracking (laserscanner)
- great tolerances due to prior work
- large variety of workpieces

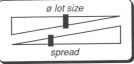

ø lot size

spread

source: Reis

Fig. 14.3 Plasma cutting with a robot

welding source.

This relatively simple and robust process has the advantage of requiring no space for a sensor in its strictest sense on the burner, while nevertheless providing fully satisfactory results in several situations. For the purposes of quality improvement, parameter adaption via observation of the weld pool or similar would be desirable. Such systems are under development, but are not yet ready for launching on the market. A decisive additional problem here is the size of the sensor, which sometimes drastically restricts the freedom of movement when complex components are involved. The simultaneous coordinated movement of robot axes and supplementary axes is also desirable in many cases, as this enables improved welding of certain geometries.

Fig. 14.2 shows a current robot welding plant for dredging shovels. Dredging shovels of up to 1 t in weight can be welded in a maximum of five layers on this plant. The plant incorporates an automatic burner-changing system and three supplementary axes which can be moved simultaneously with the robot axes in path-controlled mode. The plant is able to compensate tolerances on parts using contact tube and arc sensors.

(c) Flame cutting

Flame cutting is also a very widespread process in the field of production engineering, being employed in particular in the preliminary machining of flat metal sheeting. Whereas the 2D area is covered by the established numerical control systems with 3 axes, robots are sometimes employed to prepare the edges of parts to be welded, prior to tacking.

As in the case of path welding, the relatively high tolerances of the components to be machined represent a problem area. Edge-tracking sensors are practically indispensible. Although the high workload for the operator in the course of flame cutting makes automation of this process desirable, the frequently small batch sizes, the complex programming operations and the tolerances constitute great problems for the economical deployment of robots, as a result of which automation has so far been carried out for special applications only.

Fig. 14.3 shows a plasma-cutting plant with a robot, which is employed to prepare the edges for the subsequent welding process. For the purposes of seam tracking, the plant possesses a laser scanner, which measures the edge in front of the burner and is thus able to correct the movements of the robot.

(d) Handling and fitting

Applications in the area of the handling, particularly of tools, and the fitting, in particular of small components, are now widespread. In contrast to the very simple handling tasks which were automated at the beginning of the development process, in the form of simple pick-and-place systems, substantially higher requirements now apply. Today, the areas of application range from the robot-loading of presses through tool handling to the installation of spare wheels or batteries in automobiles. In these areas, relatively high requirements apply with regard to programming and optimization, e.g. for the purposes of collision prevention in restricted working spaces (engine compartments, etc.), and the movements required for handling purposes may be quite complex. Consequently, the kinematic systems employed for the robots usually possess 6 or more axes.

In order to compensate tolerances, handling robots can be equipped with sensors. However, as the complexity and flexibility of the gripper systems increase, so does the number of sensors used, thereby complicating the problem of integrating the sensors into the gripper and of analyzing and conditioning their signals in a suitable manner (cf. Section 3.4). As the installation of sensors still requires a substantial amount of space and causes considerable costs, it is attempted to avoid the use of

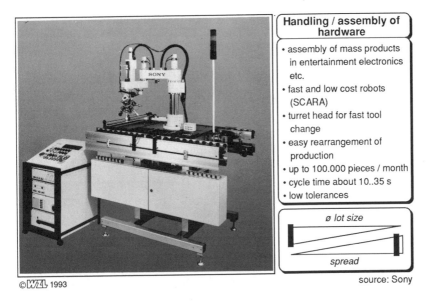

Handling / assembly of hardware

• assembly of mass products in entertainment electronics etc.
• fast and low cost robots (SCARA)
• turret head for fast tool change
• easy rearrangement of production
• up to 100.000 pieces / month
• cycle time about 10..35 s
• low tolerances

ø lot size

spread

©WZL 1993

source: Sony

Fig. 14.4 Component of an assembly system for small parts

sensors wherever possible and either to reduced the tolerances which occur or to make the grippers more tolerant towards deviations in the positions of the components. This approach may make sound economic sense and be in keeping with the trend towards "lean" concepts. In many cases, however, sensors are indispensible, or the omission of sensors leads to high costs, e.g. for measures to effect the more accurate infeed of components, which exceed the savings gained by omitting the sensors. Great potential for development still exists, therefore, in the area of sensor technology.

As the complexity of the tasks concerned increases, it becomes more difficult to distinguish between handling and assembly, and the development of these two areas is also barely separable. In this context, Section 2.2.3 should also be noted, therefore.

In the area of the fitting of small parts, in which more robots are now deployed than in the area of spot welding in the old states of the FRG, fast SCARA robots with 4 axes predominate, as the joining operations for many tasks can be carried out in vertical configuration. The design of the SCARAs makes them particularly suitable for these operations and also enables very high working speeds. Output levels of up to 100,000 pieces per month are quite common, for example, whereby the cycle times may range between 10 and 35 s. In view of the small component dimensions and the high accuracy of the parts to be fitted, tolerances are generally of secondary importance. The problems in the area of the fitting of small parts thus relate less to control of the handling process and more to attainment of the highest possible speed and the simplified adaptability of entire assembly lines to new products. These are particular important factors in the face of falling product lifetimes and increasing product diversity combined with decreasing batch sizes.

Fig. 14.4 shows a component of an assembly system for small parts which has been designed for monthly output rates of between 35,000 and 100,000 parts, and which ensures a high level of flexibility with regard to the adaption of production operations. This is achieved via the use of pallets for small parts and standardized feed and transport systems. According to the manufacturer, the level of costs for adapting production lines amounts to between 2% and 20% of the original investment [2].

(e) Coating

In this connection, the term "coating" refers primarily to lacquering and enamelling. A plant for enamel spraying using robots was first installed in Germany in 1970. The area of coating was particularly suitable for robot automation, as the levels of accuracy which require to be maintained are relatively low and the very low weight of the spray guns, etc., to be held by the robots, including the appurtenant tubing, avoided any serious strain on the robots. Furthermore, programming by means of the so-called play-back process, whereby the operator guides the robot or a kinematic model through the required motions by hand and these motions are then recorded by a controller, is relatively simple. At the same time, the operator's know-how regarding optimal handling of the spraying nozzle is utilized.

Automation of the spraying process is also desirable on account of the health risk to the sprayer, and in the interests of reducing overspraying (losses resulting from lacquer sprayed past the component) and of attaining uniform, reproducible and optimizable surface quality. To enable further improvement of the stated points and a further reduction in the coating thickness (improved appearance, reduced lacquer consumption) simulation systems are now available on the market which, in addition to enabling programming and simulation of the robot movements, are also able to provide information on the thickness of the coating on the component surface. These systems render effective off-line programming of the coating process possible for the first time, which means that only final optimization of the programmes is required in the machining cell. Sensors

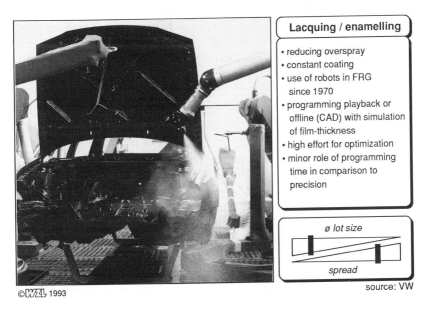

Fig. 14.5 Lacquering with industrial robots

are totally unnecessary, as the maximum required accuracy levels are in the range of a few millimetres.

Current development work is aimed above all at further reducing overspraying, e.g. by controlling the geometry of the spray jet and optimizing the switching functions of the gun. Lacquering with industrial robots is now broadly established in the automobile industry (Fig. 14.5).

14.2.2 More recent fields of application

The following sections present applications which have been implemented for the first time in recent years and which cannot be related to the areas of application discussed above. The relatively late automation of these areas by means of industrial robots is due partly to the high requirements relating to the automation components (e.g. assembly of flexible parts) and partly to fact that the employed process technology (e.g. lasers) has only recently attained an appropriately advanced level of development. Overall, experience in the production environment has shown that the majority of the problems involved in deploying robots in new areas of application are accountable to the need for the further development of process technology and to difficulties regarding the interaction between robot and periphery. The investment which is thus required to develop a marketable product can generally be justified only when the systems are subsequently employed in mass production operations, as firstly a relatively high investment volume is customary for the overall plants required for mass production operations, while secondly such plants do not require too high a level of flexibility - which would further increase the costs. For this reason, it is in particular the automobile and its component suppliers which are playing a pioneering role in the development and deployment of robots in new areas of application.

(a) Deburring, belt grinding, cleaning of castings

The aims of deburring may range from the elimination of a risk of injury through to the removal of burrs which hinder the correct functioning or assembly of components [3]. The application area of deburring is generally considered relatively difficult to automate, on account of the undefined nature of burrs, as a result of which it has rarely been automated at foundries in the past. However, as deburring is a very labour- and cost-intensive task, and is furthermore unattractive to the operator on account of the health risk, further automation of these activities would be desirable.

The undefined nature of burrs and the relatively high tolerances of the workpieces to be machined require sensors to be employed or the robot to be fitted with compliant tool-holders or special tools [4, 5, 6]. The burrs on castings can be very thick and pronounced to varying degrees. In this case, milling with robots is highly problematic, due to the low rigidity, and the deburring of such workpieces should, if possible, be carried out in the course of subsequent machining of the parts on the NC system. Apart from the restrictions regarding the range of workpieces, which result from the fact that the available sensor technology is not optimal, the high investment costs are a further contributory factor to the low degree of automation. Nevertheless, there are applications for which particularly high accuracy and quality requirements not only justify automation but actually render it virtually indispensible. An example of such an application is provided by the deburring of turbine rotors (Fig. 14.6).

Belt grinding essentially pursues the same aim as other deburring tools, with the difference that the burr is very thin and can be reliably removed with the grinding belt. However, whereas deburring with milling tools involves the robot moving the tool along the fixed workpiece, in the case of belt grinding the robot moves the workpiece along the stationary

Deburring, belt grinding, finishing

- manuelly: labour intensive and unpopular task (= high costs)
- variing bur
- individual tools and pliable tool holdings necessary
- use of sensors often indispensable
- partially modifications of peripheral equipment (e.g. belt grinding machine) necessary

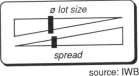

©*IWZL* 1993   source: IWB

Fig. 14.6 Deburring of turbine rotors

grinding belt. In this case, adjustments are sometimes necessary on the belt-grinding machine, in order to control the deburring process and thus ensure good results for the machining operations, i.e. complete removal of the burr without damaging the workpiece itself. Such a belt-grinding machine suitable for automation purposes is, for example, able to ensure various pressure forces by adjusting the guide rollers of the grinding belt, or to compensate positional errors of the robot and tolerances in the component via a floating working point [7]. The problems involved here result from the as yet inadequate knowledge of the variables determining the processes.

(b) Application and extrusion operations

The application of adhesives is an area which has recently become established on a widespread basis, in particular in the area of the automobile industry. For aesthetic and aerodynamic reasons, it is becoming increasingly common practice to cement the fixed automobile window panes into the car body flush with the shell of the body. This development became economically viable only when robots were deployed, which apply the adhesive evenly to the panes.

The required precision and evenness of application are relatively high,

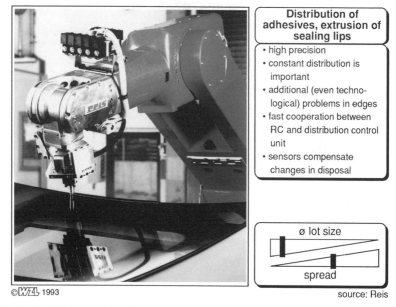

Distribution of adhesives, extrusion of sealing lips

- high precision
- constant distribution is important
- additional (even technological) problems in edges
- fast cooperation between RC and distribution control unit
- sensors compensate changes in disposal

ø lot size

spread

©*WZL* 1993

source: Reis

Fig. 14.7 Adhesion and coating of automobile window panes

imposing very high requirements on the path performance and accuracy of the robot. There is also the additional problem of controlling the emitted quantity of adhesive very precisely in accordance with the relative speed between the pane and the outlet nozzle, in order to avoid accumulations of adhesive in narrow radii. In addition to the high demands placed on the robot system, high requirements are also imposed on the dosing system (e.g. additionally integrated sensors to compensate feed tolerances) and on the interaction between these two systems. However, the widespread deployment of such plants in the automobile manufacturing and automobile components industries (Fig. 14.7) shows that the majority of the attendant problems have been solved.

On the basis of the above-described course of development, there are even plants in operation which, apart from applying adhesives, also enable the extrusion of polyurethane sealing lips [8]. The polyurethane material requires to be applied to the pane in a specified geometry, which imposes even higher demands on the system than the application of adhesives. As it is not possible to apply such a sealing lip by hand with the required uniformity, this may be regarded as an area of operations which it is possible to implement with the necessary flexibility only on account of the

Assembly

- great demand of automation
- tolerances
- use of sensors necessary or
- assembling aid (e.g. RCC)
- great number of different
  joint types
- (screw, press fit, clip, rivet,
  glueing...)
- often specially designed
  tools for handling and
  assembly
- optimized design for
  assembly as precondition

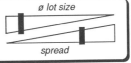

ø lot size

spread

©*WZL* 1993

source: BMW

Fig. 14.8 Automated window pane assembly in the automobile industry

developments in the field of robotics.

(c) Assembly

Assembly is an activity which is required for virtually all mechanical engineering products, and which is highly labour-intensive. The area of assembly referred to here differs from the fitting of small parts (e.g. electronic components or hifi and domestic appliances) with regard to the substantially more complex joining process.

While primarily linear assembly movements are sufficient in the area of small parts, to fit electronic components to p.c. boards or to connect two components via simple snap-on connections, the tasks involved in the area of assembly under consideration here are much more complex. Even the insertion of a pin into a borehole with only minimal oversize or even undersize cannot be carried out without the use of sensors or special assembly strategies, due to the unavoidable tolerances. Because of the resultant requirement for development work in this area, automated industrial assembly operations are currently carried out only in certain special areas, despite the recognizable potential which exists. The automobile industry clearly plays a pioneering role here - as it does in the deployment of robotics as a whole. In this industry, the automated assembly of window panes, doors and gearboxes and automated decking (joining of body and chassis) are widely established as standard practices (Fig. 14.8).

Tolerances constitute a problem throughout the area of automation. In the area of assembly, however, the situation is further aggravated by the fact that existing tolerances may not only lead to unsatisfactory results, but generally result in the failure of the complete assembly process. Here again, the simple example of "pin in borehole" serves to illustrate this problem. Consequently, the use of sensor technology (e.g. scanners, video cameras, tactile sensors, etc.) or suitable passive assembly aids is essential in practically all areas of assembly operations. The number of commonly deployed sensors and the attendant costs are clearly shown in Fig. 8.

An example of a passive assembly aid is a remote centre compliance (RCC) gripper, which has been developed especially for the above-mentioned pin-in-borehole problem [10, 11]. The mechanical design of the gripper is such that the pin to be fitted, which is provided with a suitable chamfer to facilitate insertion, can be displaced vertically to the direction of insertion and also tilted. It is of decisive importance here that a force vertical to the direction of insertion - such as will occur if the pin is inserted off-centre - results in displacement of the pin in this direction only, and does not cause canting. Canting of the pin is effected only in the event of a moment occurring in the course of further insertion of the pin, due to two-point contact between pin and borehole. The described movements are carried out without any additional controller and with the robot performing a linear assembly movement, which is why the system is referred to as passive.

Apart from the sometimes very high requirements imposed on the accuracy of the robot system in conjunction with small permissible tolerances, the position and orientation of the part to be assembled are often also of decisive importance for many assembly operations. In the case of automatic gearbox assembly, for example, which is now in industrial use [9], perfect meshing of the gearwheels is necessary. Here again, the positions of the gearwheels can be precisely determined using sensors or, alternatively, the assembly process can be made possible by means of suitable assembly strategies. The question as to which of these two approaches ultimately achieves the desired aim in a more economical manner must be decided on an individual basis for each application. A more "ingenious" configuration, avoiding the use of complex sensors, is often superior to a sensor-assisted configuration, though it does require

**assembly of
flexible parts**

- decreasing cycle
  times by picking
  up multiple mats
  simultaneously

- difficult motions
  for seperating
  and laying down
  the mats

- restricted
  accesibility

©ⅢﾂﾑⅬ1993   source: IWB

Fig. 14.9 Installation of insulating mats in automobiles

appropriate design measures.

All in all, a component design which is in accordance with the requirements of assembly operations is an important factor, or even <u>the</u> decisive factor for economically viable robot-assisted assembly operations. By implementing appropriate design measures (e.g. chamfers to facilitate installation, uniform direction of assembly, suitable gripping points, good accessibility) in the production process, it is often possible to eliminate the need for sensors and/or to drastically reduce the requirements relating to the assembly system. The fact that the restrictions relating to automatic assembly systems differ fundamentally to those for manual assembly operations is now generally recognized.

(d) Assembly of flexible parts
To date, the assembly of flexible parts has been properly developed and deployed in production operations only in a very small number of special areas. One example of such an area is the installation of insulating mats in automobiles (Fig. 14.9).

With the exception of relatively simple tasks, such as that shown here, the principle problem with regard to flexible parts is the development of

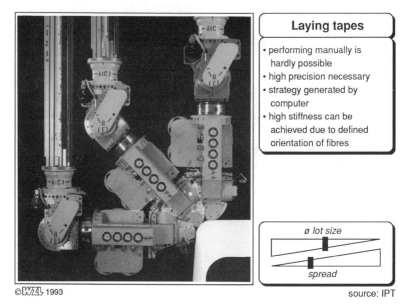

source: IPT

Fig. 14.10 Chemical-fibre-reinforced tape-laying with a gantry robot

suitable strategies for the assembly of such parts. Depending on the specific application concerned, these strategies may vary considerably and differ fundamentally from the strategies employed for rigid components. The fitting of cooling water tubes to the ends of the pipes of a cooling water pump or a cooler illustrates this problem. As the size of the tube is markedly smaller than the end of the pipe, the tube cannot be simply slid over the pipe, due to the risk of the tube kinking. In order to ensure correct assembly, procedures such as turning or twisting the tube during the fitting process require to be developed. All in all, a great deal of basic research still requires to be carried out on the automated assembly of flexible parts [12, 13, 14].

Considering that the stated problems apply in addition to those relating to the automated assembly of rigid components - some of which remain unsolved -, and in the light of the widespread use of such automated assembly operations in industry, it is clear that flexible parts possess little suitability for economical assembly by robots. At present, this area is therefore to be classified as largely unresolved and barely deployed in industry.

(e) Tape-laying

Conventional materials are being substituted to an increasing extent by non-metallic fibre composites in the production of high-tech components. The design of this material, i.e. the orientation of the heavy-duty fibres, is carried out with the aid of computer technology, and thus helps to optimize the component characteristics. The production and movement data is subsequently obtained from this design data. Fig. 14.10 shows the laying of tapes on a complex structure. To date, this process is employed with success primarily in the aviation industry. The wings of the ATR 72 regional aircraft and the bearing structures of the Boeing 777 are produced in multi-layer configuration, for example.

(f) 3D laser welding / laser cutting

In recent years, the laser has undergone a marked course of development in the direction of economic efficiency and smaller unit volumes and is thus being deployed to an increasing extent as a production tool, on account of its specific advantages, such as high energy density, freedom from wear, precision, positive characteristics of the produced seams or cuts, etc. Hand-held lasers are unfeasible, due to the normally heavy beam-guiding systems - the use of glass fibres as light guides is not possible for CO_2 lasers - and the level of accuracy required to position the focal point. Consequently, as the deployment of the laser as a tool has become more widespread, automation of the process has undergone a corresponding course of development. While 2D machining can now be largely regarded as standard, the transition to 3D contours is proving problematic. Although a number of applications do already exist here, problems often exist with regard to the path accuracy and dynamic performance of the robot, beam control, programming, and monitoring of the process by means of sensor technology.

Gantry robots are generally employed for laser welding and laser cutting, on account of their high rigidity, though the buckling arm robot is also quite suitable for laser machining purposes, at least with regard to specially selected applications [15] (Fig. 14.11). The Nd:YAG laser employed in this application has the advantage that its beam - in contrast to the CO_2 laser - can be conducted through a fibre-optic cable with virtually no dissipation, eliminating the need for complex beam guiding systems.

Further development work is required in particular in the areas of dynamic performance - above all with regard to fast changes of orientation for the hand axes - and path accuracy (minimum of +/- 0.1 required for many applications). Also, the kinematic systems of many robots do not

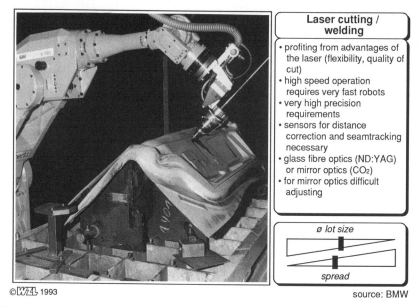

Laser cutting / welding

- profiting from advantages of the laser (flexibility, quality of cut)
- high speed operation requires very fast robots
- very high precision requirements
- sensors for distance correction and seamtracking necessary
- glass fibre optics (ND:YAG) or mirror optics (CO_2)
- for mirror optics difficult adjusting

ø lot size

spread

© *WZL* 1993

source: BMW

Fig. 14.11 Laser cutting with an 8-axis robot and Nd:YAG laser source

possess sufficient vibrational rigidity, as a result of which the accelerations which occur at high traversing speeds cause vibrations, which have a negative effect on the result of machining operations. In addition to these deficiencies regarding the robot itself, extensive development work is required in the field of sensor technology with regard to the maintenance of precise distances between nozzle and workpiece surface, and in the area of the beam-guiding system, whether this be implemented within the robot itself or externally on the robot. With regard to the use of CO_2 lasers, adjustment of the reflector in the beam-guiding system constitutes an additional labour- and cost-intensive area.

(g) Additional applications
Apart from the above-described applications, robots are also used in a large number of additional fields. Many of these are "exotic", however, i.e. they do not presently play a decisive role in industry or have yet to be developed beyond the laboratory stage, often on account of the secondary importance of the process itself within the area of production engineering. With regard to the requirements involved, the majority of these additional applications are similar to the previously discussed operations, in view of

which a more detailed discussion does not appear expedient in the context of this paper.

One of these applications which has acquired increasing importance in production operations in recent years is water-jet cutting. The requirements involved here are similar to those which apply to laser cutting, though the required accuracy is lower and the problems relating to beam guidance, etc. do not apply. Other applications are, for example, pressure joining, soldering, induction hardening, bending, etc. Details of these applications are available in [16, 17, 18, 19, 20, 21, 22, 23, 24].

14.2.3 Assessment of developments to date

The experience acquired over the past 20 years has shattered certain illusions regarding the capabilities of the robot and its areas of application. A look at the common areas of application quickly reveals that, to date, the robot is deployed primarily in the field of mass production. Although an economically viable transition to smaller batch sizes, right through to the production of individual parts, is generally regarded as desirable, only in a small number of exceptional cases is such an application in operation or at least foreseeable. In the overwhelming majority of applications, the much-praised flexibility of the robot - which initially means nothing more than the ability to guide a tool in space in any desired manner - is not utilized for a genuinely diverse range of workpieces. The reasons for this are varied, depend on the field of application concerned and are not limited to specific components. The weak points of the individual component parts of the overall robot system are examined in Chapter 3, and possible measures for improvement are presented.

To date, the flexibility of the robot is not even sufficient to enable the old robot to continue in use after switching production operations to a new product, let alone to enable utilization of the robot for totally different tasks. Although the automobile industry, for example, is now beginning to separate the plant lifetime from the product lifetime, the fact that this is emphasized as being a particularly progressive measure speaks for itself.

Understandably, the latest applications are being implemented first of all in the area of mass production, as the additional requirements regarding the flexibility of a plant are substantially lower here than when small batches are involved. Also, the costs of the extensive development work in the field of process technology and suitable programming tools are less critical when several systems are deployed.

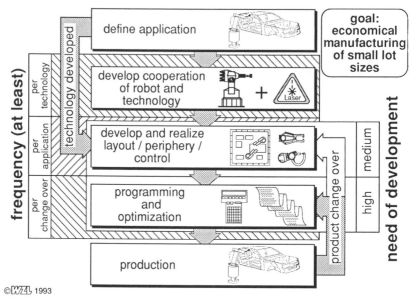

© WZL 1993

Fig. 14.12 Steps to be carried out and their frequency for the deployment of industrial robots in production operations

The deployment of robots in the most diverse fields has shown that the interaction between robot components and the components required for the process is a highly complex area which requires improvement. In many cases, linkage of the robot, as the handling device, to the actual tool and its periphery is effected via a special configuration, which is correspondingly expensive. Standard components are available only in a very small number of areas, such as welding. Consequently, a trend towards smaller batch sizes is already observable in the area of path welding, although a number of restraints nevertheless continue to apply here. As the programming time per workpiece or series is a substantially more important aspect when small batches are involved than in the case of mass production, fast and effective programming constitutes an important factor with regard to economic efficiency. Substantial development work is still required in this area.

A further fact which has been established is that the economic efficiency of robot deployment or of a development cannot be considered in isolation from the application and the attendant restrictions. Many developments in past years have revealed a degree of obsession with technical aspects, which led to unnecessarily complicated configurations and sometimes overtaxed the user. Each new development must therefore place absolute priority on the specific application for which an optimal concept is to be developed. Few developments are expedient for all applications; not even the increase in the performance capabilities of control systems resulting from the further development of microprocessors is required for all applications. High-capacity and complex systems are justified for certain applications. For other applications, however, they are often too expensive and thus uneconomical. More rigid robots, faster controllers, more complex programming languages and interfaces are desirable only when they represent an effective means to an end. Only in the context of the application concerned can it be judged whether a more rigid robot is merely more expensive or whether it represents an effective improvement (e.g. lower cycle time, higher quality). In some areas, the robot faces competition from simpler, less flexible but cheaper alternatives. The deployment of robots is expedient only when the resultant flexibility is utilized.

On the other hand, the trend towards developing robots for specific applications must not be allowed to hinder sensible efforts to standardize components, data interfaces, user interfaces, programming languages, etc. If the robot is to be deployed in an economically viable manner in new areas and for smaller batch sizes in the future, a compromise will be required between standardization and adaption to the specific application, and such a compromise can only be attained by producing all components on a modular basis. This objective is in keeping with the concept of lean production for both the user, who is provided with a very economical product, and the systems supplier, who is able to employ standard modules.

Fig. 14.12 shows the steps which require to be carried out in order to deploy a robot system in production operations. If the task concerned has not been previously implemented, the interactive system between robot and technology is first of all to be developed on the basis of the application concerned. Examples here are beam-guiding systems for lasers, new sensors for controlling the welding process (e.g. via observation of the weld pool) or suitable grippers and strategies for assembly operations. If the application concerned has already been implemented frequently in a similar form, the majority of this development work will already have been carried out. The plant layout is then planned, suitable periphery is selected, the controllers and components and their interactive functions are configured and, finally, the plant is commissioned. This is followed by programming and optimization of the system, which can also be a very time-consuming area. Subsequent switch-overs to new products will require modifications to the plant. Assuming that the fundamental aspects of the application are maintained, the plant layout must then be adapted, any required new sensors and peripheral equipment must be integrated and the plant control system must be modified, after which the plant must be programmed and optimized once again, before production can be resumed. While the changes to the plant layout and the periphery are often relatively small, complete reprogramming of the robot and subsequent optimization are generally necessary. If only variants of the same type of workpiece are to be produced, it will be sufficient merely to amend the programming for the plant.

In order to attain the objective of the economical production of small batches, development efforts are required in particular in those areas which involve labour-intensive and time-consuming measures for the purposes of robot deployment. As shown in Fig. 12, the programming and optimization phase is passed through at least each time operations are switched to a new product. This means that there is an excellent opportunity of increasing economic efficiency in this area, and there is a correspondingly urgent need for development work to this end. Design of the plant layout and the selection, development and implementation of peripheral components and the overall control system are all carried out at least once per application - i.e. before or during initial configuration of the plant. As switches to different products may necessitate additional modifications, development work is also required here, in order to facilitate such changes. The fundamental interaction between robot and technology requires to be developed only when a new technology is implemented for the first time, and thus has little effect on economic efficiency when new processes are subsequently applied.

On the basis of this situation it can be seen that the periphery, the interaction of all plant components and the extensive programming and optimization measures are the prime reasons for the often unsatisfactory level of work and costs involved in developing plants and switching production operations. Future developments must therefore focus on these areas in particular. Several suitable approaches to these problem areas are presented in the following chapter.

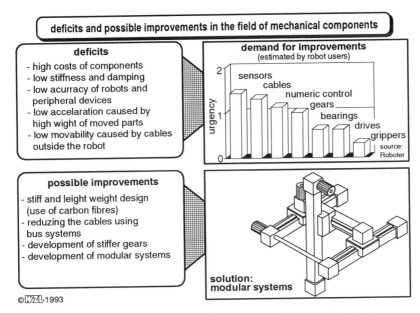

Fig. 14.13 Deficiencies and areas of development potential in the field of robot mechanics

14.3 Definition of problems for small batch sizes and suitable approaches for solving such problems

In the following sections, the robot and its environment are systematically examined in order to establish weak points with regard to specific applications, and suitable approaches for remedying these deficiencies are presented. A distinction is made between the areas of mechanics, control systems, programming, sensors/actuators, MMI and periphery.

14.3.1 Mechanics

The mechanical characteristics have a decisive influence on the manufacturing costs and performance capabilities of industrial robots, whereby the two objectives of low costs and high performance capabilities generally conflict with one another, which means that an expedient compromise requires to be found for each area of application. One method of saving development and production costs is the more widespread use of modular systems.

For simple applications, such a modular concept may involve a complete robot being assembled from externally purchased axes. Linear modules which cover a broad range or requirements with regard to size, load-bearing capacity and drive concept are becoming available to an increasing extent here. But costs can also be saved in the production of conventional gantry or joint arm robots via the increased use of modular systems for individual components, such as motors, gears or frame components.

An overview of the most important deficiencies in the area of robot mechanics and possible improvement measures is provided in Fig. 14.13. In a survey among readers of the magazine Robotertechnik [42], in which more than 6000 operators of robots took part, improvement of the cable and energy supply systems was cited as a particularly urgent requirement.

(a) Energy and signal transmission

Conventional cabling for robots involves a large number of disadvantages, the most important aspect being lost production time resulting from cable breakage. A further disadvantage is the increased danger of collision resulting from external cables and the reduced accessibility to narrow spaces, which renders the deployment of robots impossible for some applications. Not least of all, the cables on gantry robots account for a substantial proportion of the moved masses, which has a negative effect on the dynamic performance of the actuators. Additional problems arise when working media require to be supplied to the robot tool, as applies in the case of welding or lacquering, for example.

In the interests of developing remedies to these problems, new methods of energy and signal transmission have been examined at the Technical University of Aachen (RWTH), as part of the special research project "Advanced actuator and joint concepts" sponsored by the German Research Association (DFG) (Fig. 14.14). The aim of this research work was to develop a robot with unhindered and infinite rotatability of the rotary axes without subjecting the cables to mechanical strain, while at the same time drastically reducing the number of cables. The insides of the axes and joints were to remain clear, to enable the conveyance of additional media through these components.

As a basis for solving this problem, an innovative actuator concept was evolved which possessed the following special characteristics: In order to minimize the number of cables, the robot was equipped with one energy

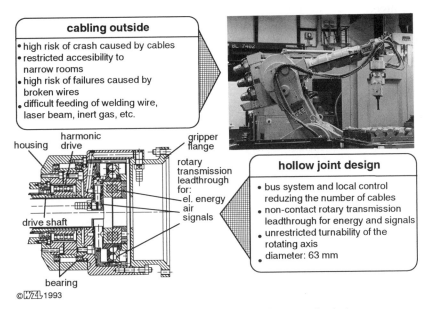

cabling outside
- high risk of crash caused by cables
- restricted accesibility to narrow rooms
- high risk of failures caused by broken wires
- difficult feeding of welding wire, laser beam, inert gas, etc.

housing   harmonic drive   gripper flange

rotary transmission leadthrough for:
- el. energy
- air
- signals

drive shaft

bearing

©WZL 1993

hollow joint design
- bus system and local control reduzing the number of cables
- non-contact rotary transmission leadthrough for energy and signals
- unrestricted turnability of the rotating axis
- diameter: 63 mm

Fig. 14.14 Cabling for industrial robots: Problems and proposed solution

bus and one signal bus, both of which were routed through the entire robot to the gripper. This concept requires all actuating components, including the power converters and controllers, to be located directly in the axial housing. The central control cabinet is no longer required, therefore.

A further important point concerns the development of non-contact rotary transformers for electrical and pneumatic energy and for electrical signals. Energy transmission is effected via rotary transformers integrated into the joints. To enable the transformer to be produced in a small size, the current is transmitted at a frequency of 25 kHz, whereby the power dissipation is relatively low at this frequency. With the employed transformers, a nominal power of 3.5 kW can be transmitted at a voltage of 650 V. At an air gap of 0.1 mm between the two halves of the transformer, the efficiency level stands at 96%.

For the purpose of non-contact data transmission a concept similar to that employed for the infrared remote control of televisions was selected. Light-emitting diodes convert the data into light pulses, which are subsequently received by photodiodes and converted into electrical signals. Due to the requirement for hollow joints, the light-emitting and receiving diodes had to be installed in two ring-shaped channels around the circumference of the joint (see Fig. 14). Each channel is employed for signal transmission in one direction, i.e. from the controller to the axes or vice-versa. To ensure that the light cones hit the receiver as required, several diodes are installed in each ring, whereby the light-emitting diodes emit the light in a tangential direction. Data can be transmitted reliably at transmission rates of 2 megabauds with this system. A system which functions in a similar manner to the energy transmitter is currently under development based on the use of inductors/capacitors. This system can also be integrated into the energy transmitter and thus requires no additional space. The system is intended to operate at data transmission rates of 4 MBd, in full duplex mode.

A further problem which required to be solved concerned the mechanical configuration of the required hollow joints. The aim of this hollow design is to provide the largest possible clear cross-section inside the axle, to enable media, such as adhesive, water, welding wire or laser energy, to be passed through these joints without impairing the mobility of the robot with external cabling. Fig. 14 provides an idea of the complexity of this task. The actuating power is supplied from a fast-running permanent-field synchronous motor via a hollow drive shaft to a harmonic-drive gear unit, the output section of which is installed on prestressed, play-free tapered roller bearings. The housing of the next axle or, as shown in the diagram, of the gripper flange, is directly connected to this output section. The rotary transmitters are integrated into the housing as a complete unit. The axial module is designed for a nominal power output of 1.2 kW and possesses a clear cross-section of 63 mm.

A complete six-axis robot was developed on the basis of this innovative robot actuation concept. Work is currently in progress to implement the concept of non-contact transmission for linear axes, as well. Such a development would eliminate the problem of trailing cables.

(b) Actuators and frame components

The accuracy attainable with a robot, the robot's static and dynamic rigidity and, not least of all, the effective service life of the device are to be regarded as criteria determining the quality of the design under specified operating conditions such as working area, effective load, number of axes and traversing speed.

Mechanical structural components   As far as the mechanics are concerned, these factors are decisively influenced by the frame components and the power trains. In [40], four gantry robots were examined with regard to these characteristics and possible improvement measures. This examination revealed in particular the positive influence of the lightweight construction resulting from the use of chemical fibre-reinforced plastics on the bridges and lifting beams of one of the four gantries. As a result, the masses to be moved in X direction were reduced from 1738 kg for the steel/aluminium design to 760 kg. Calculation via the finite element method revealed a reduction in the deformation at the tool reference point resulting from the unloaden weight from 2.21 to 0.64 mm. The use of chemical fibre-reinforced plastics also has a positive effect on inertia-induced deformation during acceleration, which was 63% lower on the examined robots with chemical fibre-reinforced plastic components than on the steel-aluminium constructions.

The static and dynamic deformation characteristics under the influence of simulated process forces were also examined in [40]. The worst measured compliance levels in horizontal direction were 3.5 μm/N for static compliance and 55 μm/N for dynamic compliance, which is more than two powers of ten higher than on milling machines. Robots are therefore totally unsuitable for cutting operations, with the exception of light deburring operations. The main weak points identified here were deformations of the crab and the Y guideways and bridge torsion. In the above-mentioned work a concept for a robot of maximum rigidity was proposed, for which mathematical calculations revealed marked improvements over conventional gantries. The most important measures involved in this proposed concept are the use of additional guide shoes in the Y and Z guideways and symmetrical suspension of the crab between two Y- beams.

With regard to the dynamic performance of the actuating systems, the investigations carried out in this area have revealed that, in the case of gantry robots at least, the motors are oversized, as a result of which the motor accounts for a large proportion of the mass inertia on an axis, which ultimately leads to a reduced level of dynamic performance. A further effective measure is the selection of a gear unit which permits the high speeds of modern three-phase servo-motors to be used as input speeds. As a result of the greater transmission ratio, the mass moments of inertia on the output side have a lesser effect on the motor. Only after exhausting the potential of these measures can further improvement be attained by means of a lightweight construction.

Gears The previously mentioned survey reveals that the users also see a need for further development in the area of gear units, whereby reduced

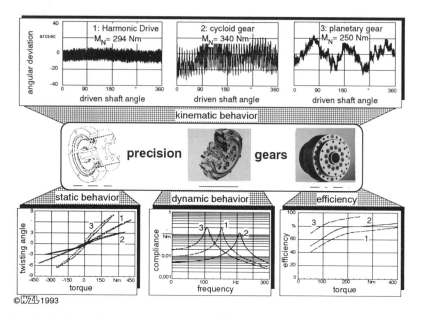

Fig. 14.15 Operating characteristics of precision gear units

mass inertia is considered of particular importance, in order to enable a higher level of dynamic performance for the drive systems. In [41], a low level of circumferential backlash, high rigidity, low fluctuation in transmission ratios and a high degree of uniformity for the transmission of motions are cited as further objectives.

These and other characteristics have been examined for 18 different gear units in the course of extensive test-stand trials at the WZL of RWTH Aachen. Fig. 14.15 shows the most important results for three modern, highly-developed gear units which are employed in robots. The three types concerned here are harmonic drive, cycloidal and planetary gear units.

With regard to comparison of the gear units it should first of all be noted that the sizes vary slightly. The measured values must therefore be considered in relation to the rated torques, which range from 250 Nm (gear 3) to 340 Nm (gear 2). The special characteristics of the respective designs nevertheless apply as described below.

The static compliance is determined by subjecting the output end of the gear unit to a rising level of torque while the driving end is fixed, thereby measuring the torsion angle. After reaching the maximum permissible load, the direction of load is reversed. A linear relationship between the load moment and the torsion angle is desirable here, with the flattest possible characteristic curve.

The cycloidal gear unit produces particularly good results here, possessing a rigidity of 118.5 Nm/arcmin at the rated torque level. The hysteresis caused by backlash and friction is minimal for all three gear units. Differences apply with regard to the linearity of the rigidity characteristics, however. Gear unit 1 reveals a marked dependence on the direction of rotation here, while in the case of the other gear units the rigidity based on the rated torque decreases disproportionately at small loads, which causes difficulties in controlling a robot drive. When the measured rigidity levels are divided by the rated torque, in order to take due account of the influence of the different sizes, the smallest gear unit (gear unit 3) produces slightly better results, though the tendencies still remain. Under dynamic load, gear unit 2 provides the best results, namely minimal dynamic compliance and the highest resonant frequency.

The efficiency level of the planetary gear (gear unit 3) is well above the levels for the other two units. This characteristic, which results in reduced heat emission, can have a positive effect on some designs of robot in the form of a reduction in temperature-induced inaccuracies. The efficiency of all models of gear units improves under load. The kinematic behaviour, that is, the uniformity of speed transmission, is plotted in the upper part of the diagram. Very good transmission characteristics at minimal speeds are revealed by gear unit 1, which possesses an angle error of only 20 arc seconds. At higher speeds, the same scale of error applies to all three gear units, however.

In addition to these measured values, additional characteristics which enable assessment of the suitability for deployment in robots can also be determined on the basis of design data. The most important factors here are required space, weight and mass moment of inertia, whereby these variables must be assessed in relation to the rated torque. Further interesting factors are the permissible input speed, the transmission ratio and overload protection (e.g. in the event of collisions). With regard to the required space and the weight, gear unit 1 is clearly superior, but at the same time it also possesses the highest mass moment of inertia, at 4.6 kgcm², compared to 0.02 kgcm² for gear unit 3. However, when these values are compared with the moment of inertia of a drive motor suitable for operation with gear units (10 kgcm²) and the inertia levels resulting from the transmission elements on the output end are taken into consideration, the influence of the gear unit on the overall mass moment of inertia appears relatively small. Finally, it can be stated that all three gear units are fully developed and highly suitable for deployment in robots. They represent the state of the art in this field. Each type has its own specific strengths and weaknesses, however, as a result of which there is no single optimal gear unit for all applications.

As a result of the swift pace of development in the area of electronics and, subsequently, in the area of control engineering, in addition to the further development of mechanical components increasing efforts are also being undertaken to improve mechanically soft designs via measures in the area of control engineering.

(c) Compensating mechanical weaknesses

Compensating processes have long been in use in the machine tool manufacturing sector, in order to correct leadscrew errors or deformations caused by the intrinsic weight of heavy components on large machines, for example [34]. Position setpoint correction processes involving the transmission of altered setpoints to the position controller are suitable for these constant or only slowly-changing errors. In the simplest cases, this intervention is carried out in the form of a regulating measure only, i.e. the disturbance variables and their effects on positional accuracy are stored in the controller in the form of correcting functions or correction tables, but the actual positional error is not recorded. However, the application of such processes requires a knowledge of the disturbance variables, which must either be measured on the finished device or approximated by calculations. Such measuring operations may require several days, however [35].

When it is possible to measure the actual deviation, e.g. the positional deviation at the tool centre point, a controlled compensation process can be employed, thereby eliminating the need to carry out measurements on the gantry. However, only in rare cases is it possible to determine the position of the TCP throughout the entire working area, using an optical measuring process, for example, as a result of which controlled compensation is only rarely applicable for robots.

The above-described process of stipulating altered setpoints for the positional control circuit operates too slowly to enable dynamic errors to be compensated. In order to compensate vibrations, efficient control processes are required whereby the vibrational movements measured on the structure of the robot are fed directly into the control process. Due to the complexity of these processes, only a small number of research projects [36-39] have so far been devoted to this area, whereby restrictions were made with regard to the number of axes to be examined (max. 2). Furthermore, the vibrations were compensated in the horizontal plane only, neglecting the influence of gravity. According to [40] the processes also require a precise knowledge of the controlled system and the disturbance variables, whereby the provision of this information by sensors and updating of the model of the controlled system in the event of changes resulting, for example, from wear, different axial positions or loads involve complex requirements. Finally, irrespective of the type of compensation process employed, compensation is possible only for errors in degrees of freedom for which corresponding axes are available on the robot. In the case of a linear gantry, for example, displacement of the TCP in Y direction can be reduced by increasing the torsional rigidity of the X bearing profile, but not by compensation. It is therefore expedient and desirable to carry out improvements of the mechanical characteristics of robots in the course of further progress in the field of control engineering.

14.3.2 Control systems

The purpose of a control system is to enable optimal utilization of the potential provided by the flexible handling device, i.e. the robot. Practical experience has shown that a number of improvements are required which it will be possible to carry out in the short term, while several long-term objectives also require to be pursued. Weaknesses which are cited particularly frequently by users include the lack of standard interfaces, e.g. for data saving, inadequate facilities for structured programming, difficulties in defining movement paths (e.g. no variable tool centre point) and robot positions, insufficient cycle times for the internal control circuits, inadequate motional and slip characteristics (for reduced wear on gears), etc. In some areas, such as programming languages (cf. section 3.3), developments have already been carried out. On the basis of the current level of development in the field of robot control systems, the long-term requirements can be divided into the four general areas of user-friendliness, modularity, interfaces and performance capabilities, as shown in Fig. 14.16.

The user-friendliness of a controller is of particular importance when frequent reprogramming, optimization or general operator intervention are necessary. This applies in particular in the case of small batch sizes. It should also be taken into account, however, that when robot cells are deployed on low-manned shifts higher requirements apply to the user

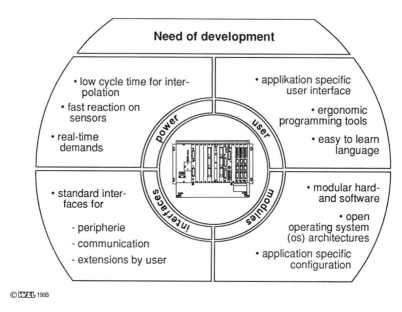

Fig. 14.16 Objectives and concepts for improvements in the area of control engineering

interface, as a large number of cells, which may be completely different in configuration, require to be monitored by a small number of operators at such times. As the personnel on low-manned shifts cannot be as experienced in handling the individual systems as an operator trained for a specific cell, more comprehensive diagnostic systems and even clearer operator guidance facilities are required, to avoid overtaxing the personnel in the event of errors. Considerable development work requires to be carried out in this area, although initial developments in the direction of improved operator guidance systems are already observable. By comparison with the existing scope of technical possibilities, which have been implemented in certain areas, such as NC technology (window-oriented user interfaces, help systems, diagnosis systems, etc.), the user guidance facilities for present-day robot control systems have been kept relatively simple. One reason for this is certainly the fact that robots are still employed primarily in the field of mass production, where switches to different products are relatively rare. It is generally argued here that a sophisticated user interface is uneconomical on account of its unavoidable additional costs, as the resultant time which can be saved in programming, operation and maintenance is of negligible significance when switches to different products are rarely carried out.

It is true that the development of a new, user-friendly user interface and provision of the necessary hardware (display with graphics capabilities) result in an initial increase in the development costs for a controller and a subsequent rise in its retail price. However, if such development work is carried out on the basis of existing user interfaces from the PC area, such as WINDOWS or OS/2 - PM, these costs can be reduced substantially. Furthermore, in addition to the stated advantages, there are additional long-term benefits, such as a reduced scope of training requirements (different controllers have the same, familiar user interface, differing only with regard to the scopes of functions and menu structures, etc.) and a higher degree of compatibility with other computer worlds. Also, in the long term the development costs for new user interfaces can be reduced drastically, as only components relating to specific control functions and applications will require to be adapted. As the degree of acceptance and deployment of such user interfaces increases in the field of control engineering, their advantages will justify the development costs which, in the long term, will be lower than presently supposed. The closely related area of programming is discussed in detail in the next section.

On the basis of current technological standards, the performance capabilities of robot controllers with regard to their processing capacities is generally satisfactory, with the exception of certain specific applications. Special requirements may apply to the computing capacity of a controller which is required to carry out various different tasks. The combination of high traversing speed and high accuracy, for example, such as applies when using a laser as a tool, requires the shortest possible interpolation cycles. The additional processing of arithmetic operations, sensor signals or user-defined interrupts imposes further demands on the controller's capacity, sometimes resulting in a marked reduction in the processing speed. This can be countered only with faster processors, multiprocessor systems or transputers, whereby this would result in substantial price increases. With regard to computing capacity, such measures are therefore economically viable only in certain cases. Furthermore, in view of the price trends for microprocessors and the vast increases in capacity which have been achieved in this area in recent years, it may be assumed that the number of applications requiring still more capacity will continue to fall in the future. Nevertheless, there will always be some applications for which the use of high-capacity processors or systems is justified.

When intelligent sensors are connected to the controller, these impose real time requirements, i.e. the controller must respond to signals from the sensor within a preset time. Real-time capability does not impose any requirements on the computing capacity of the controller, but merely requires a response to a certain event within a defined time structure. It is not necessary to respond within a certain time which is identical for all controllers; rather, the maximum possible response time must be known.

In order to reduce the substantial complexity of integrating an individual controller into a plant control system or connecting peripheral components, uniform interfaces at different levels are desirable. In 1983, an internationally standardized general guideline for the standardization of communications protocols was adopted in the form of the ISO reference model (ISO 7498). The aim of standardizing this reference model was to coordinate the development of open, i.e. vendor-neutral communications interfaces. The ISO reference model divides the tasks of data communications into seven mutually independent areas or layers beginning with the lowest layer, the physical layer, which stipulates the hardware structure for data transmission, through to the seventh layer, the application layer, which specifies the actual important task.

In 1984, the General Motors company in the United States of America began to draw up its own specific functional profile for the area of automation engineering, on the basis of the ISO reference model. This functional profile has become known by the name of MAP (Manufacturing Automation Protocol). The application of a functional profile involves selecting from the respective layers of the ISO reference model all those standards which are required for the application concerned, virtually in the form of a path. In actual fact, MAP should, therefore, stand for Manufacturing Automation Profile. As mentioned above, the protocols of the 7th layer of the ISO reference model are of importance to the MAP user, as they represent the direct user interface. In the MAP profile, these are the following protocols:

- Manufacturing Message Specification (MMS - ISO/IEC 9506) for controlling any automation devices,
- File Transfer Access and Management (FTAM - ISO 8571) for file transmission and access to the contents and attributes of files,
- Network Management (NM - ISO 9595) for the coordination, control and monitoring of resources which are employed for communications in open systems and
- Directory Services (DS - ISO 9594), for access to the network-related information stored in directory systems.

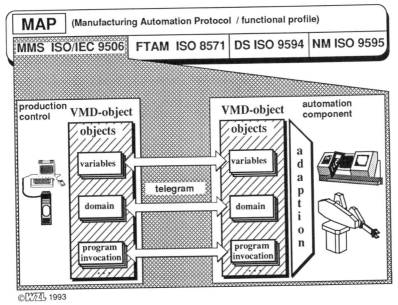

Fig. 14.17 Objects in MMS (Manufacturing Message Specification)

The Manufacturing Message Specification (MMS) has been evolved especially for the development of open systems in the production field and thus provides a basis for the implementation of open DNC communications interfaces. MMS defines the structure and contents of communications messages which require to be exchanged between the programmable automation devices in production facilities, in order to perform the required production tasks.

In addition to the MMS kernel specified in parts 1 and 2 of MMS, which defines the basic communication functions, additional device-related communication functions are defined in the MMS companion standards. MMS companion standards for numerically controlled production machines, robots and programmable controllers have been internationally standardized under the designations NC Companion Standard (ISO/IEC 9506, Part 4), RC Companion Standard (ISO/IEC 9506, Part 3) and SPS Companion Standard (ISO/IEC 9506, Part 5). For the first time in history, MMS provides an internationally standardized DNC interface. Fig. 14.17 shows the relationship between MAP and MMS, and the advantages of defined access to the internal functions of various automation devices.

At field bus level, many different protocols are currently available on the market (PROFIBUS, CAN-bus, SERCOS, Interbus-S, etc.), none of which has become firmly established as the standard option, however. Consequently, the desired advantage of such bus systems, that is, standardized communications between components from different manufacturers, is essentially lacking. Despite all the advantages and disadvantages of the systems with regard to their performance capabilities, it is to be hoped that one protocol will become firmly established in the long term.

A further requirement which has yet to be implemented is a uniform interface to enable the user to expand the control functions. Such expansion measures may include, for example, modification of the user interface, i.e. altering and extending the menu structure, adding new functions, e.g. to control peripheral components and define user-configured functional modules designed in accordance with the requirements of the specific application concerned. Systems suppliers or users should be provided with the possibility of configuring the control system in accordance with their specific requirements. The user-interface of a robot controller which is implemented by a control systems manufacturer can only provide a basic scope of commands, as the number of possible applications for a robot is too great to be covered by one single user interface. The capability to configure control systems in accordance with the specific requirements of individual applications is an essential prerequisite for economical deployment in small-batch production. In order to meet users' requirements for control systems designed in accordance with their specific requirements, some manufacturers have specialized in certain applications, such as inert-gas-shielded welding or spot welding. In order to avoid the need for such specialisation, which restricts the scope of potential customers for manufacturers, and to provide system designers with a possibility of carrying out extensive adaptions to controllers or within control systems, an open control system architecture is necessary. Such an open control system is comprised of several modules, which carry out certain functions of the RC. These modules possess defined interfaces connecting them to one another and to the basic control system, which means they can be exchanged as required. This means that users and designers are not only able to add modules, but also to replace existing modules (e.g. new sensor functions, transformation algorithms, user interfaces, etc.). This form of open architecture for the control system and the operating system thus provides maximum scope for application-oriented configuration of the control system. At the same time, the control systems manufacturers are not required to disclose the algorithms which they employ, as the respective modules are self-contained, and the development costs can be contained by means of concepts which can be expanded on a modular basis.

The idea of such a modular software architecture can also be applied to the control system hardware. In this context it is unrealistic to assume that control systems manufacturers could be prepared to disclose information on the internal functions of their controllers, in order to enable the development of component parts. The internal functions of a control system with appropriate performance capabilities may constitute a competitive advantage which no manufacturer will surrender voluntarily. A conceivable development, on the other hand, would be implementation of the previously mentioned standardized interfaces, e.g. for sensors, axial controllers, etc. Certain computer architectures also permit an additional expansion of the computing capacity via the integration of additional processors. Multiprocessor and transputer architectures, which possess a certain degree of scalability, are suitable for this purpose. For the area of robot control systems in particular, a number of developments on the basis of transputer architectures already exist [26, 27, 28]. However, the scope of measures required to attain suitable distribution of the tasks among the processors and thus for scaling purposes should not be underestimated.

In the long term, standardization in the areas of the processor, operating system, user interface, etc. would lead to a marked reduction in costs, and facilitate handling of the control system for user and designer alike. It would furthermore be possible to utilize developments from other areas of information technology (processors, compilers, MMI-APIs, etc.), thereby reducing the scope of development work while increasing the scope of functions. It must not be overlooked, of course, that in certain areas the components of a control system are subject to different requirements to those which apply, for example, in the PC sector. For example, the operating system for a controller, which represents the essential basis for a modular software architecture, must possess real-time capabilities which standard PC operating systems are unable to provide. The advantages of such a procedure prevail, however, as the resultant scope of modification is nevertheless quite justifiable. Further information on the subject of the "open controller" is provided in paper 13.

14.3.3 Programming

With regard to the desired transition to smaller batch sizes in particular, the area of programming and the time and costs involved in this area play a very important role. The time and costs involved in programme generation will rise in the future, due to the increasing use of sensors and communications with peripheral components. The increasing lengths of programmes and the transition from the pure programming of movements to the generation of programme-logic structures will impose substantially higher requirements on the programming of future robot control systems. The deficiencies and development concepts discussed in this section are to be assessed in the context of this future course of development, and are not geared simply to the requirements which commonly apply to programming today. In view of the expected course of development, the use of new processes and tools will be essential for most applications. Only in a small number of applications will the economical deployment of a robot be possible without new tools, particularly when small batch sizes are involved.

Definition of the problem areas:

An analysis of the most common present procedures for generating RC programmes reveals the most important weak points which remain in this area (Fig. 14.18). In the case of the on-line or hybrid process, the tool centre points which the robot is to approach in the programme are taught into the system and stored on a programme sheet or in data lists in a single operation. This is usually carried out by means of teaching, i.e. by moving the robot into the desired target position using the manual controller. In order to avoid using the capacity of the robot cell and the subsequent lost production time, a kinematic model of the robot is sometimes employed in a separate programming area which is identical to the operating cell. Irrespective of any such models, a programme structure, which contains all the commands necessary for correct running of the programme apart from the precise positioning information, is generated in a further step. This structure also contains technological information, such as current, voltage, feed rate, oscillating parameters, etc. for path welding. Such information is incorporated into the programme primarily by manual input.

The order in which the two steps of teaching in tool centre points and generating the programme structure are carried out is unimportant. After carrying out these two steps, the two parts of the programme must be connected, either by entering the positions recorded on the programming sheet or by integrating the data list into the programme. Finally, the programmes are tested and optimized in the operating cell.

The above-described procedure reveals a number of deficiencies which make the generation of programmes a difficult and thus time-consuming

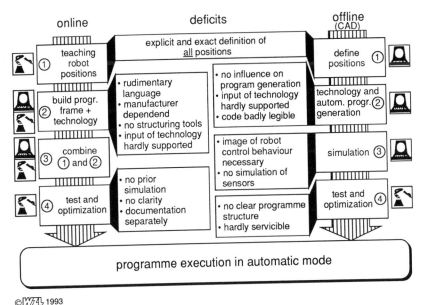

Fig. 14.18 Procedure and weak points in customary RC program generation

matter. In the course of teaching in the tool centre points, all the positions which are to be approached within the programme must be specified explicitly and with a high degree of accuracy. When it is considered that an application-oriented programming user interface, as presented in section 3.5, is able to reduce the number of points to be taught in to less than half the original number, subject to certain basic conditions being fulfilled, the potential and the importance of such a user interface become clear.

An example of such a reduction in the number of points to be taught in from the area of path welding is provided by the programming of a weld seam in an inner corner. In order to reach the inner corner point without collision, the burner must be inclined on its path into the corner in such a manner as to prevent the burner mounting from colliding with the wall of the component. As the burner welds its way out of the corner, the inclination can be reduced accordingly. When the classical procedure is applied, each intermediate point at which a change of orientation is to take place must be explicitly taught in. As the specification of three points including orientation is often sufficient to enable automatic generation of the intermediate points, the scope of work required is reduced accordingly when using an application-oriented user interface with appropriate macros.

With regard to generation of the programme structure, the problems relate primarily to the programming language and the employed programming systems. Many robot control systems incorporate very rudimentary programming languages which possess few of the characteristics of high-level languages. The definition of parameterizable functions, the free definition and naming of data types and variables, the execution of common computing operations, etc. are far from standard facilities here. Furthermore, each robot control system which is available on the market possesses its own programming language, necessitating highly extensive retraining when switching to a different control system.

Even when a high-level-type programming language is available which enables structuring of the programmes, there is a lack of tools to provide further assistance in structuring the programmes, leaving the user to carry out this work. Also, there are virtually no facilities to support the input of technological data.

With regard to testing and optimization of the generated programmes, there is a lack of facilities for preliminary simulation of the programme, to enable the identification of programming errors prior to carrying out full-scale testing, and the lack of clear structures for the programmes, particularly when simple programming languages with virtually no structuring capabilities are employed, makes reading and correction of the programmes more difficult. Apart from individual comments in the programme, programme documentation can only be carried out separately from the programme text, in the form of flowcharts or textual explanations.

Programme generation with the aid of CAD systems involves - apart from input of the motional and positional information in the simulation system - the input of technological data and other statements (e.g. gripper open/closed). After complete specification of the machining task, programme generation is carried out automatically and enables simulation of the generated programme to be carried out directly afterwards. In the operating cell itself the programme undergoes final testing and the motional sequences are optimized. Although this method of programme generation is relatively user-friendly, this procedure nevertheless has several weak points, apart from the high acquisition costs for the programming and simulation system. A general problem is the tolerances which inevitably apply between the CAD model and reality, which make the use of sensors essential.

As in the case of on-line or hybrid programming, all the robot positions must be specified, i.e. essentially the teaching-in procedure has merely been shifted to the computer. The need for development work in the direction of application-oriented programming interfaces which is mentioned above therefore applies equally here. The procedure for the input of technological parameters also corresponds to the procedure on the on-line side and thus requires improvement.

Automatic programme generation certainly offers the advantage of providing programmes free of syntax errors. However, the generated code is difficult to read, as it is barely structured, which means that during final optimization in the cell it is expedient to modify the motional behaviour, but not the programme structure itself. This means that direct maintenance and editing of the programmes is practically impossible. An additional problem of automatic programme generation is that the user cannot directly influence the programme generator. This restricts the programme to a specific controller, plant configuration, sensor system, etc. Any changes on the plant require reprogramming of the programme generator, which can be carried out only by programmers from the manufacturing company. Simple adaption by the user is not possible.

With regard to the development of simulation systems, the implementation of an exact replica of the control functions and, in particular, of the transformation algorithms, has proven a very complex matter, as the manner of treatment for exceptions is not always obvious. The control systems manufacturers display very little willingness to disclose detailed information on the algorithms which they employ, as they constitute the major part of the development work which is invested in a control system. Furthermore, there are no facilities for extensive simulation of the employed sensor technology, which means that only limited testing can be carried out on programmes employing sensors.

Concepts for potential solutions:

A programming language to be employed in the area of robot control systems is essentially subject to two contradictory requirements. On the one hand, the programming process should be as simple and clear as possible for the operator, so as to minimize the scope of training required when new controllers are introduced. On the other hand, however, high-capacity programming languages are required which are based on high-level languages from the field of computer science, e.g. Pascal, and thus enable the generation of highly complex but also high-capacity programmes. These apparently complementary requirements can be fulfilled in diverse ways. One possibility is strict separation between the generation of complex programmes in a high-level language with all its possibility by trained personnel, e.g. in the operations scheduling department, and programming on the machine, which is carried out at a very low level, with the machine operator having little influence on the

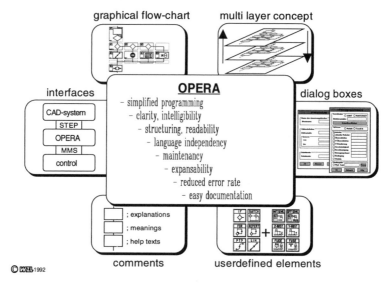

Fig. 14.19 Principles and advantages of the OPERA programming system

programme. This results in the problem of different programming languages being used at the same company, leading to communication problems between machine operator and off-line programmer. Also, an additional compiler is necessary. And retranslation from the low-level into the high-level language is not generally possible by simple means. Another alternative would be to use the same language at both levels, but restricting the scope of commands for the machine operator to a subset of all the functions. The machine operator could be denied any possibility of generating functions, for example, while still being able to activate functions. The definition of such a subset represents a sensible alternative and has been implemented in the new DIN standard 66312, the IRL robot programming language (Industrial Robot Language), which is based on Pascal.

Current robot control systems generally incorporate relatively simple programming languages, though control systems which employ high-level languages for programming are also available on the market [29]. To further simplify programme generation, the user is often assisted by menus and masks for entering the parameters required for a command. The advantages of this procedure are firstly the reduced frequency of errors and secondly the fact that the programmer does not need to know all the possible parameters for a command and its precise syntax.

But with this procedure the problem still remains that although a complex programming language provides the possibility of structured programming, it is left to the programmers to implement this possibility on the basis of their abilities. No direct support is provided for a structured approach to programme generation. This is a particularly critical factor in view of the fact that programme developers generally receive training in the field of programming in the course of their vocational training, as it is not possible to impart a sound knowledge of the methods and procedures involved in structured programming in the course of such training. Supporting programming tools are essential in conjunction with complex programming languages, in order to ensure that the generated programmes nevertheless possess good readability and maintainability.

A possible tool to provide the programmer with better support in generating programme-logic structures is the graphic, structure-oriented OPERA programming system (Open Programming Environment for Robot Applications) which has been developed at WZL to generate programmes in the form of programme flowcharts, as are established for the purpose of separate documentation. Fig. 14.19 shows the essential principles of this system. The basic element of the programming system is the graphic programme flowchart, consisting of symbols supplemented by comments, which specify the respective functions to be carried out. The basic functional scope of this system is based on the IRL (Industrial Robot Language) programming language, i.e. all the commands present in IRL can be incorporated into the flowchart in graphic form. The programme code which is generated automatically by the system is in accordance with the IRL standard and can be translated into ICR (Intermediate Code for Robots) without any problem whatsoever. The flowchart enables programme structures, such as branches or loops, to be presented in a clear form. In order to ensure a clear overview of the programme, the flowchart can be spread over as many successive layers as are required. Each lower-level layer thus contains the contents of an appropriate element of the higher-level layer.

It is thus possible to group any number of elements from one layer into a block and to transfer this block to a lower-level layer. These elements are then no longer visible in the original layer and can be displayed only by switching to the lower-level layer for the block element. This corresponds to the human procedure of grouping unclear elements together and enables programming to be carried out in accordance with the principle of top-down design. Starting with a very abstract view and division of the task

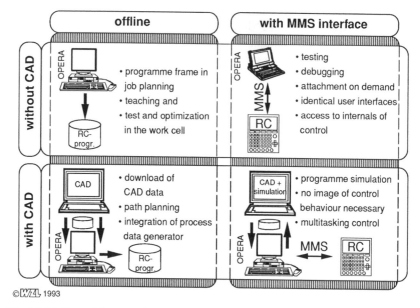

Fig. 14.20 Potential areas of application for the OPERA programming system

concerned, this process involves continually refining the modules of a programme until the actual programming language level is eventually reached. High-level layers contain only increasingly more comprehensive functional blocks. This method of starting with a global approach to the task concerned and subsequently incorporating an ever-increasing level of detail automatically leads programmers to adopt a more systematic procedure and to structure their programmes in a more systematic manner.

The parameters required for the individual command elements are entered via dialogue boxes, which eliminate the need for the user to learn all the possible activation parameters for the command and their syntax and reduce the scope for errors. Furthermore, the users can add any number of their own elements to the system by storing them as functions, allocating them graphic symbols which they have created themselves and entering them into the system. There is then no difference whatsoever between these elements and those already incorporated in the system. In order to document such user-generated functions together with their activation parameters and the overall programme, programmers can provide programmes, functions, variables, etc. with comments and even draw up their own help texts for user-generated functions, and integrate

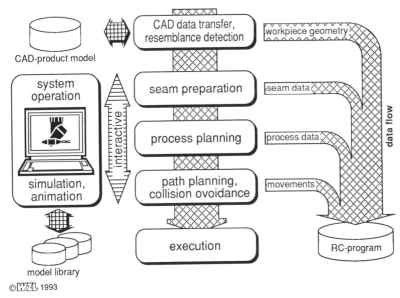

CAD-product model

system operation

interactive

simulation, animation

model library

©*WZL* 1993

CAD data transfer, resemblance detection

seam preparation

process planning

path planning, collision ovoidance

execution

workpiece geometry

seam data

process data

movements

data flow

RC-program

Fig. 14.21 Integrated system for the off-line programming of welding robots

these into the OPERA help system. Interfaces for connection to a higher-level CAD / planning system and to a robot control system (MMS standard) are also under development. Among other things, the MMS interface will enable debugging operations to be carried out by tracing the programme run in the control system on the programming system and monitoring the contents of variables, the state of inputs/outputs, etc.

The programming system can be deployed in four different configurations (Fig. 14.20). In the case of purely off-line operation, the system can be used without connection to a CAD system to generate the programme structure at the operations scheduling stage. The further procedure corresponds to the left half of Fig. 14.18. When a CAD / planning system is employed, information on robot positions, technological data, etc. can be input and incorporated into the RC programme.

When an MMS-interface is employed, the system can additionally be used as a mobile debugger in the form of a laptop, for example, which is connected to the control system via MMS for the purpose of programme testing or debugging. An advantage of this combination is that the system is deployed only when required, as a result of which not every control system is subject to the attendant costs. Furthermore, when carrying out direct debugging, the same user interface as has been used previously for programme generation is now available for operation and programming, and full access to the internal functions of the control system is provided via MMS.

The final configuration requires a simulation system and an MMS-interface, so that the programme can be run on the control system without moving the robot and displayed on the simulation system. It is not necessary to map the control system in the simulation system, as the control system itself is employed. In this case, the simulation system simply displays the robot position read from the control system in graphic form. To enable effective deployment of this combination, a control system with multitasking capability is required, which continues to control the production cell during simulation of the programme to be tested. Although no such control systems are currently available, they may be expected to become available in the medium term, in the wake of the further development of the performance capabilities of processors and the deployment of modern operating systems.

With regard to CAD-assisted programming with automatic programme code generation, Fig. 18 presented weak points relating to the specification of tool centre points via a method similar to teaching and the lack of support for technological programming. These deficiencies can be avoided by means of a programming and simulation system configured especially for the task concerned, which interacts closely with the CAD system used for design purposes. An example of such a configuration is presented below in the form of a system developed for programming welding robots in the shipbuilding industry [30]. The key feature of this system is the integrated system for generating programmes, which extends from CAD design through to simulation of the fully planned machining sequence (Fig. 14.21).

The data taken from a CAD system designed specifically to meet the requirements of the shipbuilding sector provides the starting point for further planning of the machining sequence. On the basis of this data, the existing workpiece model is supplemented and expanded by the addition of geometrical and technological data for the weld seams in the next planning stage (seam processing). The resultant working model provides the basis for the next planning stages. In the following tool planning phase, the suitable burner tools are selected on the basis of the working model, and collision-free operation of these tools is configured. The result of this planning stage is a path model which contains the data relating to the

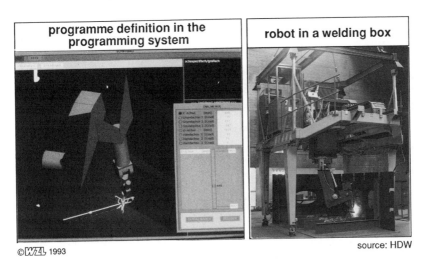

programme definition in the programming system	robot in a welding box

©*WZL* 1993   source: HDW

Fig. 14.22 Off-line programming system for welding robots in the shipbuilding sector

welding process for all burner paths. The final planning phase involves incorporation of the employed machines, in this case the robot welding systems. In this phase, the welding, transfer and auxiliary paths are defined and the feasibility of the individual burner operations and all additional movements in the context of the overall system are checked and verified. All the information required for subsequent generation of the programme code is then available in the internal data structure.

The system also incorporates a module for similarity identification, which compares the current working model with the working models of previous planning procedures and establishes any similarities. Similar components may differ, for example, with regard to the absence of a few seams or minor dimensional differences. In such cases it is often possible to use most of the completed planning measures for a previous component in unchanged or only slightly modified form. The automatic similarity identification function establishes the appropriate previous planning measures and incorporates them into the new planning.

Due to its specific configuration for the field of path welding in the shipbuilding sector, this system, which is shown in Fig. 14.22, is able to provide facilities for more effective planning than is possible with standard systems for general applications. Thus there is also a need for the development of modules for specific applications in the area of CAD-

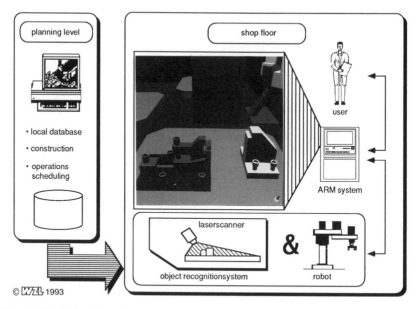

Fig. 14.23 Task-oriented motion planning and simulation system, ARM

based programming systems, similar to the previously stated need for such specific configurations in the area of the user interfaces of control systems in the RC area (cf. section 3.2).

Other approaches involve carrying out programming at an even higher level. In the case of the task- oriented programming of industrial robots, a user is no longer required to specify *how* a machining task is to be performed, but simply enters general commands, such as "Move component from A to B". All sub-tasks are generated automatically by the system and are incorporated into the generation of an executable robot programme. The problems which arise in connection with the differences between the model in the simulation system and the real operating cell are minimized via the integration of sensors. The data in the simulation system is updated before and after a machining task, thereby guaranteeing collision-free running of the programmes (Fig. 14.23).

The potential areas of application for such a system are in the production of small batches and individual products, where the ratio of programming time to execution time in relation to the quantities produced becomes uneconomical when conventional systems are employed [31, 32].

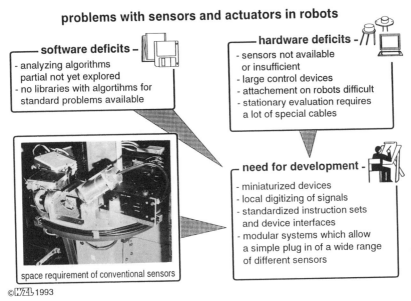

Fig. 14.24 Problems involved in the deployment of sensors/actuators in robots

14.3.4 Sensors / actuators

(a) Problems involved in the deployment of sensors and actuators in robot systems

The areas of sensor and actuator technology are of fundamental importance in expanding the scope of application for industrial robots. The current level of technological development in these areas involves substantial problems with regard both to the hardware and to the software required to analyze the sensor signals, whereby the hardware-related difficulties - in the case of sensors, at least - are probably the more easily surmountable problems. Fig. 14.24 provides a summary of this problem area.

A primary deficiency with regard to hardware is that there is a lack of adequate sensors for several human sensory capacities which would facilitate the deployment of robots. An example here is the human hand's sense of touch. Should it be possible to simulate this sense in a robot gripper, thereby enabling it to perceive the distribution of pressure in the fingers of the gripper and to detect the slipping of a gripped object, this would represent a great step in the direction of a flexible universal gripper. Such a development would also require a suitable mechanical design for the gripper, however, similar to the simulation of the human hand described in [43].

A problem which appears more simple to solve, however, is that of the considerable space required by conventional sensors and actuators and their appurtenant analysis and control units, an example of which is shown in the photograph in the middle of Fig. 24. The photograph shows a collision-prevention system for robots, which is discussed in more detail at a later juncture. This system consists of a laser delay-time sensor, the beam of which is deflected via a reflector which can be moved in three axes. To analyze the sensor signals and control the drive motors, an extensive range of electronic components is required. These electronic components can either be fitted on the robot or installed on the floor.

With the systems which are currently available on the market, these two alternative methods of installation are subject to problems, at least in the case of gantry robots. As the devices are not designed for robot applications, they normally require a 240 V or 110 V a.c. power supply, which is not necessarily available on the lifting beam of a gantry. This voltage is internally converted into a d.c. voltage of 12 V or less, using power supply units. The power supply units sometimes require fans, as a result of which these two components account for by far the largest proportion of the space required by electronic devices. By supplying the required power to the devices via a low voltage power supply available on the robot (usually 24 volt), the size of the electronic equipment could be reduced and the energy supply problem solved. The scale of this problem is illustrated by the example of stepping motor controllers which are currently available, where the ratio of the space required for the controller to that required for the motor is well over 20:1.

Stationary installation of the devices involves different problems. As most sensors and actuators require special cables, a large number of cables require to be drawn through the entire robot, whereby cable lengths of over 70 m may be required for large gantries. In addition to the high costs and the previously described problem of mechanical strain on the cables or even the destruction of cables, a further problem with regard to the transmission of sensor signals is the increasing susceptibility to interference when the length of the transmission path increases.

To remedy this situation, in addition to miniaturizing all the components it would also be expedient to carry out the local preprocessing of all signals and then to feed these preprocessed signals into a digital bus system. Increased flexibility of the analysis and control units would also be desirable with regard to the connection of various sensors or actuators.

With the technology which is currently available, it is virtually impossible to integrate sensors into an application without a knowledge of electronics. Devices are required, however, which enable the simple insertion of various sensors and actuators. A level of user-friendliness must be attained here which is at least comparable to that which is available in the area of personal computers where, according to the degree of proficiency which they possess, it is relatively simple for users to install external components, such as screens or keyboards, or even internal components, such as plug-in cards or hard disks. A certain degree of standardization is necessary here on the software side, however, at least with regard to the primary functions, such as the reading-in of signals or the configuration of input/output channels. A further expedient development would be the provision of software libraries containing frequently required basic functions relating, for example, to the integration, differentiation or Fourier transformation of signals. Here again, the PC serves as an example: software designers are not expected to programme routines for accessing storage media themselves, irrespective of whether the medium involved is a hard disk, diskette or tape drive.

The greatest problem, however, for which no solution is foreseeable, concerns the inability to transfer the capabilities of human intelligence to computers to any satisfactory degree. This deficiency becomes manifest in sensor analysis operations, in particular when abstraction and the identification of relationships and patterns are required, as is often the case. The high initial expectations for the so-called AI sciences have long given way to a more sober assessment of the potential which is available in this area. Certain degrees of success have been attained recently with fuzzy logic and, in particular, in the field of computer "vision" using neural networks. A potential application for image processing in the area of flexible production is described below.

(b) CAD-assisted image processing

Image-processing systems are being employed to an increasing extent in automated assembly operations for the purposes of robot guidance, workpiece identification and process control. For workpiece identification and controlling tasks, the systems require input data which is able to describe possible workpieces or the required state of an assembled unit. The data determined from the image of the current state is then compared with these so-called scenic models [49], to provide the desired information on the type of workpieces involved or the success of an assembly operation.

In a large number of industrial applications, these models are stored in the programme itself, that is, the image-processing software is adapted to the specific task concerned. Increased flexibility is attained with systems which enable a so-called teaching process. This involves taking photographs of sample workpieces or sample scenes with a camera, on the basis of which outstanding features are then determined via the same processes which are to be subsequently employed to assess the actual scenes. Parameters which influence the quality of the results of the image-processing operations can be adapted to the specific application concerned, in some cases in interactive mode. From the features which are established by this method, the user is then able, to a certain degree, to select those which are to be employed for the purposes of the subsequent comparison operations. As teaching-in of the scenic models requires manual intervention or at least an interruption in production operations for the teaching process, this procedure is unsuitable for the automated production of small batches.

CAD systems are being used to an increasing extent in the planning and simulation of assembly operations, however, and it is therefore practical to use the information which is stored in such systems on the product to be assembled and the assembly environment for the purpose of designing the scenic model, thereby avoiding the need for the teaching process. To this end, concepts which enable the more flexible utilization of image processing to monitor automatic assembly processes when small batch sizes are involved have been developed and implemented in the research project "CAD-Ankopplung von Bild-verarbeitungsystemen" ("Integration of CAD and Image-Processing Systems"), sponsored by the German Research Association (DFG).

The industrial processing of grey-scale images is generally concerned with various characteristics, such as surface quality (texture), the size and form of a constant grey-scale value, grey-scale value dispersion, etc [45]. However, the values for these characteristics cannot necessarily be derived directly from the CAD data, in addition to which they are highly dependent on the light exposure. For the application concerned here, the edges of the objects contained in the image were therefore employed as characteristic features, as edges are purely geometrical features, the position of which is known from the CAD model. However, the lighting conditions also influence whether the edges of the workpiece can be identified in the grey-scale values recorded by the camera. It is practically impossible, therefore, to assess on the CAD screen which edges are suitable for the purposes of workpiece identification and which are not.

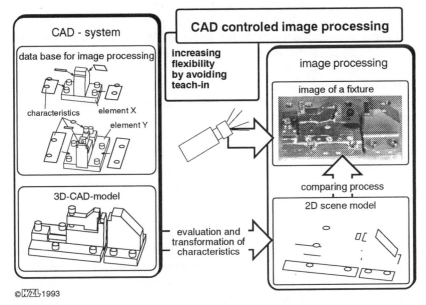

Fig. 14.25 Integration of CAD into an image-processing system

For this reason, the research project was limited to products in modular design. With such products, it is possible first of all to observe all the components contained in the module through the camera at different angles and under different lighting conditions, in order to determine edges which are particularly rich in contrast. These identification features, the values of which have proven particularly invariant under varying lighting conditions and at different angles of vision, and which are referred to below as edges relevant to image processing, can then be stored in a file and thus made accessible to the CAD system.

The robot-assisted assembly of modular holding devices developed at WZL provided the reference basis for the research work [44]. This assembly process involves screwing the individual elements of the device to a machine pallet to form an operational holding device. The device can be adapted to the specific machining task concerned by selecting, arranging and adjusting the elements, which means that each different configuration will produce a different image when monitoring is carried out using a CCD camera. Teaching-in of the scenic model is therefore unsuitable here.

When a holding device is designed on the CAD screen, the positions of all the edges present in the CAD model which are relevant to image processing can be determined, transformed into a two-dimensional description in accordance with the angle of vision and converted into camera coordinates, by means of a software package developed in the course of the project. The image-processing software can then compare this reference image with the current image of the real scene photographed by the camera, in order to assess whether the correct elements are in the correct positions.

Fig. 14.25 is intended to illustrate this process of designing the scenic model on the basis of the CAD data. The top left of the illustration shows two holding elements with the edges which are to serve as identification features and, below this, the 3D CAD model of a simple holding device comprised of these elements. On the basis of this information, a three-dimensional description of all the features to be checked by the image-processing computer is first of all generated. This representation is then projected into the two-dimensional camera coordinate system, taking due account of the angle of vision and the distance of the camera from the object, whereby concealed edges must be masked out. On the basis of this reference model, the image-processing system is able to carry out monitoring and control functions without any need for prior teaching.

In order to carry out this procedure, the CAD model and the camera image must be made congruent, however. For this purpose, the object-to-image ratio of the camera and the position of the camera coordinate system relative to the so-called world coordinate system of the CAD system must be established by means of a camera calibration process. The exact camera position is determined in two stages, whereby the attainable accuracy has a direct influence on the accuracy of the image analysis process.

In the first stage of the camera calibration process, points are sought in the camera image whose three dimensional coordinates in space are known. In the application concerned here, the centre points of the mounting boreholes in the base plate of the device are employed for this purpose, as their exact three-dimensional coordinates are known from the CAD model of the pallet. In the second stage, the two-dimensional coordinates which the image-processing system determines from the camera image and the appurtenant three-dimensional world coordinates are incorporated into a general form of the camera function of representation.

The mathematical description of the image-mapping process can be divided into three-dimensional transformation of the coordinates of the world coordinate system into the camera coordinate system and subsequent perspective projection into the two-dimensional image plane. The first part of the mapping process involves unknown quantities in the form of the six degrees of freedom of the camera in space, while the perspective projection stage involves only the focal length as an additional unknown, when an ideal camera free of optical distortion is employed. The Radial Alignment Constraint process described in [49] enables the calculation of all unknown quantities from five known calibration points. The accuracy of the calibration process can be improved by using more than five points, as the optimal solution then requires to be found for a redundant linear system of equations.

The prior knowledge of the expected image which is provided by the CAD data can also be used to find weak edges, thereby enabling the use of an image-processing system for complex workpieces and under conditions which are in compliance with practical requirements. For this purpose, model-controlled contour-finding algorithms have been developed which make use of this prior knowledge to enable fast and reliable image analysis. So-called sub-pixeling techniques are also employed, whereby the resolution of the camera is increased by means of interpolation processes. These techniques enable the position of a holding element on a pallet measuring 500 x 630 mm to be determined to an accuracy of approx. 0.2 mm.

Both the software connected to the CAD system for the purpose of calculating the input data for image processing and the image-processing and camera-calibration software have been tested under realistic practical conditions in the "Integrated Production and Assembly System" at WZL. As neither the device elements nor the assembly environment were modified for improved identifiability, the analyzed image material imposed high requirements on the reliability and stability of the algorithms, due to the complex forms and the metallic surfaces involved. In spite of these conditions, incorrectly assembled elements are identified at a reliability factor of more than 99%, which means that the automatic monitoring of assembly operations without teaching is possible and also expedient for batch sizes of "one".

Whereas image-processing systems are suitable primarily for so-called in-plane measurements, as distances vertical to the image plane can be determined only indirectly, the laser system described in the following section is an out-of-plane measuring system. Although this system is able to measure the distances from objects to the sensor directly, measurements vertical to the laser beam can only be calculated indirectly, on the basis of changes in distance.

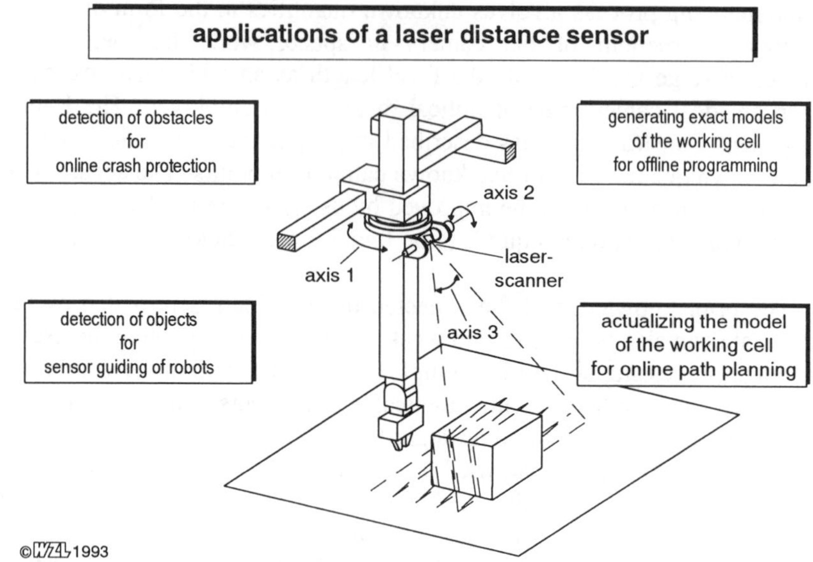

applications of a laser distance sensor

detection of obstacles
for
online crash protection

generating exact models
of the working cell
for offline programming

axis 2

laser-
scanner

axis 1

detection of objects
for
sensor guiding of robots

axis 3

actualizing the model
of the working cell
for online path planning

© WZL 1993

Fig. 14.26 Potential applications for a laser ranging sensor

(c) Potential applications for a laser measuring system

The system shown in Fig. 14.26 has been developed to protect a gantry robot from collisions, and is suitable for applications in diverse areas. To avoid restricting the gantry's freedom of movement, the sensor equipment (see also Fig. 24) has been mounted on the crab. To protect the gripper of the robot from collisions with obstacles located in the working space, the height profile of the robot cell is determined and compared with the current height of the gripper. When the gripper is in its lowest position, the distance between the bottom edge of a gripped workpiece and the sensor is approx. 2.5 m, as a result of which the sensor to be deployed here had to be designed for this measuring range. As fast and stable measuring operations were also required, a laser delay-time sensor was selected, which scans the surface on a point-by-point basis at a maximum frequency of 10.7 kHz.

A collision can occur only when the colliding objects are located in the path of movement. The path of movement is therefore scanned on-line and checked for obstacles. In order to carry out continuous scanning of the entire movement path, the laser beam (measuring beam) must be emitted at a high frequency at right angles to the robot path. To this end, the measuring beam which is emitted from a fixed optical system is directed onto a pivoted reflector (axis 3), from which it is diverted into the surrounding area. As the galvanometer which is employed to move the reflector has a resolution of 0.005 degrees, the laser beam can be positioned to an accuracy of 0.17 mm, even at a distance of 2 m from the reflector. Via an analogue input on the controller of the galvanometer it is furthermore possible to set the pivoting angle and thus the scanning width, which must correspond to the width of the movement path. A balanced, surface-coated aluminium reflector is employed to divert the laser beam.

At the same time as the height profile of the movement path is measured relative to the crab, the data from the shaft encoders is read out. On the basis of the values for the Z axis it can be calculated how far the gripper is extended into the working space. The two values (measured value from the laser and position of the Z axis) are compared. If the measured value from the laser is smaller than the lowest position of the Z axis, including the gripper, the robot will be stopped.

An important feature of the collision prevention system developed here is that the movement path is measured and checked for obstacles irrespective of the programmed path or the direction in which the robot is moved by hand. For this purpose, the entire sensor unit is rotated around the Z axis of the gantry robot via axis 1 and positioned in the direction of travel of the robot. The necessary drive power is provided by a servo motor mounted on the crab of the robot, which transmits the motive energy via a toothed belt to the rotating housing on which the entire scanning unit is mounted. The rotating housing consists of a split, stationary inner steel ring on which a toothed aluminium outer ring is mounted in rotatable configuration. The servo module is controlled via a control card in a PC.

As the braking distance is dependent on the speed of the robot, the lead distance of the laser beam must be adapted to the robot speed via axis 2. As the mechanics and controller of the second axis rotate around the Z axis of the robot (axis 1 of the collision-prevention system), particular importance was attached to a lightweight design. Additionally, a high level of resolution is required for positioning purposes, as at a distance of 2.5 m a turn of the second axis by only 1 degree alters the lead distance by 43.6 mm. Consequently, a stepping motor with 400 steps per revolution and a harmonic drive gear unit with a transmission ratio of 100:1 were employed. This results in a resolution of 40000 steps per revolution for the second axis.

For the purpose of collision prevention, the movement path of the robot is presently scanned on a point-by-point basis at a frequency of 500 Hz.

The gantry robot is stopped only when a measured value and thus an object in the movement path is higher than the robot's gripper. Once the robot is stationary, an attempt is made to circumvent the obstacle, by means of a two-stage procedure.

The simplest method of evasion for a gantry robot is to pass over the top of the obstacle. Once the robot is stationary, the first stage thus involves exact scanning of the movement path in front of the gripper. On the basis of the resultant height profile, the maximum height of the object obstructing the movement path can be determined. If it is not possible to pass over the top of the object, in the second stage the collision-prevention system scans the areas to the left and right of the movement path, in order to locate a possible evasion route.

Apart from this system to protect the robot from collisions, software modules have also been developed which enable the use of sensor guidance, e.g. when gripping workpieces. The collision-prevention system is currently being developed so as to enable automatic scanning of the entire robot cell. The image of the cell generated in the robot coordinate system is subsequently transferred to a simulation system. The model of the robot cell can then be updated in the simulation system. In this way, modelling errors can be corrected, to enable the programmes generated off-line to carry out collision monitoring and control on the basis of the real image of the cell. In this way, the reliability of robot programmes generated off-line can be increased substantially, with only a minimal scope of additional operations. Provided that sufficiently fast algorithms were available, continuous updating of the cell model could be used to carry out the automatic on-line planning of collision-free paths, leaving only a small number of important basic aspects of the programme to be specified in the course of off-line manual programming.

14.3.5 Man-machine interface

In order to generate an executable welding programme, the programming procedures which are employed today in the field of industrial production operations require the use of the welding robot, to record the points and contours which the robot is to approach and move along. This method of programming is firstly extremely complex and time-consuming, as the robot can only be moved indirectly using the customary hand-held programming devices, while secondly the method cannot be carried out in parallel with production operations, which means that the robot is not

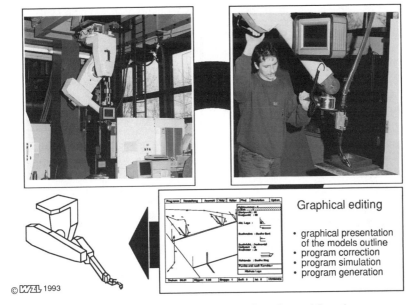

© WZL 1993

Graphical editing

- graphical presentation
  of the models outline
- program correction
- program simulation
- program generation

Fig. 14.27 Shopfloor-oriented programming system for robot welding plants

available for productive operations during programming [33].

The primary objective of a project concerning the ergonomic deployment of a CNC robot system for welding which was carried out at WZL in conjunction with the industrial sector was to develop a plant for application in the area of job production which provides the user with an ergonomically ideal environment for generating even the most complex welding programmes (Fig. 14.27).

The primary components developed in the course of this project are

- the programming robot, optimized in accordance with ergonomic requirements,
- a hand-held programming terminal with a sensitive screen surface,
- a wireless voice data entry facility for effective user assistance,
- a graphic programming system for processing and checking the generated pseudo-code and for generating the RC control programmes, and
- macro and sensor technology to support programming operations (Fig. 14.28).

As a result of the integrated concept for all the functions required to generate a new welding programme, the ratio of arc-machining time to

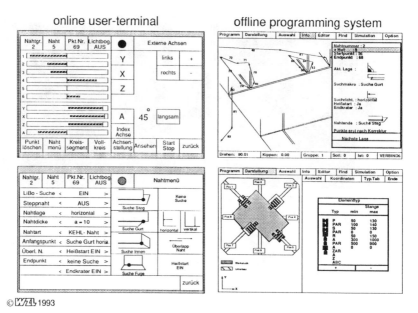

© **WZL** 1993

Fig. 14.28 Operator interface of the shopfloor-oriented programming system

programming time is 1:4, compared with ratios between 1:40 and 1:100 for conventional programming employing teach-in procedures.

14.3.6 Periphery

The periphery accounts for a large proportion of the costs relating to a robot system. According to [48], handling devices are initially only restricted moving devices, which acquire flexibility only in conjunction with the periphery. Such periphery is implemented in the form of material-flow systems, which are used to link individual production or assembly islands. Extensive modular systems for the transportation and intermediate storage of workpieces have long been established for the mass production of small parts. Virtually any configuration of production lines can be implemented via the combination of longitudinal and transverse transport paths, lifting and turning units. Particularly widespread are twin-belt systems, whereby standardized workpiece-carrying pallets rest on belts on either side and are transported by these belts via friction. Such systems require adaption to the specific application concerned only with regard to the workpiece holder on the upper side of the pallet, which means that a large number of components can be reused when switching to a different product. More recently, systems have been introduced with a transport line consisting of passive rail elements and self-propelled pallets fitted with battery-powered drive units [49]. As each pallet possesses its own computer controller, a decentralized control system can be implemented here. At the start of operations, the pallet is provided with a description of its route, which can be updated at defined information points. Between these points, the pallet controls its route independently by activating switching points. The advantage of this concept lies in increased flexibility in controlling the flow of material. Also, the scope of cabling required for the system is reduced.

Bunkers and vibratory conveyors are additional components which are often required in robot plants, as a result of which a broad range of these components is available on the market. The same applies, with certain qualifications, to press-fitting, rivetting and screwing stations. Grippers, fixing and aligning devices often require substantial development and design work, however, particularly when implementing new assembly operations. Standard concepts are also lacking in the area of transportation and storage systems for large parts, such as are often required in the automobile industry.

14.4 Summary and prospects

Critical assessment of the experience acquired over the past 20 years in the area of the robot deployment in production operations has shown that extensive development work still requires to be carried out in many areas. It is also evident that the robot and its periphery must always be considered in conjunction with the field of application in which they are to be deployed, in order to obtain a meaningful assessment of the economic efficiency in particular of specific developments. Future developments must not move exclusively in the direction of increased capabilities via complex configurations, but must take due account of the need for economical deployment.

Despite the desire for "lean" concepts, a great deal of development work requires to be carried out in many fields of application, in order to attain the aim of economical production operations for small batches in particular. But in the long run, increased performance capabilities can only be attained at a viable level of expenditure via the extensive modularization and standardization of individual components. Along with the need for more user-friendly operation and programming, this requirement may well represent the primary challenge for the manufacturers of control systems and robots in the coming years.

Although the scepticism with regard to the deployment of robots in production operations, which results primarily from economic considerations, is often justified, the numbers of robots in operation will nevertheless continue to increase in the future. On the one hand, the robot is practically irreplaceable in many areas for quality reasons, while on the other hand the ongoing development process will result in continuously falling overall costs for the deployment of robots. Where necessary, the performance capabilities of the systems will be increased. This paper has provided some practical ideas as to how the overall system summarized under the name "robot" could be optimized.

Part Five

ENVIRONMENT:
ECOLOGICAL ASPECTS IN
PRODUCTION TECHNOLOGY

15 Cooling lubricants: the ecological challenge facing production engineering
16 Strategic cost/benefit evaluation of product development, production and waste disposal

15

COOLING LUBRICANTS - THE ECOLOGICAL CHALLENGE FACING PRODUCTION ENGINEERING

15.1 Introduction

Manufacturing companies and factories are affected by environmental issues in every conceivable area of their business. This applies particularly to manufacturing. Their mission, to manufacture a product by adding process materials and energy to raw materials, inevitably runs into conflict with arguments about the resultant emissions and waste. While these tended in the past to be regarded as a `necessary evil', increasing environmental awareness and pressure to reduce costs are compelling companies to rethink their strategies. It makes sense from both ethical and economic angles to involve ecology on an equal footing with other factors in planning and evaluating products and production methods [1].

The ecological behaviour of a company is influenced to a considerable extent by external factors such as state legislation and public opinion (Fig. 15.1). Both of these can have a substantial economic impact.

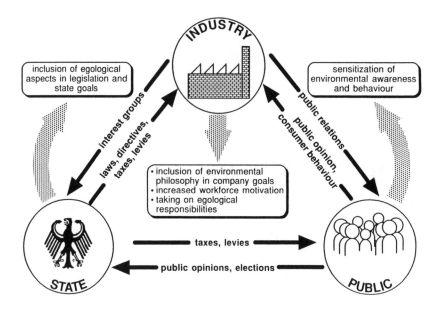

Fig. 15.1 Influencing ecological behaviour

Environment-friendly products and advertisements extolling the virtues of environment-friendly manufacturing processes exert a positive influence on the readiness to buy and clearly help companies to gain a competitive advantage. The state, however, also influences the line adopted towards the environment by industry. This can be direct, via environmental laws and directives or indirect via a system of rewards and penalties. An example of this would be the increased taxes and levies for ecologically unsound materials and the tax breaks for investment in the environment [2].

It therefore makes sense not only from the point of view of social responsibility to integrate ecological considerations into corporate targets. More ecological awareness as regards products and manufacturing methods brings with it a number of advantages which are likely to increase still further in the long-term.

- reduced energy, water and raw material costs
- lower waste disposal costs
- tax breaks for environment-friendly products and manufacturing methods
- improved market opportunities with ecologically sound products

- less liability risk in case of damage
- decreased health risk for employees
- increased staff motivation

all of which contribute to an enhancement of the image of the company concerned.

One of the main sources of ecological problems is manufacturing as it is here that material and energy turnover is highest. The step towards `clean manufacturing' requires thorough understanding of chemical and physical reactions and their impact on humans and on the environment. Knowledge in this complex field is still patchy. Additionally, ecology-oriented optimization is normally associated with investment which does not bear fruit immediately. This may cause some difficulty especially for small and medium sized companies.

Ecologically sound behaviour is frequently held to mean observing the law governing environmental protection and industrial safety. Insular solutions are worked out or the payment of increased sums is accepted as the `lesser evil'. However it makes more sense and is more beneficial in the long term for a company to develop a comprehensive ecological concept. Fig. 15.2 illustrates what course the development towards becoming a more ecologically oriented manufacturing enterprise might take [3].

The initial step must be to identify environmental priorities. With the aid of a systematic itemization of type and quantity of input and output variables, a materials and energy balance can be drawn up for individual manufacturing methods or products. This can form the basis for the allocation of problematic materials to the appropriate method or product. Previously unaccounted for loss, caused for example by leakage or adhesion to the workpiece can thus be localized and appropriately assigned.

An inspection of the factory, viewing and hearing about sources of difficulty as well as discussion with those involved permits problem areas to be defined subjectively. Accidents and occupational diseases should likewise be recorded and their causes identified [2].

The direct allocation of costs, the comparison of emissions with legal limits and recommended values in conjunction with evaluation of the risk potential indicates where the priorities lie thus permitting measures to be drawn up accordingly. These will typically include long-term and short-term measures depending on the time and level of investment required and may be organizational changes which can take effect immediately or adaptive measures such as encapsulation, filter systems etc. They may also include the complete avoidance or adaptation of ecologically unsound

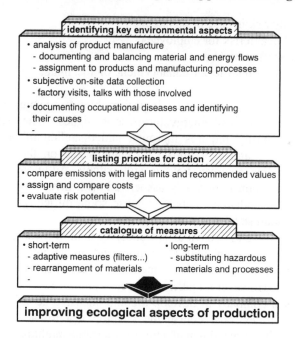

Fig. 15.2 Improving ecological aspects of production

methods which require long-term planning. Parallel to such commitment, an environmental concept for the planning of future products and manufacturing methods should be drawn up. Ideally this should involve all areas of production.

15.2 Environmental problems occurring within the work process

From the totality of production sequences, attention will now focus on the machining process itself (Fig. 15.3). Cutting with a geometrically defined cutting edge, cutting with a geometrically undefined cutting edge and massive forming will now be more closely examined. Tool, workpiece, process materials and energy are regarded as process input variables. The work process, or rather its design results on one hand in a product which demonstrates the characteristics required in terms of quality, economic efficiency and, increasingly, recyclability. On the other hand, unwanted residual materials and emissions are also produced. Their characteristics in terms of type, quantity and state are influenced by the process.

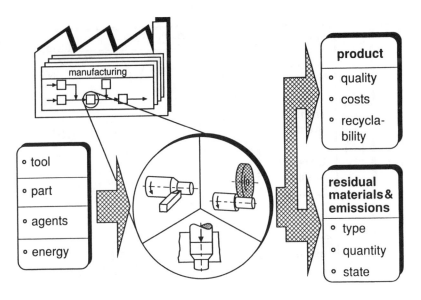

Fig. 15.3 The manufacturing process as a scope of balance

After defining the scope of balance under review, the manufacturing process must be analysed as a whole, rather than in terms of individual operations. In general it can be said that basic operations are carried out and mechanisms take effect, describing the relationship between tool and workpiece. The influence they exert on the process result and the process environment must also be taken into account.

The discussion of the physical, chemical and biological operations, interrelationships and their actual functions forms the basis for any understanding of the process (Fig. 15.4).
– physical reactions: (variables): speed, acceleration, temperature, pressure, force, structural transformation, diffusion...
– chemical reactions: corrosion, oxidation, pyrolisis, ...
– biological reactions: formation of bacteria

Various reactions during the machining process can cause a number of residual materials and emissions. Tools frequently shed used material or material which has become unsuitable for further use and can also result in emissions. Other residual material may contain chips or sludge, spent cooling lubricant and newly formed chemicals or reaction products, worn workpiece material and process material. Emissions include undesired

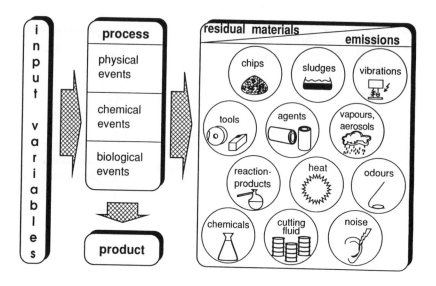

Fig. 15.4 The process influences residual materials and emissions

smoke, mist and vapours in addition to noise, heat, odours and vibrations.

The greatest problems arise nowadays as a result of the deployment of cooling lubricants. One problem is the machine workload. The handling and disposal of cooling lubricants is increasingly giving rise to problems which contradict the principles of economically and ecologically viable production..

The vast majority of manufacturing processes exploit the good cooling and lubricating characteristics of fluids such as oil, emulsion and solutions. The approach so often adopted in the past of using the various types of cooling lubricants unhesitatingly and, in many cases without any sense of responsibility is now coming in for growing criticism. Large industrial companies which have developed and applied environment-friendly concepts for the deployment of cooling lubricants tend to lead the way. Small and medium sized companies are frequently far from any such step. Rising costs for waste disposal are however now forcing all companies to implement cooling lubricant strategies adapted to suit their own manufacturing structure.

The pie charts in Fig. 15.5 showing the amount of cooling lubricant used annually in Germany illustrate the high potential for improvement. In 1991, the total quantity of all water-miscible and water-immiscible cooling

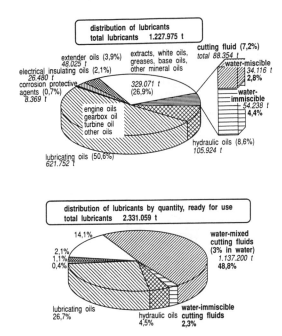

Fig. 15.5 Distribution of deployment of cooling lubricants

lubricants was almost 90,000 t. The consumption of approx. 34,000 t of water-miscible cooling lubricant is fairly low in comparison with the total consumption of cooling lubricant. Mixed with water as shown in the lower part of the diagram, the 1.1 m tons represent an enormous amount which is very significant in terms of work safety and waste disposal.

15.3 Problems related to cooling lubricants

The fluid used in metal working operations is viewed now more critically than ever before. Efforts to increase the performance level of cooling lubricants are now required to take equal account of the demands for environment and work friendly operating conditions expressed by society and legislator as well as by trade associations and federations.

Various problems are connected with the use of cooling lubricants. Attention focuses on the ecological and economic aspects of waste disposal and on contact with it at the place of work (Fig. 15.6). The first part of the diagram shows that most of the spent cooling lubricant, which

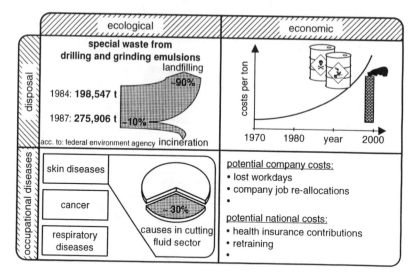

Fig. 15.6 Ecological and economic problems surrounding the use of cooling lubricants

has been designated as hazardous waste, is simply dumped and that due to the high costs of incinerating such waste, only a small quantity of it is in fact subsequently incinerated. There is, however, only limited dumping capacity. Both means of disposal are leading to an increasing quantity of hazardous waste and are using up raw material. The economic pressure imposed on companies as a result of disposal is increasing progressively and can already reach up to 20 % of the part costs [5].

The increasing incidence of occupational diseases is putting further pressure on both company and national budgets [6].

15.3.1 Uses and characteristics of cooling lubricants

Cooling lubricants have two principle uses in cutting and forming operations: cooling and lubrication. The priority regarding their functions and the demands they are required to meet varies widely depending on the operation concerned and determines the characteristics they are required to demonstrate.

The grinding process generates, for example, considerable heat. This is caused by the undefined cutting conditions prevailing around the grinding grain and the high relative speed between grinding wheel and workpiece. Much of the volume to be removed from the workpiece is not cut; it is, rather, displaced by the usually extremely negative rake angle and subjected to plastic deformation. The energy generated by this deformation enters the workpiece and tool as heat and therefore demands effective cooling from the cooling lubricant. Efforts must be made at the process design stage to prevent the generation of heat by reducing friction between grinding grain and workpiece. This calls for effective lubrication. Whether the required effect in grinding processes is lubrication or cooling depends on the grinding method and the process parameters selected.

When cutting with a geometrically defined cutting edge, the priorities are different. Here the chip, or rather the workpiece slides over a larger tool contact area at much lower relative speeds. While grinding is carried out at speeds of 30 to 250 m/s, speeds reached in turning operations, for example, are lower by a factor of 50 to 100. In cutting processes with geometrically defined cutting edge, up to 80 % of the heat is dissipated via the chips, 10 % to 20 % is absorbed by the tool and approx. 5 % to 10 % by the workpiece [7]. The primary function of the cooling lubricant is to cool tool and workpiece at high cutting speeds and cutting temperatures, and to reduce adhesion and abrasion at low cutting speeds by providing lubrication.

Cooling lubricants must meet further requirements in cutting operations which are necessary either for the process itself or are indispensable if machining is to be economically viable (Fig. 15. 7). Cooling lubricant must be regarded as a medium which interacts with machine, workpiece and tool. It must, therefore, demonstrate a number of different characteristics enabling it to satisfy a wide variety of requirements.

As shown in Fig. 15.7, the characteristics required for a given operation are determined largely by technological demands taking account of environmental and handling factors and industrial law.

Forming operations lie within a different range in terms of relative speed and contact zone than cutting operations. In cold and hot forming operations, temperatures of up to 300° C and 1200° C respectively are reached. Due to the generally low relative speed, the heat generated in the workpiece does not present any difficulties. The tool may require cooling only when a high-speed press is being used or in hot forming operations. In order to lower the forming forces and tool load and to prevent bonding, lubricants must frequently be used to reduce friction.

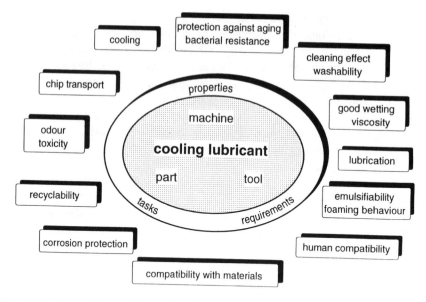

Fig. 15. 7 The task in hand determines the characteristics of cooling lubricant

15.3.2 Composition

According to German Standard (DIN) 51 385 [8], cooling lubricants are classified as being water-immiscible, water-miscible or water mixed cooling lubricants. Water-miscible cooling lubricants in the state in which they are applied are termed water-mixed cooling lubricants (Fig. 15.8).

In accordance with the special requirements of the various methods applied and of the cutting and workpiece materials used, cooling lubricants have a wide variety of compositions. Additionally, their composition and thus the characteristics of these fluids change during use. As regards material composition, a distinction must be drawn between primary materials present in the cooling lubricant on delivery and secondary materials which form during use or which are carried over into the lubricant during use.

In the state in which cooling lubricant is delivered and in its state when ready to use, it consists of base materials and additives. Base or primary materials are natural hydrocarbons (solvent raffinates, hydrocracking oil) synthetic hydrocarbons (poly-alfa-olefins), synthetic esters (Bi, Tri and Tetra ester), vegetable esters (from vegetable oils such as rape seed oil, sunflower oil), polyglycols (e.g. water-miscible or oil-soluble polymers or

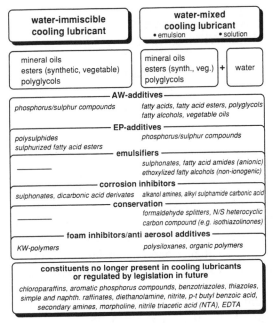

Fig. 15.8 Composition of cooling lubricants for cutting operations

compounds containing these materials. Water is added to water-mixed cooling lubricants as an additional basic constituent [9 -13].

Primary oils based on mineral oils are the main constituent parts of the water-immiscible cooling lubricants and a major constituent of the water-miscible cooling lubricants. They consist of a number of hydrocarbons whose parent substance can have paraffin, naphthenic or aromatic structure [10 - 12]. Solvent raffinates are in general extracted and deparaffinized mineral oils. These base oils are characterized by high ageing stability [14] as well as by a relatively low aromatic hydrocarbon content. In addition to the solvent raffinates, hydrocrack oils and synthetically manufactured oils such as poly-alfa-olefines or synthetic ester are increasing steadily in importance. Hydrocrack oils which are manufactured using the catalytic hydrocrack method from which they take their name likewise have a low proportion of aromatic hydrocarbons. The aromatic hydrocarbon content has a positive effect on the capacity of the base oils. However their use should be minimized due to their irritant effect on human skin and should not exceed 5 % [13]. Synthetically manufactured base oils generally have a higher viscosity index than conventional base oils. They are also less liable to evaporate, have longer service life and contribute to enhanced work hygiene [13]. In recent years, vegetable and synthetic ester and polyglycols have established themselves as environment-friendly, biodegradable base fluids.

Water is used in cooling lubricant emulsions and cooling lubricant solutions. The stability of emulsions depends to a considerable extent on the quality of the water [11]. The nitrate content of the water must not exceed 50 mg/l [12, 15].

In order to adapt the cooling lubricants as well as possible to the use for which they are required, agents are added which change both the physical and the chemical characteristics of the base fluid. Groups of substances added to cooling lubricants are: antiwear additives which form a lubricating film (AW additives), high pressure additives, so-called EP additives (Extreme Pressure), anticorrosion additives, foam retarding agents, antifog materials, dispersing agents and surface-active substances. Emulsifying agents, solutizing agents, biocides, odorous substances and colouring agents can also be added to water-miscible cooling lubricants [10].

Antiwear additives, so-called physical additives improve the wettability and adhesive capacity of the mineral oils on the surface of the metal. Fatty acids, vegetable oils and synthetic colouring agents such as carbon acid esters belong to this group. They react at room temperature with the metal surface and form a lubricating film of metallic soap [12 - 16].

EP additives are chemically active high pressure additives which form protective films with high levels of pressure and temperature resistance. Chemical compounds containing phosphorous and sulphur are used as EP additives. The effect of the EP additives is attributable to the formation of chemical reaction layers in the limit friction range on solid state contact.

Cooling lubricants containing chloroparaffins as EP additives should no longer be used. Additives containing chlorine pose a considerable threat to human life and to the environment. Contact with cooling lubricants containing chlorine can lead to chlorine acne. The incineration of cooling lubricants containing chlorine at insufficiently high temperatures can generate dangerous pyrolyzates. Cooling lubricants with EP additives containing chlorine are difficult to dispose of. The cooling lubricant is therefore classified after use as hazardous waste. It is, therefore, recommended that chlorinated substances should not be added to cooling lubricants [12].

For water-miscible cooling lubricants, emulsifying agents are the most important group of additives in terms of function and quantity. They are used to form stable emulsions. Emulsifying agents are surface-active

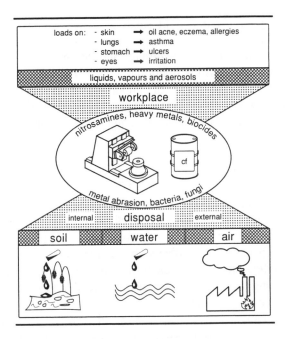

Fig. 15.9 Environmental impact caused by cooling lubricant

bipolar substances. They reduce the surface tension of the oil droplets suspended in the water and prevent these from coalescing and rising to the water surface to form a layer of oil there. Combinations of anionic and unionized emulsifying agents are generally used in cooling lubricant emulsions [16, 17].

A number of chemical compounds of which the most important are alkanolamines and carbonic acids and their salts are used as corrosion retardants. The effect of these additives is due to a chemical reaction with the metallic surfaces [17]. Sodium nitrite used to be the most important corrosion retardant. Nitrite compounds may however no longer be added to cooling lubricants because of the inherent risk that nitrosamines could form [12].

The growth of bacteria and fungi in emulsion is kept under control by the use of biocides, A number of preserving agents are available, some of which are already contained in the cooling lubricant and some of which can be added to the emulsion if required. Formaldehyde splitters and N/S heterocyclic compounds are characteristic of this group of substances.

Biocides are generally added in small amounts. Some of them can be highly toxic [11, 12, 17].

The quality of the base materials and of the substances added to enhance their characteristics is normally such that they contain traces of other materials in the form of impurities. The type, quantity and composition of these foreign substances depends on the manufacturing process. Specifications are obtainable only from the manufacturer [11, 12].

15.3.3 Ecological Aspects

The use of cooling lubricants is potentially dangerous from two points of view. Humans are involved directly at the place of work and indirectly via the disposal of the different forms of cooling lubricant (Fig. 15.9).

Machine operators are affected by contact with various substances within the cooling lubricant. Their health can be impaired by skin contact or by swallowing the substance but also by breathing it in or by the irritant effect of oil mist and vapours.

Measures aimed at reducing the health risk at work are machine encapsulation and extraction systems as well as the use of low evaporation cooling lubricants. Skin disease caused by cooling lubricants can be largely avoided by thorough skin cleansing and care [15].

The cooling lubricants can be disposed of after use either within the company or externally through a waste disposal company. Depending on the quality and mode of disposal, the result is ground, water and / or air pollution.

Cooling lubricants change during use in machining. When new, cooling lubricant represents only a slight risk. This changes in the course of its service life. Secondary substances which form or find their way into cooling and lubricating fluids during use include reaction products, foreign bodies and micro-organisms. In principle, a variety of different compounds are conceivable, particularly in view of the contact of the cooling lubricant with catalytically effective metals at increased temperature. A considerable number of these products occur in such low concentrations that they are not detectable. The reaction products most frequently associated with cooling lubricants are: nitrosamines, polycyclical aromatic hydrocarbons, decomposition products of the mineral oils, breakdown products (usually acids) of the microorganisms, metals, metal oxides and metal salts [11].

Even the use of nitrite-free water-miscible cooling lubricants cannot guarantee that nitrite compounds will not occur in the cooling lubricant or

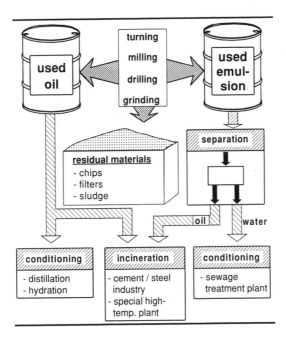

Fig. 15.10 Disposal of cooling lubricant in cutting technology

will find their way into the cooling lubricant circulation. Nitrite can form as a result of microbial metabolic reaction of nitrate. When alkanolamines are used as corrosion-retardants, the formation or introduction of nitrites can trigger a reaction to nitrosamines. Nitrites or nitrosating agents can get into the cooling lubricant via service water, nitrogen oxide from combustion engines (gas or diesel driven fork-lift trucks, cigarettes or other tobacco smoke), tempering salt or antirust oil containing nitrite (11, 12, 15).

Substances which find their way into the cooling lubricant while it is in use are regarded as extraneous material. A distinction can be drawn between unintentional introduction (lubricants or hydraulic fluids, metal dust, corrosion retardant washed from the surface of the workpiece to be machined, cleaning fluid, airborne substances from other sources of emissions) and substances which are added deliberately (concentrated cooling lubricant, water, additives to make up for loss, preserving agents, system cleaning agents).

Microorganisms prefer water-mixed cooling lubricants. At temperatures of 20 - 40°C, which are ideal for most microorganisms, emulsions offer good conditions for growth as the microorganisms can

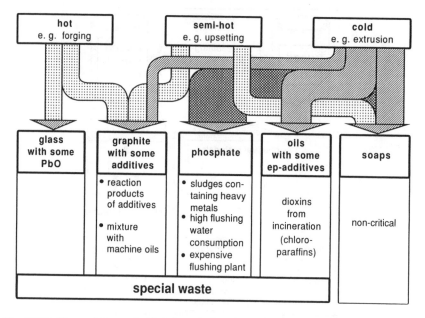

Fig. 15.11 Use and disposal of lubricant in forming technology

utilize all constituent parts of the cooling lubricant which are biologically useful. While the cooling lubricant concentrates have a relatively low nuclei formation rate, microorganisms and nutrients which enter via workpiece, hands, air and impurities (dregs of drinks, cigarette ends etc) during operation, result in increasingly rapid production of microorganisms. Bacteria and fungi are the most common culprits [11].

Clearly, the material composition of the cooling lubricant is extremely complex. The users of cooling lubricants however often find it impossible to obtain information about the composition of the products used and about any health-impairing characteristics. Synoptical tables or safety data sheets and product specifications are generally of little help to users striving to fulfil their obligations with regard to the German Dangerous Substances Directive (duty to determine contents). The information required can only be obtained from the manufacturer [11].

Specialist disposal of waste oil and emulsions is becoming increasingly more important. There are now various options open to companies interested in conditioning these materials, regardless of whether this is conducted within the company or externally (Fig. 15.10).

With sufficient care, water-immiscible cooling lubricants need not be replaced. The refill quantity necessitated by loss via chips and workpieces is sufficient to ensure that the qualities are largely maintained. Sometimes, however, replacement is essential when, for example there is an unavoidable increase in extraneous material (chips, water), substantial quantities of extraneous oils (machine or hydraulic oil) or decomposition products as a result of aging. Depending on the state it is in, used oil is either taken to be reprocessed or incinerated.

Water-mixed cooling lubricants cause much greater problems. Here the emulsion in the machine must be changed regularly which means that there is an enormous amount of fluid to be disposed of. In order to recycle the water, the water and oil phases are first separated. It is advisable to use ultrafiltration as other methods do not, as a rule, reach the boundary limiting values now set. Once the emulsion has been pre-cleaned, the cooling lubricant is fed under pressure past semipermeable filter areas in the ultra filtration operation. Water, dissolved salt and smaller organic molecules penetrate the membrane and form the permeate (filtrate) while molecular materials with larger molecules, such as hydrocarbons, are retained and enriched (residue).

The permeate is passed on for further reprocessing, to the sewage treatment plant, for example in order to separate heavy metals, nitrites, cyanide, chromate and other dangerous substances using thermal and chemical methods. The residue, i.e. the oil phase is subsequently incinerated.

Various types of residue, the majority of which consists of chips, filters and sludge are left after cooling lubricants have been conditioned. An investigation should be carried out to determine whether these substances can be de-oiled or desiccated in a reconditioning operation prior to disposal. The authorities responsible must be notified of the composition and, if necessary, a reconditioning operation must be undertaken. Then, provided the authorities and management of a waste incineration plant or dump agree, it can be incinerated or dumped. Otherwise, the regulations governing waste disposal must be observed.

The diversity of forming operations with their wide range of differences in terms of temperature, surface topography and pressure in the gap between tool and workpiece has resulted in the development of a large number of lubricants. Fig. 15.11 shows a classification of the lubricants used depending on the operation in hand.

In cold forming operations such as extrusion, the workpieces are generally phosphated (exception: rolling etc). The resultant sludge, some of

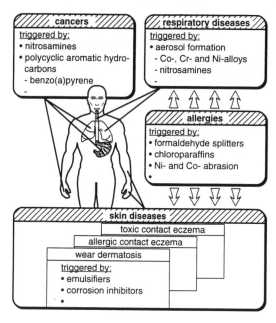

Fig. 15.12 Potential risk of health impairment

which contains heavy metals is classified as hazardous waste. The considerable consumption of flushing water is a further ecological problem in spite of the sophisticated machinery. In this application, oil is used as a lubricant as well as the relatively innocuous soaps and graphite. Due to the development of dioxins, oils containing EP additives pose a major threat to the environment when they burn. For this reason, they are seldom used.

In semi-hot and hot forming operations such as forging, graphite, usually with water is the most widely-used lubricating and/or cooling medium. Graphite itself is relatively innocuous. The additives however can undergo a chemical transformation at high process temperatures. Due to the unavoidable mixture with the lubricants in the machine, the reaction products are insufficiently documented. An additional, less widely used lubricant is glass powder. When PbO is added to the glass, the resultant waste produced requires special disposal.

Daily contact with lubricants involves a number of health risks which can arise from individual constituent parts or from impurities introduced either in the course of the machining operation or which find their way into the lubrication system. Contaminants can affect the body either through direct skin contact, through the respiratory system or mouth. This is intensified when lubricant spraying and turbulence results in the formation of aerosol spray. The type and extent of the risk posed to human health can only be estimated, such is the complexity of the interaction system. A further factor increases the difficulty of assessing the danger; personal variables such as the sensitivity of the individuals concerned, whether or not they are smokers etc.

Among the recognized occupational diseases, skin diseases (occupational dermatosis), respiratory illness and cancer are those which occur most frequently in connection with work with cooling lubricants. The triggers for these diseases can be individual components or reaction products and other substances which have found their way into the system (Fig. 15.12).

In Germany, the diseases which crop up most frequently in association with work with cooling lubricants are skin diseases. In some factories in the metal industry, up to 20 % of the work force is affected at any one time [18]. The increase reported in recent years is particularly high. In 1980, trade associations reported just under 11,000 cases of serious or chronic skin disease, by 1990 the number of reported cases had jumped to 18,750. Surveys undertaken by the trade associations of the mechanical engineering and small iron industry show that approx. one third of these skin diseases stem from work with cooling lubricants, with the groups lathe operator, grinder, miller and driller being particularly hard hit [12].

A distinction must be drawn between different types of skin disease, some of which occur together but are attributable to different causes [6, 12, 19].

One such disease is toxic contact eczema which is caused by instantaneous irritants (alkaline compounds etc) and which affects all humans. These illnesses are generally attributable to careless handling of cooling lubricants, particularly emulsions and are avoidable provided the substances are treated with care and appropriate safety measures are taken.

Wear dermatoses are the most frequent skin complaints. They occur mainly following long-term direct skin contact with water-miscible cooling lubricants. They are caused by the emulsifying, anti-rust flushing and washing agents which are added and which attack the skin's natural protection. Direct contact with oil can also cause lasting damage by making the skin excessively dry.

Allergic contact eczema is caused by a special reaction of the cellular immune response to allergen. The proportion of reported cases of this serious form of occupational skin disease is growing steadily in relation to other skin disease. The reasons for this are germicidal agents such as formaldehyde splitters or rust retardants containing amines.

All of these skin diseases can, however be minimized by exercising care and by avoiding use of certain additives.

The greatest problem facing industrial medicine in recent years has been the formation of nitrosamines in cooling lubricant [12, 19, 20]. Nitrosamines are carcinogenic substances and can be absorbed via skin and respiratory tract. They develop from nitrite and amines. The nitrite does not even have to be contained in the cooling lubricant; it can find its way there directly or as a nitrate. Bacterial impurities then convert the latter into nitrite. In contrast, amines are basic constituent parts of most cooling lubricants. A distinction is drawn between primary, secondary and tertiary amines. Only the stable nitrosatable secondary amines are involved in the development of nitrosamines. Simply ceasing to use secondary amines as has already been done is, however, no guarantee that nitrosamines will not form. Like nitrite, amine can be introduced into the circulation as an impurity from outside.

The polycyclic aromatic hydrocarbons (the most important of which is benzo(a)pyrene) are also designated as carcinogenic. The use of highly refined mineral oils has reduced considerably the proportion of benzo(a)pyrenes in water-miscible cooling lubricants. However, additional PAHs can form in water immiscible cooling lubricants particularly after a long service life.

Chlorinated hydrocarbons, contained as preserving agents and EP additives in cooling lubricants are a major health hazard. Chloroparaffins in particular are regarded as highly carcinogenic. These are no longer present in most recently developed cooling lubricants.

Some substances present in operating emulsion are liable to cause obstructive diseases of the respiratory tract. There is, however, very little documentary evidence concerning the long-term effects of the absorption of vapours and aerosol gases by the lung. It can be assumed that the toxic effect of inhalation of mists and vapours on the whole organism has previously been underestimated [18, 21].

The proportion of court cases in which compensation has been awarded is, at 3 %, relatively low. This must not be allowed to blind companies to the considerable expense incurred as a result of each individual case of skin disease occurring among the work force. Besides absenteeism, it must be remembered that moving employees around so that those affected can occupy more suitable work places can be both inconvenient and costly in several ways [6, 12].

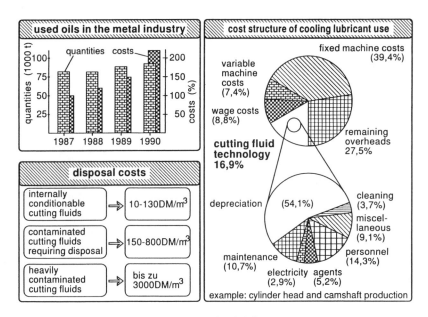

Fig. 15.13 Costs involved in the use of cooling lubricants

15.3.4 Legislation and directives

Legislators have already reacted to the discovery that cooling lubricants can represent a serious danger to health. Nitrosamines, for example have been classified as carcinogenic industrial substances and in 1992 a limit TRK value was set. This means that cooling lubricants are now covered by the German Dangerous Materials Act. As a result, work places must now be monitored regularly to ensure that the amount of nitrosamines in the air does not exceed the permitted level [10].

Further directives which can be implemented only after a great deal of reorganization are expected to take effect in 1993, thus increasing the strain on the already overstretched budgets of the companies concerned.

Examples of the changes mooted are weekly measurements of the nitrite concentration in cooling lubricant in conjunction with the obligation to replace the cooling lubricant when the limit concentration of 20 ppm nitrite has been reached. At the present state of development in cooling lubricant technology, nitrite concentrations of 20 ppm can be reached at any time. As the reasons for this are yet to be discovered, there is no means of effectively preventing the development of nitrite.

Companies working decentrally with machines with only one operator will, in future, be required to provide additional personnel to monitor the cooling lubricant. Added to this will be a number of preventive medical check-ups to be carried out within the scope of industrial medicine for the personnel at the machine. This trend towards tighter legislation is expected to become even more pronounced in the future. Cooling lubricant will thus become an increasingly important factor in terms of the economic efficiency of manufacturing plants.

These directives place responsibility for and obligations towards the work force firmly in the hands of the employer who is required, under the terms of the Dangerous Materials Act, to identify materials which may pose a threat to the work force and to investigate the possibility of using substances or compounds which may be less dangerous. The employers are also required to carry out investigations in order to determine whether or not the limit values for dangerous substances are being observed and, if required, to ensure that safety measures are implemented. Official instructions must also be issued regularly regarding the safe handling of dangerous substances and informing staff of the risks posed by the substances and compounds.

15.3.5 Economic aspects

The obligations imposed by legislation and directives are forcing companies to re-think the financial aspects of the use of cooling lubricant. As illustrated in Fig. 15.13, whereas the quantity of used oil has risen only slightly, the financial cost of disposing of used oil and emulsion has doubled rapidly.

When reconditioning is carried out within the company, costs fall to DM 130 per m3. In the case of heavily contaminated cooling lubricants which have to be designated as waste requiring special disposal and incinerated, this figure rises to as much as DM 3000 per m3.

The right-hand side of the diagram provides a detailed breakdown of the cost structure within companies. In the example of the manufacture of serially produced parts for cars, the cylinder head and the camshaft, costs connected with cooling lubricant technology account for almost 17 %. They are much higher than the total tool costs (approx. 3 %). A detailed view of this share shows that as well as the high depreciation level, personnel and maintenance costs are included at 25 %.

The example illustrates the enormous financial reserves which will be required to fund work with cooling lubricants. Against this background and

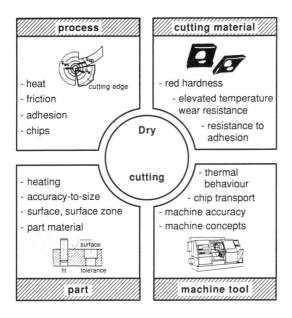

Fig. 15.14 Dry machining - requirements and boundary conditions

taking account of all the risks they pose to the environment, existing manufacturing structures will have to be re-thought and improved.

15.4 Approaches to reducing environmental pollution

Nowadays, industrial safety and environmental protection rank alongside the technically optimized manufacturing process. There is a wish to see harmonization of ecology and economy within economically viable production. For the metal working industry this means using cooling lubricant in technologically sound applications while, at the same time, implementing and refining maintenance and recycling concepts.

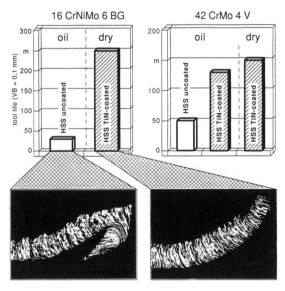

cutting conditions: v_c = 10 m/min, h_{cu} = 0,1 mm

Fig. 15.15 Dry broaching using coated HSS

15.4.1 Avoidance of cooling lubricants

The most logical step which can be taken to avoid the problems associated with the use of cooling lubricants is dry machining. A number of machining operations currently use vast quantities of cooling lubricant which is not necessary from a technological point of view. The question must be raised as to whether the use of cooling lubricant is essential for existing or future machining operations. In order to answer this question, an analysis of the prevailing boundary conditions is required in conjunction with detailed knowledge of the complex interactions linking the process, cutting material, part and machine tools with one another (Fig. 15.14). On this basis, conclusions can be drawn as to measures and approaches which can be adopted in order to implement dry machining.

Dry machining operations do not involve any of the use of cooling lubricants. When designing a process without cooling lubricants, measures must be adopted which permit functions normally undertaken by cooling lubricant to be assumed by some other substance.

The factors which exert most influence on tool wear are adhesion and abrasion at low cutting speeds and diffusion and oxidation at high cutting

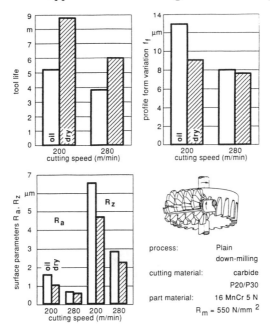

Fig. 15.16 Comparative performances for dry and wet plain milling

speed and, therefore, high cutting temperatures. Consequently, cutting material must demonstrate little adhesive tendency to the material of the workpiece as well as high levels of hardness and wear resistance at elevated temperatures. These requirements are satisfied to differing degrees by the cutting materials already available [8].

Coated tools are examples of such cutting materials. With their high levels of hardness and chemical stability, a number of cemented carbides meet all requirements even at high cutting speeds. The variety of characteristics of the cemented carbides also permit the development of layers whose characteristics can be adapted to suit the demands imposed by dry machining [22].

Tools which have been coated permit dry machining to be extended to areas in which cooling lubricants are generally regarded as essential. An impressive example of the efficiency of a PVD-coated HSS tool in dry machining operations is broaching, carried out on soft and tempered steel materials (Fig. 15.15). The generally low cutting speeds of 1 - 25 m/min are characteristic of this broaching process. The formation of build-up on the cutting edge is characteristic of the machining of steel materials at

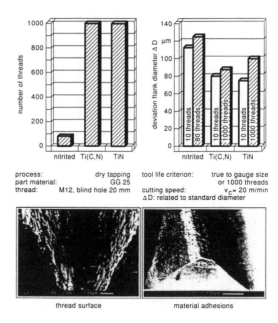

process: dry tapping
part material: GG 25
thread: M12, blind hole 20 mm

tool life criterion: true to gauge size
or 1000 threads
cutting speed: v_c = 20 m/min
ΔD: related to standard diameter

thread surface   material adhesions

Fig. 15.17 Dry tapping

these low cutting speeds. Broaching oil is generally used to reduce this.

As the chip longitudinal cross section in Fig. 15.15 shows, the formation of build-up on the cutting edge is almost completely suppressed by the use of coated HSS tools due to the reduced adhesion between the receding chip and the cutting material. The reduction in build-up on the cutting edge represents a significant improvement in terms of both surface quality and tool life. As comparison with a broaching operation conducted on tempered 42 CrMo 4 shows, this considerable improvement in performance is maintained in dry cutting operations due to coating [23, 24]. Care must, however, be taken to ensure that chips can be removed safely from the chip space of the broaching tools under dry cutting conditions.

The enhanced performance achieved in dry cutting operations as compared with wet cutting in gear hobbing with uncoated metal carbide also is attributable to a different mechanism (Fig. 15.16). The greater thermal shock effect results in micro-spalling in carbides and thus to more rapid wear. In dry cutting operations, longer tool life can be achieved. Dry machining also has advantages in terms of tooth quality. Both surface characteristics and profile form deviation are better than they are when cooling lubricant is used. Similarly positive results have been observed in gear shaping operations [25].

In general, it can be said that the cutting materials available meet the requirements for dry machining. This applies to use at high cutting speeds which demand the elevated temperature hardness and wear resistance demonstrated by coated or uncoated carbides, cermets, cutting ceramics or CBN as well as at low cutting speeds which rely on the low adhesive tendency of hard coating on HSS or carbide tools.

Casting materials, steel, aluminium alloys and non-ferrous heavy metal can be dry turned and milled. By virtue of their cutting temperatures which are considerably lower that those of steel, materials for casting are particularly suitable. An additional reason is that in cutting operations using emulsion, the casting dust tends to mix with the emulsion forming sludge which, in conjunction with chips, results in deposits in the working area of a machine and in the cooling lubricant system and can cause corrosive damage there. For dry machining operations, the machine must be encapsulated and must have an extraction system in order to protect the work force from casting dust.

While the materials listed rarely require cooling lubricant in turning and milling processes, conditions in drilling, reaming, tapping and thread milling are generally more difficult. In drilling operations, heat occurs not only at the cutting point but also as a result of friction between the heel of the twist drill against the bore-wall. Additional heat is generated in the tool and workpiece via the chips. In tapping operations, the tools are subjected to very high mechanical and thermal stress by crushing, friction and adhesion. The option of dry machining is very limited in these operations.

As the examples of broaching and gear hobbing show, developments in the field of cutting materials are contributing to the avoidance of cooling lubricants even in the case of cutting operations which are regarded as extremely difficult due to complex tool geometry and / or process kinematics when process control is adapted accordingly. This is made possible above all by the use of coated HSS or carbide tools. Machining task, workpiece material, tool and cutting conditions must, however be adapted to suit one another.

One example of this is tapping in grey cast iron without cooling lubricant (Fig. 15.17). In the machining case presented, 1000 threads were manufactured using TiN and Ti(C,N) coated taps without reaching the tool performance limit. In comparison, using an uncoated nitrited tool under the same cutting conditions, only 80 threads were manufactured before reaching the tool performance limit. Material adhesion to the tool flank is characteristic of thread cutting without cooling lubricant. The high performance of the coated tools is attributable to the lower adhesion levels between the carbide layer and workpiece material, which in the case of uncoated tools results in heavier wear and a significantly larger flank diameter.

One topic of interest which is closely related to the subject of 'phasing out cooling lubricants' is the substitution of grinding by machining processes with geometrically defined cutting edges. While the use of cooling lubricant is essential in almost all grinding operations, parts made of hardened steel can also be dry machined in turning operations using tools made of cutting ceramic or CBN. Due to their high level of elevated temperature hardness and chemical stability, these cutting materials do not require to be cooled. As regards process consistency and the surface qualities achievable, cooling proves disadvantageous in the case of these brittle cutting materials as temperature stress can result in micro-spalling and cutting edge chipping. The use of cooling lubricants is also associated with shorter tool life and less favourable process results as a result of adverse influence exerted on the surface zone.

Turning hardened parts in a dry cutting mode permits not only the cooling lubricant and energy consumption required by grinding to be dispensed with but, more importantly the disposal of additional grinding sludge. This substitution is a very far-reaching step. As a rule, it necessitates or permits the reorganization of the entire previous manufacturing sequence. This, in turn makes way for economies in terms of energy and raw materials, extending far beyond the savings which stand to be made by dispensing with the use of cooling lubricant [26, 27].

An important aspect which arises in connection with dry machining is the consistent adherence to part tolerances. Particularly exacting dimensional and shape tolerances can represent a severe restriction for dry machining and can necessitate special measures.

The mechanical energy introduced into the cutting process is transformed almost entirely into heat. While most of the machining heat is absorbed by the cooling lubricant and carried off in wet machining operations, dry machining results in tool, workpiece and machine tool all being subjected to a higher level of thermal stress. The consequence can be dimensional and form deviation of the part. The design of a dry machining process must take adequate account of this aspect.

The level of part accuracy achieved under dry cutting conditions depends primarily on the quantity of heat entering the part and the geometric part dimensions. It is essential to design the cutting process in such a way as to minimize the amount of heat being transferred to the workpiece. This can be achieved by minimizing cutting and by exerting influence on the heat distribution. There are several approaches to this. Cutting can, for example, be minimized by selecting the largest feeds and positive cutting edge geometries possible while the heat distribution can be influenced beneficially by selecting high cutting speeds. The volume of material to be removed is, however, the most significant factor.

In general it can be said that a dry machining operation is always possible when the part is not required to meet any exacting standards of accuracy. This also applies to pre-machining of workpieces which must undergo a further machining stage before they reach their final contour and level of accuracy.

Parts which are finish-machined in a dry operation should have as small and uniform as possible over-measures which can be removed in one cut. Near-Net-Shape-Parts, whose form is already close to the final contour required after casting, forging or extrusion operations are ideal.

High levels of part accuracy can be achieved when hardened workpieces are turned in a dry cutting operation. The reasons for this are the very small part overdimensions to be removed and the heat distribution which is better than in grinding operations. While 60 - 80 % of the machining heat flows into the workpiece, this proportion is only approximately 20 % in hard turning operations. The reduced machining time, which is lower by a factor of 10 to 20 than in grinding operations, means that the quantity of heat flowing into the workpiece is only 1 to 2 % of that in grinding operations [28 - 30]. This illustrates the necessity of cooling during grinding operations and, at the same time, focuses attention on the opportunity to carry out dry hard turning operations. This has been confirmed in laboratory trials. When finish turning operations were carried out on the inner and outer surface of cylindrical rings (outer diameter 130 mm) without cooling lubricant, after a total machining time of 9 minutes, the part temperature was found to have risen by only 4 degrees. The specifications of the IT 5 quality required had been met.

A further factor which exerts influence on the accuracy of the parts is the behaviour of the machine tool when no cooling lubricant is used. In addition to flushing away chips and cleaning guide elements, cooling lubricant also moderates the temperature of the machine components, thus ensuring that the conditions required for the manufacture of precise parts are in place. These functions are not performed in dry machining operations. The heat passed to the machine via workpiece, tool and hot chips can result in individual machine components heating and thus lead to

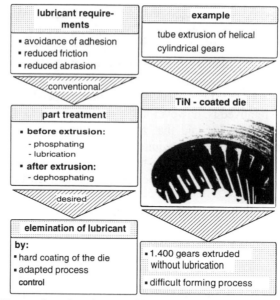

Fig. 15.18 Cold extrusion of steel gears without lubrication

form and dimensional error. Special measures are needed to ensure that the hot chips are removed quickly and reliably from the machining area, that the heat introduced into the machine is compensated and that the spindle and attachments are cooled. This presents a challenge to machine tool manufacturers to develop a machine concept adapted to suit the special needs of dry machining. Users intending to invest in a new machine should add dry machining capability to the specifications which must be met by the manufacturer.

The fact that in practice, dry machining has not realized its potential significance in view of the efficiency of the cutting materials available, is due to a number of factors. One of these is certainly that in many companies a very wide range of parts and materials are machined on the machines available. Another is that the cooling lubricant in most machine tools can be used regardless of the material, cutting material and machining method. Although cooling lubricant is in many cases, not technically necessary and in interrupted cutting has an adverse effect, it is frequently useful for secondary functions such as washing out the chips. When this is the case, it is not possible to implement a consistent policy of dry machining.

Economic and ecological gains can be made in connection with dry

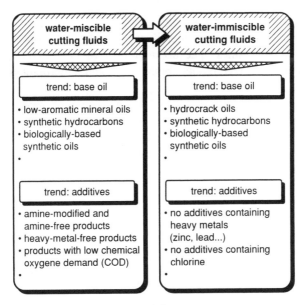

Fig. 15.19 Trends in the field of cooling lubricants

machining only when the appropriate machine tools are never filled with any cooling lubricant at all. This, however means that the necessary conditions in terms both of machine tools and process planning must be in place. A start might be to divide the manufacturing process into dry and wet machining. In concrete terms this would mean ensuring that parts which can be dry-machined were machined only on machines specifically for this purpose and parts or machining steps requiring cooling lubricant would be performed on machines designated as being for wet-machining.

As has been demonstrated, there are a number of ways of introducing dry machining into companies in which it will be impossible to do without cooling lubricants completely, due to the wide range of conditions imposed by the workpiece materials, operation and part. The phrase 'wet machining where necessary, dry machining where possible' may express a possible compromise between technological and economic considerations.

In the course of the development of new coating materials and methods, the question presents itself as to whether forming manufacture might also be possible without cooling lubricant. To date, there have been no clear findings although the handling of phosphate acids and lubricants is becoming increasingly difficult in extrusion operations.

The high level of investment required for large-scale flushing machinery, encapsulated phosphatizing machines and specialist disposal of the sludges containing heavy metals are causing problems for small and medium-sized companies. The abandonment of lubrication altogether in forming operations can reduce both costs and environmental pollution substantially.

Laboratory tests carried out on the extrusion of gear wheels have shown that gear wheels can be cold forged without the use of lubricants (Fig. 15.18). Using the TiN-PVD-coated die, 1400 helical cylindrical gears were extruded without reaching the end of tool life in this very difficult forming process.

15.4.2 Modify cooling lubricants

It is only possible in certain cases to dispense with cooling lubricants. This means that in order to improve the manufacturing process in ecological terms, products which are quickly biologically degradable, and physiologically and ecotoxologically innocuous will require to be developed and used. There is a demand for new base oils and additives as well as modification of water- miscible and water-immiscible cooling lubricants while reducing the contents. Development must be directed at bringing about increased use of water-immiscible cooling lubricants as a substitute for emulsions and at developing environment-oriented water-immiscible and water-miscible lubricants. These could be offered as a compatible comprehensive system along with the corresponding hydraulic, gear and guide-way oils [31]. Cooling lubricants, particularly oils should continue to be removable by washing using watery cleaning systems as this would make the ecologically dire process of subsequent de-oiling using chlorinated hydrocarbons redundant.

The basic trend in the application of cooling lubricant is moving away from water-miscible towards water-immiscible products (Fig. 15. 19). A survey of only the base oils of both groups shows that attention is focusing on reducing the use of aromatic compounds in the development of superior oils [32]. A further priority is the easing of the problems connected with disposal, i.e. less risk to water, biological degradability at appropriate dilution etc. Here, for example, a compromise must be found between degradability and the economic demand for longer tool life.

The deployment of so-called hydrocrack oils (HC oils) and low-aromatic mineral oils is an attempt to take account of the PAH problems. At the same time, hydrocrack oils reduce oil mist and the tendency to evaporate.

The development of synthetic oils or oils based on `native' base substances is a more far-reaching step in this direction. Synthetic hydrocarbons, as representatives of the first group are PAH free, low emission and free of aromatic compounds.

Low viscosity synthetic water-immiscible cooling lubricants are used as an alternative to emulsions and are based not on mineral oils but on industrially manufactured fatty acid esters. This very wide term covers a wide range of base substances with different characteristics, qualities and price levels. Their components (carbonic acids and alcohol) can be mainly or partly natural in origin or completely synthetic. Their degrading rates, which range from very fast to very slow depend on their structure. Their toxic characteristics are similar to those of vegetable oil. However they exceed these by far in their application characteristics at low and high temperatures. In comparison with equiviscose mineral oils, they demonstrate much lower evaporation rates and higher flashpoints thus causing less environmental pollution and a higher level of operating safety. As there are products around with a quality superior to that of mineral oil products, increases in the quantities of lubricants and ester-based functional fluids are to be expected, not only for conservation reasons but also due to the rising requirements the machines are having to satisfy [31].

Water-immiscible cooling lubricants also require environment-friendly additives. These can, for example, be polymer additives with shearing stability for the reduction of oil mist. Demands for zinc-free base oils are growing increasingly loud. Other additives too, such as barium or chlorine likewise have no place any longer in cooling lubricants due to their eco-toxicity [33].

Going on from the base substances listed above, the development of water-miscible materials should progress according to the following principles [19, 34]:

- Nitrogen components not nitrosatable (development of nitrosamines is avoided)
- Boric acid derivative free cooling lubricant (environment and
- Tensides which pose no physiological risk (derived from the
- EP/AW agents (e.g. ester-based) which pose no physiological risk)
- Biostatics without boric acid compounds
- Lubricating effectiveness enhancers without chlorine, sulphur or phosphorous

An additional aspect to be considered with regard to additives is avoidance of the use of complexing agents and thus of water-soluble heavy

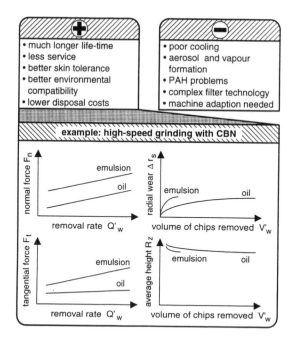

Fig. 15.20 Advantages and disadvantages of using oil instead of emulsion

metal compound in order to permit waste disposal without burdening the waste water unduly. Additionally, the number of additives used should be kept as low as possible and should be adapted to suit the needs of the application in hand.

Complete avoidance of amines of any type in emulsions, as should be demanded for ecological reasons, would entail considerable difficulty. It is likely that problems would arise in connection with stability which, in turn, would necessitate stronger preserving agents with all the attendant disadvantages. There remains, therefore, a slight risk which should be minimized as far as possible by taking care accordingly [20].

The product group of water-miscible biostable or easy-care cooling lubricants takes particular account of the microbiological problems which severely restrict the economically efficient use of water-miscible cooling lubricants. They consist of mineral oil and an emulsifying /anti-corrosion agent. These products are usually bacteriocide-free which makes them more compatible. Due to the slow bacterial growth rate they keep for much longer [35].

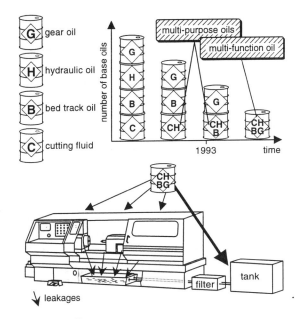

Fig. 15.21 Multifunction oil

There is a perceptible trend away from water-miscible products towards low-viscosity water-immiscible cooling lubricants. A comparison of the characteristics entered in Fig. 15.20 for water-immiscible and water-miscible cooling lubricants shows water-immiscible cooling lubricants to be much more favourable. Their higher viscosity in comparison with emulsions can be balanced out by using low viscosity cooling lubricant which still have a high flash point provided special base oils are used [33].

A major advantage of the water-immiscible products lies in their considerably longer service life. Provided they are well cared for, service life can extend over a period of years. This reduces the cost of care and maintenance as well as the quantities to be disposed of. These respond better to recycling. From the point of view of industrial medicine, the advantage that they are tolerated better by human skin is offset by the disadvantages caused by mist and vapour formation and by the problems posed by PAH. The latter, however, can be countered by imposing the appropriate demands on the base oil [34].

The change from emulsion to oil is not without problems as regards the machine. Increased viscosity and the tendency to mist and vapour formation demand that the machine and the filters be adapted to meet the requirements of the cooling lubricant system. The machining area must be encapsulated and fitted with a corresponding extraction system.

A change-over to water-immiscible cooling lubricant can be more beneficial not only for ecological reasons but also from a technological point of view as shown in the example of high performance grinding in Fig. 15.20. The use of oil instead of an emulsion results in lower machining forces and, more importantly, in a prolongation of tool life by a factor of 8.

The demand for a reduction in the materials used, is naturally resulting in the development and use of products with uniform structures which meet the specifications of all the cooling and lubricating agents required by the machine. These products are the so-called multifunction oils (Fig. 15.21). In addition to a cooling lubricant, normally, a gear oil, a hydraulic oil, and a bed track oil, all of which differ both in terms of viscosity and additives are needed. Material lost through leakage which always occurs gets into the cooling lubricant and causes impurities there. Multifunction oils which can be used both as hydraulic gear oil and cooling lubricant have been available for years. These products are suitable for cutting and forming processes under low to medium EP conditions such as turning, drilling, milling, sawing, grinding and thread grinding for steel up to medium strength, iron and nonferrous heavy metal [32, 33].

Oils which meet the requirements of cooling lubricant, bed track oil and hydraulic oil are currently being tested. It is planned to include gear oil in a subsequent development step. The advantages of a Multifunction oil lie in the reduction of the number of substances used and in the lower level of maintenance required. As shown in Fig. 15.21, loss through leakage is absorbed by the cooling lubricant without any difficulty.

A change of this nature demands that the machines be retooled accordingly. In this case, alternatives must be considered to the individual oil circulation systems. Seals, pumps and filter systems must be adapted to accommodate the changed characteristics of the media with which they come into contact. The materials used to build hydraulic systems and transmissions rule out the use of active sulphuric compounds.

On the whole, the trend regarding cooling lubricants is moving away from water-miscible products wherever this is technically possible and towards water-immiscible products with a view to eventually developing a product capable of carrying out all the cooling and lubricating functions required for the smooth running of a machine.

The refinement of the individual components is aimed at prolonging service life while at the same time reducing the ecological risk. These are exemplified by synthetic cooling lubricants containing biological base oils

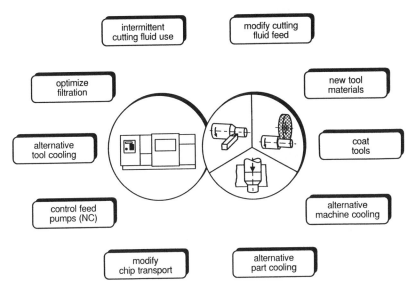

Fig. 22 Tool and machine related measures

which are, however, still four times more expensive. Efforts to optimize them are hampered by conflicting technological and ecological demands.

15.4.3 Reduce the quantity of cooling lubricant

Many machining operations rely on the use of cooling and lubricating fluids. For these cases too, efforts are already under way to improve production in ecological terms. One option is to reduce the quantity of cooling lubricant used, i.e. only the volume of cooling lubricant really necessary is supplied. Measures concentrating on the tool and machine are shown in Fig. 15.22.

In the case of the machine tool, there are various options open. Chips, frequently carried away by whatever fluid is used, can be removed by different means. Sloping bed tracks, transport systems, special chip guiding systems and alternative media such as compressed air can all be used. The machine tool temperature, which is important for form and dimensional accuracy, can be kept constant via closed cooling circulation systems, for example. In the field of filter technology, there are a number of methods of maintaining the quality of the lubricating agent at a high level for a long

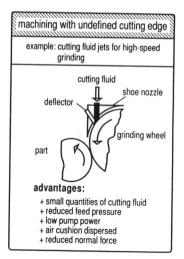

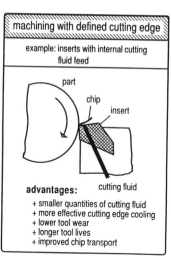

Fig. 15.23 Selective cooling lubricant feed

cooling strategies for tool and workpiece. A number of ways present themselves, depending on the machining operation selected, of reducing the quantity of cooling lubricant to a minimum (Fig. 15.23). Large amounts of cooling lubricant are generally used for cutting with a geometrically undefined cutting edge. This applies particularly to high speed machining. Investigations show that flow of water-immiscible cooling lubricant can be reduced from 200 l/min to approx. 25 l/min when the shoe nozzles are adapted [36]. At the same time, feed pressure is lowered from 30 to 1,5 bar. This is made possible by reducing the air cushion on the peripheral time. This permits the service life to be extended to such a degree that the fluid no longer requires to be changed so frequently. Concrete details are given in Chapter 4.4.

The cooling lubricant supply is normally designed in such a way as to ensure that one or more working areas are kept wet continuously. However a supply system which has been synchronized with the machining process is quite adequate to permit the cooling lubricant to work efficiently. The cooling of workpiece and tool and chip transport are frequently independent processes. Intermittent cooling lubricant supply and the use of cooling lubricant pumps regulated by the machine control system reduce not only the total volume of cooling lubricant but also the amount of energy required. The dimensions of the cooling lubricant system can be kept smaller.

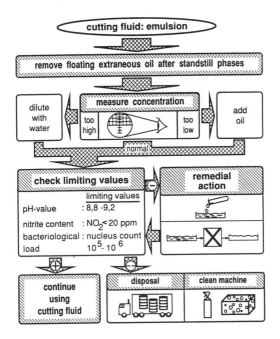

Fig. 15.24 Cooling lubricants: checking and maintenance

Loss through leakage should also be mentioned at this point. Fluid escapes through the base of the machine, by evaporation and vaporization as well as along with the workpiece when it is removed. Closed systems, encapsulation and cleaning systems promise improvement.

The tools themselves offer a variety of options as regards reduction of the volume of cooling fluid used. Cutting materials with high resistance to hot wear and coated tools are very well suited for use in dry machining operations, or, if this is not possible, for use with substantially lower quantities of cooling lubricant.

Consideration is also being given to the development of alternative area of the grinding wheel. This has the fortunate side-effect of reducing the normal force, thus increasing dimensional and form accuracy during the process.

Examples from the area of cutting with a geometrically defined cutting edge are internally cooled tools. While drills with cooling ducts have been in common use for some time now, inserts with internal cutting fluid feed for plunge and slice turning are an innovative development. The cooling lubricant is fed at high pressure directly to the machining point via channels in the carrier tool and in the insert. The more effective cutting edge cooling system reduces tool wear considerably. In addition to reducing the quantity of cooling lubricant required, this approach permits tool life to be doubled when alloyed steel is machined and increased by a factor of four when nickel base alloys are machined. Chip removal from the channel is also better than is the case when conventional feed systems are used, thus enhancing the surface quality of the workpiece [37, 38].

The sole purpose of using cooling lubricant is frequently to moderate the temperature of workpiece and machine during the cutting operation. In such cases, the question must be whether or not alternative media which are not required to lubricate could be used. It is conceivable that fluids such as pure water containing only environment-friendly anti-rust agents might be suitable. If the only requirement is chip removal and / or that machine components are cleaned, compressed air is a viable alternative.

15.4.4 Treat cooling lubricants, concepts for recycling

Maintenance is of central importance in the handling of cooling lubricants. A program of continuous maintenance is essential particularly in the case of modern cooling lubricants, if they are to remain in good condition over a longer period of time. This applies particularly to emulsions, whose service life is directly related to the amount of maintenance they receive. Emulsions require to be checked and maintained regularly in order to retain their efficiency and to reduce the risk of illness (Fig. 15.24). Service manuals should be consulted for instructions [10, 12].

The first step involves the removal of floating extraneous oil. This is normally syphoned or skimmed off after a machine standstill phase. In the second step, the concentration is determined using a refractometer. Depending on the result, the emulsion is either replenished with base oil or diluted with water. Weekly checks on limiting values of the cooling lubricant determine pH value, nitrite content and level of bacteria present. When the values deviate from standard, remedial action corresponding to the composition of the emulsion must be taken to restore the chemical and biological limiting values. If the deviation from the standard value is too high, the emulsion has to be disposed of.

Before refilling, all relevant parts of the machine must be cleaned thoroughly. This cleaning process is usually very difficult both in the case of simple machines and in the case of the complex machining centres which are in common use. Reservoirs, pumps, chip transport systems etc

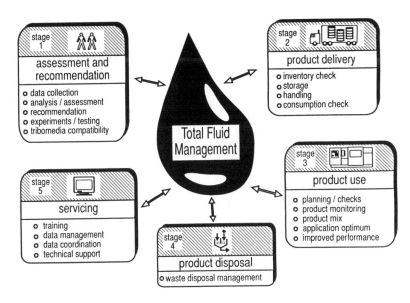

Fig. 15.25 Total Fluid Management (taken from Fuchs Petrolub)

are frequently inaccessible and therefore require costly dismantling work. Inaccessible spots are an ideal breeding ground for yeasts, fungi and bacteria which are almost impossible to clean. The machine tool building industry is called upon to consider the need for a machine tool with a cooling lubricant system which is easily accessible when new machines are at the design stage.

Small and medium sized companies do not, as a rule, have the personnel and laboratory capacity necessary to obtain the necessary data concerning the state of the cooling lubricant. Increasingly, companies are approaching the supplier of their cooling lubricant for help. One concept, called 'Total Fluid Management', which has already been put into practice, is shown in Fig. 15.25. This system comprises all the stages involved in handling cooling lubricant. It begins with the assessment of the machining process and the conclusions drawn regarding recommendation of the appropriate cooling lubricant. The product is delivered with information regarding correct storage and treatment. When the product is in use, attention focuses on product monitoring and care as well as on the optimization of application. Waste disposal of the used products is likewise the responsibility of the cooling lubricant supplier. The supplier is also responsible for service, taking over data management in the field of

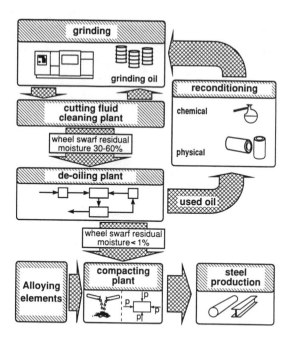

Fig. 15.26 Recycling concept for the disposal of grinding sludge

logistics and carrying out in-company personnel training.

Recycling is a further step in the reduction of environmental pollution by cooling lubricants. As shown in Fig. 15.10 there are a number of subsequent uses for the individual fluid phases depending in the type of cooling lubricant and degree of contamination. There is also considerable potential for disposing of the solid phases. While it is relatively easy to re-use chips from cutting with a geometrically defined cutting edge, large quantities of waste are left after grinding operations, especially those carried out using high speed technology. This waste consists of grinding chips, grinding oil and filtering aids which must be disposed of so as not to cause unnecessary pollution. The law, as it stands at the moment, states that it depends largely on the composition and consistency of the grinding sludge whether it is designated as hazardous waste [13]. Degree of contamination and residual moisture are important variables.

Fig. 15.26 presents a concept for re-using grinding sludge. Grinding sludges emerging from conventional filter machines generally have a residual moisture of content of 30 % to 60 %. The main prerequisite for an environment-friendly solution is that these sludges are de-oiled in external de-oiling plants until their residual oil content is below 1 %. The used oil which has been filtered out can then be returned to the mineral oil manufacturer. There it can undergo physical and chemical reprocessing and returned to the machining process. The grinding dust is analysed in order to determine its constituent parts and alloying elements are added to it in accordance with the specifications of the steel manufacturer. It is then compressed and pellets or briquettes are produced, which can be used to produce steel.The concept is one of circulation, preserving in exemplary fashion both the environment and raw materials which are rapidly being depleted.

15.5 Recapitulation and prospects

Cooling lubricants are among the most commonly used process materials in the metal working industry. They are necessary primarily for the process itself in order to meet the technical conditions required for the operation in hand. However, various problems arise in connection with their use both in the direct work environment and with their disposal. Due to the increasing number of laws and directives governing industrial safety and environmental protection, the use of cooling lubricants is putting intense economic pressure on manufacturing companies.

Analysis and discussion of this subject shows that in accordance with the model shown in Fig. 15.2, the problems arising from the use of cooling lubricant should be tackled in the following stages (Fig. 15.27).

The first step is to carry out an analysis of the present situation, setting out the type, quantity and circulation time. All costs incurred in connection with the use of cooling lubricant such as purchasing, maintenance and disposal are then balanced and allocated to the appropriate manufacturing operation and product. Costs, technological efficiency and risk, which can in come cases only be assessed subjectively, must be compared on the basis of appropriate criteria and priorities must be set for future action.

Improved care, monitoring and maintenance of the cooling lubricant are short-term measures which can be implemented immediately in the manfacturing department. Additional steps concentrate on reducing the quantity of cooling lubricant used and on modifying the cooling lubricant medium. These measures require thorough understanding of the process. At this point, the question should be raised as to whether it might be possible to do without cooling lubricant altogether [40].

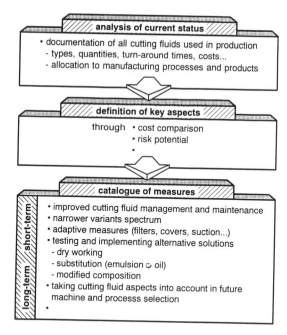

Fig. 15.27 Stages in improving the use of cooling lubricants in ecological terms

A study of the whole subject of cooling lubricant shows clearly that it is essential to re-think the manufacturing process. It is no longer sufficient simply to direct all efforts at optimizing the efficiency of a process. The ecological aspects must now be accorded equal consideration. Researchers and individual companies are called upon to assume special responsibility for developing and implementing innovative environment-friendly manufacturing processes.

16

STRATEGIC COST/BENEFIT EVALUATION OF PRODUCT DEVELOPMENT, PRODUCTION AND WASTE DISPOSAL

16.1 Introduction

The contributions to this book demonstrate the increasing complexity of the decision-support process. The attainment of objectives such as 'Competitive strategies for a global market' (Chapter 1), `Quality Management as a corporate strategy' (Chapter 2) and `Strategies for the introduction of a quality management system' (Chapter 3), is dependent on the availability of the appropriate facts and figures.

Environmental protection as an additional objective of the engineering branch is discussed separately in the last chapters in which the ecological challenge is integrated into the broader aspects of production engineering [1]. The following contribution offers a perspective for ways in which managerial staff can turn the various demands with which they are confronted to the use of their companies in order to consolidate their competitive advantage through reliable decisions and efficient management

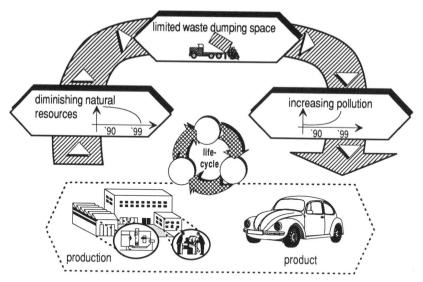

Fig. 16.1 Initial situation

in product development, production and waste disposal (Fig. 16.1).

The current state of evolving social awareness is coupled with the knowledge that natural resources are limited in a variety of ways. Many natural resources will soon cease to be available in sufficient quantities [2]. At the same time, capacity for the disposal of production and product waste is dwindling rapidly [3]. There seems to be no easy solution in the form of additional dumping capacity in sight.

The challenge facing manufacturing enterprises particularly in Germany is, therefore, the reduction of all forms of threat to the environment. This applies both to the design of production processes and to product design. In this context, two main tendencies have come to the fore. Regulations ensure that due account is taken of new criteria regarding product and process design; at the same time, the time factor involved in technical decisions is being extended to cover the entire product life cycle.

Responsibility for the production process itself nowadays goes hand-in-hand with responsibility for the eventual disposal of the products. It is, therefore, essential to extend the terms of reference (Fig. 16.2). Legal provisions and directives governing waste disposal laws are forcing more and more companies to take back their own products once these have served their purpose. Initially, in Germany a packaging law came into force

Fig. 16.2 Extended scope of balance

in 1991, regulations on electronic waste are expected to follow in 1994 and further regulations, e.g. for scrap cars are on the cards.

Product development and the design of production processes remain, however the most important elements in the effort to meet all corporate targets including the protection of the environment. A comprehensive approach to evaluation is essential if meaningful projections are to be made at the product development stage about the entire product life cycle.

16.2 Systems of objectives in changing times

16.2.1 Economic and ecological boundary conditions

The prevailing climate of evolving social and economic views is taking concrete shape as an `socio-ecological market economy' with new boundary conditions for manufacturing companies [4]. The term itself is an expression of the dual influence exerted on the company. Ecology-oriented `framework' legislation - the mechanisms of which are comparable to those of social legislation - in conjunction with the rules of the market defines the

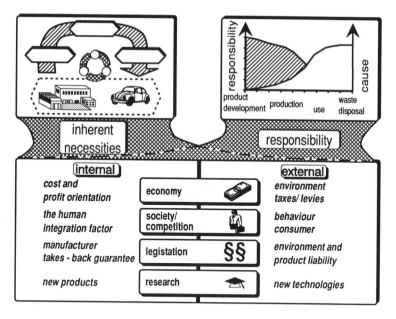

Fig. 16.3 Factors influencing corporate strategies

responsibility of the entrepreneur (Fig. 16.3).

Direct market demands arise from the changing buying behaviour on the part of the consumer which, in the case of public customers is explicitly formulated as environment-friendly procurement policy [5]. The effects of a system of environment dues and taxes integrating external effects into company cost accounting are at variance with current market thinking. An example of this is the discussion about raising exhaust gas tax on CO_2 released into the air.

Direct statutory restrictions and regulations, which do not conform to market trends, especially those with a national character, require to be evaluated differently. Restrictions imposed for the protection of the environment stipulate, for example that filters be installed and in the case of tank farms, collection basins built. An operating license is issued only when these conditions, which are reflected in capital and operating expenditure, have been met.

In terms of internal significance and external responsibility, changing influences exerted on the company can be divided into four categories: economy, society/competition, law and research. Companies can do little to influence the economy as a whole. The effects on costs and profit must

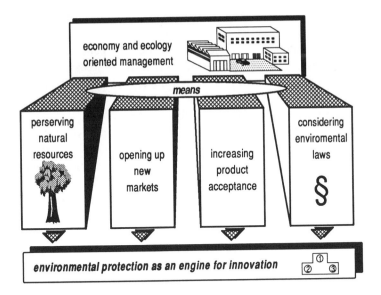

Fig. 16.4 Economy and ecology oriented management

be considered separately and must be taken into account in other deliberations. Environment taxes and levies must now figure in any calculation as new cost elements.

Companies are having to rethink both their processes and their products and to adapt to changing market and boundary conditions. In this sense, evolving boundary conditions can be regarded as exerting a positive influence as an innovative factor within a company.

16.2.2 Environmental protection as an innovative factor

This innovative factor should be exploited by revamped economy and ecology-oriented management taking account of both short and long term aspects (Fig. 16.4).

Nonrenewable resources can be preserved only by the selective implementation of innovations. The protection of the human race through either bio-technological or environmental considerations is just as important as the conservation of natural resources [6]. Even air and water, long considered inexhaustible, are now being regarded with growing respect.

Companies, of course, have largely the same objectives as before: diversification of the product range in conjunction with opening up new markets, thereby increasing product acceptance. It is particularly important at the present time for companies to corner new areas of the market concerned with the protection of the environment. This, in turn, encourages product acceptance on the part of the consumer. The tag `environment-friendly' has become a powerful marketing argument.

This time-oriented approach is, however, out of tune with the decision-support procedures common in manufacturing companies today. It would therefore make sense to ease companies into the process of considering the long-term impact of industrial production by passing appropriate laws.

Given the statutory requirements, foresighted ecology management is essential if the risks of corporate action are to be limited effectively and economic management is to be successful in the long-term.

In contrast, the influence exerted on the markets by growing environmental awareness is clearly tangible. In this context, two phenomena are emerging. On one hand, product acceptance can be increased by careful implementation of suitable measures in conjunction with `eco-marketing'. It is hoped that the introduction of, for example, the German environment angel or the European environment-friendly logo as sales-boosting product certificates will go some way towards promoting this strategy.

On the other hand, completely new markets are being captured by substituting products and technologies or by developing new products to suit changing consumer behaviour.

16.2.3 Extended objectives

The examples and tendencies outlined, show that only an evolving system of objectives guarantees the long-term future of a company and success on the market. The changing requirements addressed in Chapter 1.1 `Competitive strategies for a global market' necessitates that the ecological aspect be extended. Previously known monetary objectives must now be accompanied by consideration of `ecology' and `environmental cost' (Fig. 16.5).

Additionally, however, the structure of the objectives is changing. Whereas the decision-support process condensed conventional indicators in a hierarchy - e.g. according to Du Pont - until an unambiguous solution was reached, this is frequently no longer possible. The new system of

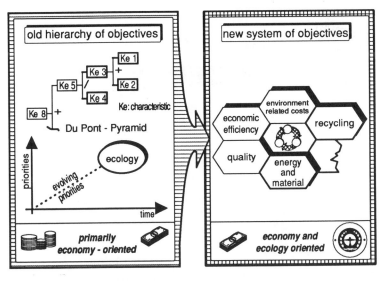

Fig. 16.5 System of objectives change with the times

objectives requires a structure to be found which allows for a suitable compromise between seemingly contradictory characteristics - e.g. one balancing production costs and quality. A simple hierarchy is insufficient and must therefore be replaced by more efficient decision-support mechanisms.

The development and deployment of these more helpful mechanisms and the provision of the information required is dependent on the continuing development of information technology. The growing availability of high-performance computer components is resulting in technology-led changes in product and process design [6]. In future, due to the multiplicity of criteria and to the increasingly complex system of objectives, efficient decision-support processes will require the cooperation of a number of experts from various areas of a company.

The process of selecting the material to be used, previously decided on the basis of the productivity of the production process and improvement of the cost structure in order to maximize profits, is now extended to accord equal importance to additional criteria such as quality management and waste disposal requirements (Fig. 16.6). The decision regarding the selection of plastic or special metals for a product or assembly group is becoming more and more complex.

The results of the Delphi Survey (Fig. 16.7) show that the people who

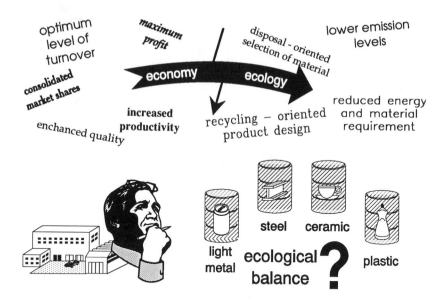

Fig. 16.6 System of objectivs in changing times - example: selection of material

bear responsibility within a company are aware of the situation and that they expect to extend the scope of their evaluation strategies. Ecological and energy-related aspects in particular will become increasingly important.

The trend towards more comprehensive evaluation is also expected to continue. Whereas attention used to focus on the evaluation of production and revenue, evaluation of the total life cycle of a product is increasingly being regarded as essential by the companies questioned in the Delphi Survey.

Stricter framework conditions such as the obligation to take back packaging waste are forcing companies to include ecological and energy-related considerations as well as financial aspects in their evaluations.

A similar development is expected with regard to the use of the product. The form of evaluation planned, taking account of strategic and energy-related considerations, indicates that a more customer-oriented view is gaining ground. It is essential that the question of energy consumption, particularly of energy-driven products (e.g. household appliances, vehicles) be given more consideration, if energy consumption is to be reduced still further.

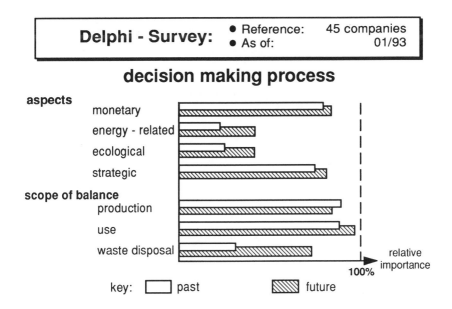

Fig. 16.7 Trends in evaluation

In manufacturing, where the potential for rationalization is greatest, energy-related and ecological issues are coming to the fore along with the classical financial factors. Measures based on ecological and energy-related evaluation and aimed at environmental protection can thus help to cut production costs.

16.3 Strategies for evaluation: Tools for target-oriented decision support

Although the survey carried out shows that industry is becoming generally more sensitive to ecological issues, it also reveals that there are still serious shortcomings in the process of actually translating changing priorities into deeds. No suitable technical alternatives have yet been discovered for a number of ecologically unsound products, processes and materials. In areas in which technical alternatives already exist, it can be a complex, if not impossible task to determine which of the myriad of solutions on offer is best in ecological terms.

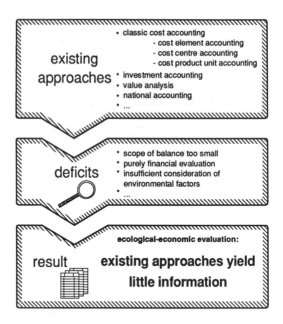

Fig. 16.8 Weak points of existing approaches to evaluation

The experiments required to find the optimum solution can only be carried out on a few products or manufacturing processes of particular industrial importance. Suitable planning tools are required to evaluate material, design or manufacturing alternatives objectively. When appropriate evaluation strategies are implemented, the weak points of existing technical solutions to be identified and rectified at an early stage by adopting alternative approaches.

One general criticism of existing evaluation methods is that they are frequently too complex and that not all relevant aspects are considered, or rather, account is taken only of the financial aspects (Fig. 16.8). The survey confirms that no satisfactory approach to evaluation has yet been developed which characterizes product or process specific environmental behaviour with any clarity.

One of the restrictions of the modern ecology-oriented approaches to evaluation is that the results are often difficult to reproduce. The comparison of supposedly characteristic environmental indicators comprising environmental quantities which do not bear comparison, frequently leads to over-simplification. A critical inspection of these approaches reveals that this procedure merely shifts the deficits in

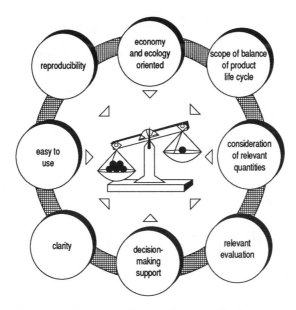

Fig. 16.9 Requirements to be met by a future-oriented evaluation system

reproducibility from evaluation in its narrower sense to the deduction of environmental characteristics [7].

16.3.1 Requirements to be met by an economic/ environmental evaluation system

Attempts to ensure that practice-oriented systems of evaluation meet certain demands inevitably run up against the weak points identified. The Delphi Survey confirms that users of such systems are keen to make further use of results obtained in evaluations. The companies questioned said that their primary need was for a system of evaluation which they could use to aid day-to-day decision-making with regard to, for example, environment-friendly and at the same time economic production optimization (Fig. 16.9).

An ideal evaluation system would meet the following demands. It must be capable of being used to optimize details of individual processes as well as to select alternative product / production concepts. It must also be able to look beyond short-term financial aspects and take all relevant

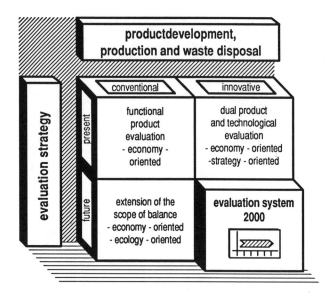

Fig. 16.10 Evaluation portfolio

environmental factors into account. Environmental factors related to the later phases of a product cycle, such as the actual use and subsequent waste disposal, call for careful consideration.

The latter demands especially are in direct contrast to the concern that the evaluation system should not be excessively complicated. The system must be easy to use, if it is not simply to be wheeled out on special occasions but is to make a broadly based and active contribution to the more rational use of our natural resources.

16.3.2 Approach to a solution: Evaluation system 2000

The weak points already identified in existing approaches to evaluation and the development requirements defined by the companies are used in the following as a basis on which to demonstrate the concept of a solution which, it is hoped, can respond to the dilemma. The target and actual situations as described by the companies concerned have been shown here in the form of a portfolio (Fig. 16.10).

Interrelationships between product development, production or waste disposal are currently evaluated primarily in terms of economic aspects. It

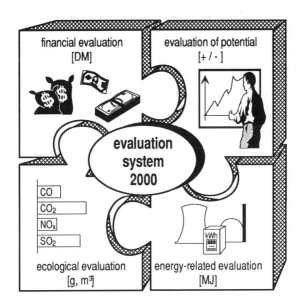

Fig. 16.11 Future-oriented product and technology planning

is only in the case of product or production design of long-term interest that factors which are financially unquantifiable are included in the decision-support process. Consequently, demands are now being heard for a much more broadly-based approach to be adopted for the future evaluation of innovative product or production concepts.

Future-oriented product and technology planning must be geared to dealing with the increasing number of overlapping cross-references and demands. The customers' requirements must be met, manufacturing costs must be kept to a minimum and environmental issues must not be neglected. Reciprocal dependencies intersect existing departmental boundaries between design, process planning, process control, resource management, manufacture and assembly. The challenge facing science is to provide a tool capable of taking control of this complexity. A comprehensive system of evaluation permitting monetary, strategic, energy-related and ecological components to be considered simultaneously is required (Fig. 16.11).

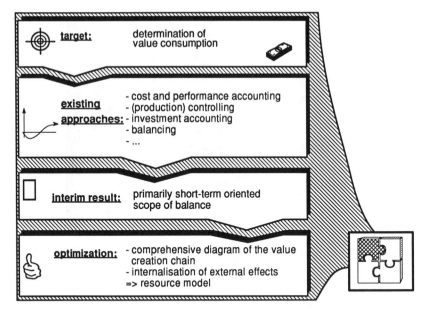

Fig. 16.12 Monetary evaluation procedure

16.3.3 Potential application

A multidimensional evaluation concept will be developed in the following by examining the potential of various existing evaluation strategies for adaptation and subsequent integration in the system required. Research approaches to meeting demands which have not previously been fulfilled by any of the existing evaluation strategies will first be presented. The established evaluation strategies will be examined in terms of `cost', `potential', `energy' and `ecology'.

(a) Monetary strategies for evaluation
The objective of conventional evaluation strategies, as everyone who has a passing acquaintance with business administration knows, is the determination of actively-based cost accounting for a defined area under examination. This can be either a product, a manufacturing process or a technology (Fig. 16.12). Within the scope of production controlling, alternative technological applications are evaluated with a view to determining which one incurs the lowest level of cost. Capital investment accounting is similar; the area under examination in this case is the

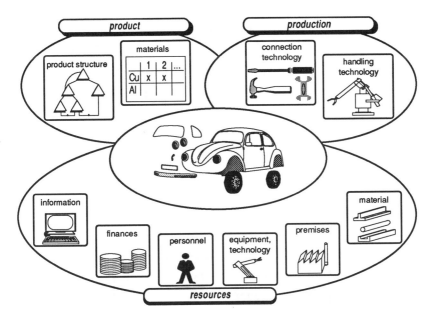

Fig. 16.13 Corporate structures

consideration of the amount of capital to be invested and the anticipated income. Comparison on the basis of cost permits the selection of the specific optimum cost and offers relevant decision support.

The statutory accounting system demands that assets be listed on a set date; like cost and output accounting and production controlling, the time to which the balance refers is primarily in the past. It is only in the case of investment accounting that the period of time involved extends into the future, although the time frame seldom goes beyond the medium-term.

If the quality and therefore the clarity of monetary evaluation are to be improved, a more detailed diagram of the chain of added values must be drawn up. All relevant quantities including those which are long-term must be included. It is essential to extend the time frame if aspects such as product waste disposal, still regarded as being of secondary importance, are to be given due consideration. Recommendations regarding decisions can shift as a result of a consistent move towards the `internalization of external effects', i.e. application-oriented consideration of resources which were previously free of charge (Fig. 16.13).

A strategy aimed at more actively-based evaluation of variant and assembly costs has been developed by the Laboratory for Machine Tools

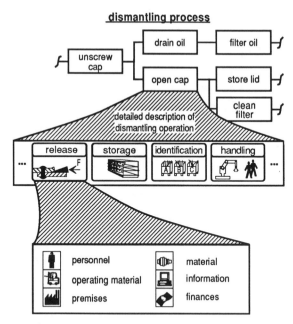

Fig. 16.14 Approach to evaluating a dismantling operation

and Industrial Management (WZL) of the Technical University of Aachen (RWTH) [8, 9]. With the `internalization of external effects' in mind, synergetic effects were used to develop a method of evaluating dismantling costs. Initial experience in practice indicates that this approach has considerable potential.

The use of descriptive models can be of assistance in drawing up a detailed diagram of the chain of added values. Computer-aided linkage of the information contained in the model about the product, the production process and the resources required can enable a comprehensive evaluation of the individual factors contributing to a dismantling process to be drawn up. The evaluation, therefore, is the analysis of information regarding a certain criterion or a certain decision: e.g. environmental compatibility of a product and the process of dismantling it [10].

An approach to the evaluation of a dismantling process is given as an example in Fig. 16.14. It is necessary to divide the dismantling process into individual steps and to define which resources are required. The level of resource consumption is determined by evaluating the product characteristics and the degree to which they contribute to the stresses and

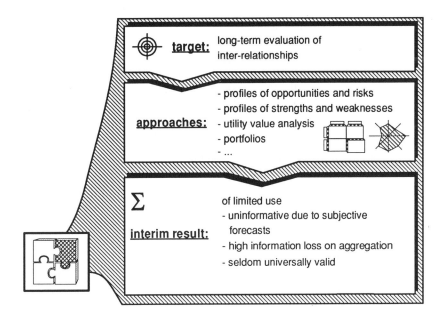

Fig. 16.15 Evaluation of potential

strains to which the resource is subjected. These values are subsequently used to quantify the dismantling cost.

(b) Strategies for evaluating potential

The approaches to evaluation which quantify in monetary terms are in conflict with those assessing the potential uses. Approaches which assess the potential uses are geared to the interpretation of descriptive quantities which cannot be quantified easily, if at all [11]. In general, the potential of this method can profitably be brought to bear when complex relationships are to be evaluated from a long-term point of view. Conclusions can be drawn as to suitable technological or corporate strategies from the results (Fig. 16.15).

Methodical approaches are, e.g. the opportunity-risk or strength-weakness analysis, the utility value analysis, the portfolio technique. One characteristic of these methods is that planners often introduce their own influence at the stage during which information is concentrated and therefore contribute consciously or not to the subjectivity of the outcome. Added to this is the fact that the concentration of information itself and the abstraction of general statements inevitably entail the loss of detailed information. The usefulness of such methods in concrete, individual cases is, therefore, restricted.

The currently voguish attempt to use a strengths-weaknesses analysis to define `the' ideal material from which cars should be built is more likely to harbour risks for the environment than to open up new potential. Generalized unqualified statements for or against one group of materials precludes exploitation of the specific potential offered by the material for use in individual cases. As the determination of a more environment-friendly alternative is, however, dependent on the concrete boundary conditions, general recommendations are not necessarily suitable for individual cases.

(c) Energy-related strategies for evaluation

Increasing sensitivity with regard to environmental issues, the painful realization of the relationship between pollution and energy consumption and, not least, steadily increasing raw material and energy prices have helped to change attitudes in future-oriented companies. These are beginning to accept that energy consumption must become an important factor in production planning [12]. Already, even although the planned energy taxes have yet to take effect, many companies viewing their energy requirement anxiously. In the capital investment industry, almost 40 % of the manufacturing costs are directly or indirectly attributable to energy and material consumption - and the trend is upward [13].

The demands of company management for cost reductions coupled with the demands of economists for restraint with regard to the depletion of resources is compelling manufacturing companies to treat as a matter of urgency the need for optimization of both energy and material consumption. The availability of suitable tools is a prerequisite for the development or at least recognition of alternative production processes which make more rational use of energy.

One of the requirements which such tools will have to meet is, for example, that energy-related quantities should be considered not in terms of changing reference values e.g. prices in units of currency, but in objectively measurable units such as energy requirement in megajoules. Additionally, the tools must be universally applicable, yet still show the interrelationships between specific energy and material consumption. Account must be taken of both the energy used in the manufacture of semi-finished goods required for the production process in question and the resources required to use and subsequently dispose of a product. One approach which meets those demands is based on the evaluation of

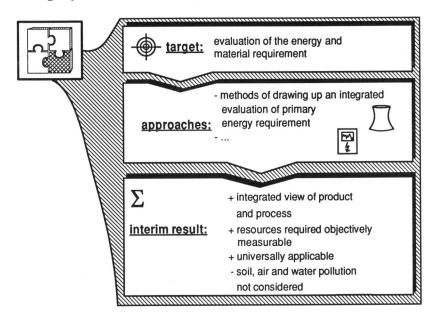

Fig. 16.16 Energy-related evaluation

product-specific primary energy consumption [14]. This is energy requirement-oriented product line analysis (Fig. 16.16).

A tool like this has been developed at the Fraunhofer-Institute of Production Technology (IPT). One result of research work undertaken at the IPT is the development of a method of evaluating comprehensively energy and material consumption. This tool enables potential reductions in energy and material consumption to be recognized at an early stage and to be evaluated [15, 16].

The terms of reference for this evaluation encompass all relevant energy and material flows (Fig. 16.17). If a product or a production alternative is to be evaluated comprehensively and objectively, the scope of the balance must be extended to include the entire product life cycle. As previously mentioned, an energy-related evaluation must not, for example, be restricted to the analysis of that area of production which is of primary interest to the production engineer. The energy and material consumption during the phases of use and waste disposal must also be taken into account [14].

Depending on whether or not material consumption is directly linked to one product, the energy used can be characterized as indirect or direct

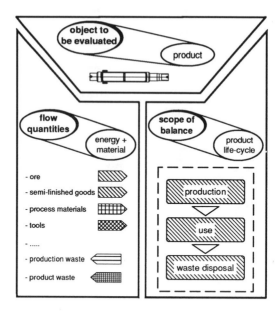

Fig. 16.17 Comprehensive evaluation of energy and material consumption

expense.

The term indirect material consumption covers, for example expense incurred in the preparation of the material or process materials; the term direct material consumption refers to expense in connection with tool and machine use. In terms of energy consumption, this means that the transformation upstream of the available primary energy into the form of energy required must be included in the calculation. Energy consumption can subsequently be classified according to the same principles as material; either as direct or indirect use.

The numerical evaluation of energy is based on the following sequence of steps: 'collect-balance-evaluate' (Fig. 16.18). Collection involves the quantification of the above-mentioned energy-related power demand. This entails the measurement or calculation of the amount of energy and material required for each individual stage in the production/operation. It is essential to take account of the conditions under which the product in question is deployed when assessing the expenditure involved. Calculations or measurements of efficiency can be used to quantify expenditure. Records of the cost of waste disposal must include the calculation of whether the production or product waste accumulating

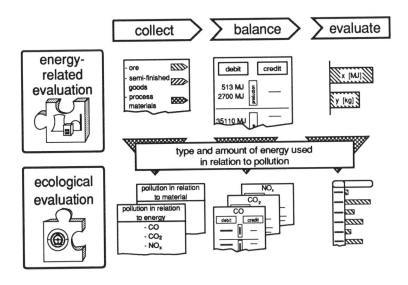

Fig. 16.18 Development of an ecology-oriented system of evaluation

during the product life cycle can be recycled or whether it must be dumped. If recycling is technically feasible, investigations must be carried out to determine whether or not it is viable from the point of view of energy consumption. When the cost of recycling exceeds the accumulated energy content of the primary raw material plus terminal storage cost this is obviously not the case.

For evaluation purposes, dumping will be considered to entail additional energy expenditure for collection, compression, transport etc. In cases in which recycling is considered economical from the point of view of energy consumption, the evaluation will be based on recycling. Instead of energy and material consumption for waste disposal, an energy `credit note', or rather a residual energy value will be awarded for the secondary raw material produced. This algorithm permits energy savings achieved by using circulating materials to be recorded precisely. At the same time, due regard can be given to the question as to whether the secondary raw materials produced in the recycling process are of the same or lower quality as the initial material [16].

This evaluation strategy permits all product-specific relevant energy and material flows to be recorded and condensed in two characteristic quantities: primary energy requirement and material requirement.

(d) Ecological evaluation strategies

The type and scope of the energy and material consumption and the degree of resultant pollution are directly dependent on one another. Investigations have, therefore, been undertaken to determine to what extent the energy-related evaluation strategy developed can also be used to evaluate product-specific pollution. Since all of the energy and material relevant to the process has already been recorded in the energy-related method of evaluation, it can now be confirmed that there is, indeed, a considerable potential for adaptation: the current method of evaluation will now be extended to embrace a comprehensive method of analysis which takes account of both ecological and economic aspects. This procedure can build upon many of the research results previously obtained. This will not only permit a considerable amount of the knowledge already available to be re-used, but will also allow tried and tested scientific methods to be used [16, 17].

The same sequence of steps presents itself for the operative development of ecology-oriented and for energy-oriented evaluation. A distinction is drawn in the collection phase between pollution caused primarily by the energy requirement and pollution caused primarily by the material requirement. Pollution arising mainly as a result of energy conversion is regarded as energy dependent; CO_2, SO_2, NO_2 etc. depend on the type of energy used. Cooling lubricants, emulsions, waste etc. are regarded as environmental quantities which are dependent on the material requirement. Individual material balance sheets are drawn up on the basis of these records. The balance sheets can then be used as a base on which to develop a product-specific environment profile.

All values in the environment profile are shown as physical units; in addition to the absolute evaluation, this depiction can be used for an ecology-oriented comparison of production concepts.

In the case of complete agreement or non-agreement of two environment profiles, the derivation of recommendations is unambiguous. When there is partial agreement between the environment profiles, an initial approach may be adopted which condenses the environment profile to a descriptive parameter [11]. Each form of pollution is different and cannot easily or objectively be compared with other forms. Interdisciplinary development of an extended evaluation strategy has yet to materialize.

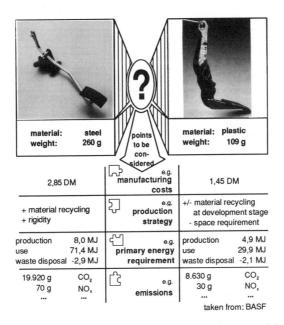

	e.g. manufacturing costs	
2,85 DM		1,45 DM
+ material recycling + rigidity	e.g. production strategy	+/- material recycling at development stage - space requirement
production 8,0 MJ use 71,4 MJ waste disposal -2,9 MJ	e.g. primary energy requirement	production 4,9 MJ use 29,9 MJ waste disposal -2,1 MJ
19.920 g CO₂ 70 g NOₓ ...	e.g. emissions	8.630 g CO₂ 30 g NOₓ ... taken from: BASF

Table material: steel, weight: 260 g | points to be considered | material: plastic, weight: 109 g

Fig. 16.19 Multidimensional evaluation - example: car accelerator pedal

16.4 Case studies

Although detailed work still requires to be undertaken on the multidimensional system of evaluation under development, the existing strategies will be capable of use in the near future to assist corporate bearers of responsibility to reach decisions objectively. The applications currently possible will be demonstrated on the basis of the following case studies.

The `substitution of materials' is a subject of discussion in more boardrooms than ever before - not only in the automotive industry. Light metal, steel, plastic or ceramic? Which of these materials will be the right one for which application in the future? The answer to these questions is becoming ever more complex. On one hand, existing materials are constantly being refined, resulting in the development of a number of new material variants. On the other hand, the number of criteria to be satisfied is increasing continuously. The decision as to which material is best suited to the application in hand cannot, therefore, be made on the basis of abstract recommendations. Account must be taken of specific boundary conditions [12].

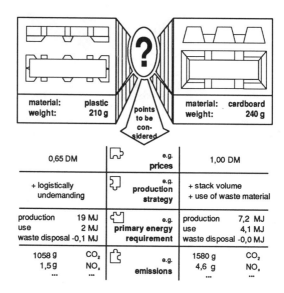

0,65 DM		e.g. **prices**	1,00 DM	
+ logistically undemanding		e.g. **production strategy**	+ stack volume + use of waste material	
production	19 MJ	e.g. **primary energy requirement**	production	7,2 MJ
use	2 MJ		use	4,1 MJ
waste disposal -0,1 MJ			waste disposal -0,0 MJ	
1058 g	CO₂	e.g. **emissions**	1580 g	CO₂
1,5 g	NOₓ		4,6 g	NOₓ
...	...		...	...

taken from: TCE VIDEO EUROPE

Fig. 16.20 Multidimensional evaluation - example: video recorder packaging

An example of the preparation which must be carried out on the quantities involved in the decision is given in Fig. 16.19. The preparation required for the decision as to which of two passenger car pedals should be selected involves more than simply a comparison of the manufacturing costs. The energy requirements and similar parameters must be known. As previously indicated, all of these quantities must be considered if an optimum selection process is to be guaranteed [18].

Systematic preparation of the absolute descriptive quantities suggests that technical innovations can present the potential for saving not only in monetary terms but also in terms of energy and ecological resources. The degree to which the different forms of rationalisation reinforce or cancel one another out depends on the part or product in question.

The application of the multidimensional evaluation system will help to make discussions which were previously based mainly on emotions more objective. As demonstrated in Fig. 16.20, the evaluation concept already developed is universally applicable.

The application of the concept is not limited in any way to selection processes involving only material. It can be used, for example, to prepare the way for decisions regarding logistics, transport or packaging concepts.

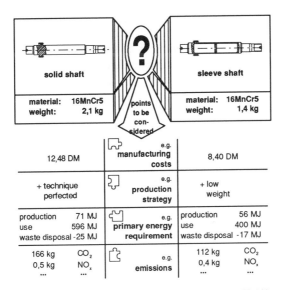

	e.g. manufacturing costs	
12,48 DM		8,40 DM
+ technique perfected	e.g. production strategy	+ low weight
production 71 MJ use 596 MJ waste disposal -25 MJ	e.g. primary energy requirement	production 56 MJ use 400 MJ waste disposal -17 MJ
166 kg CO₂ 0,5 kg NOₓ ...	e.g. emissions	112 kg CO₂ 0,4 kg NOₓ ...

taken from: SFB 144

Fig. 16.21 Multidimensional evaluation - example transmission shaft

In accordance with the procedure described above, decision-making criteria concerning video recorder packaging were selected and prepared systematically. In order to take due account of the consumption of all resources, quantities characterizing the logistics involved were selected. These values, presented in condensed form, aid management in making strategic decisions.

In addition to lending rather more objectivity to decisions regarding selection, the evaluation method has considerable potential in terms of the systemization of technical measures to optimize the manufacturing process.

A shaft like those used in conventional passenger car transmissions is shown on the left hand side of Fig. 16.21. The results of a comprehensive investigation of the resources used were used as an information bank on which to base attempts to come up with a more rational alternative product. Investigations conducted into the conventional requirements showed that any reduction in energy-related and ecological values consumed, would require the mass to be lower. Guided by the evaluation of the potential of innovative shaping and jointing technologies, a hollow sleeve shaft has been developed, with 30 % less mass and which is capable of the same functions as the solid shaft.

Evaluations carried out parallel to product development have enabled the resources required for the realization of new shaft designs to be defined at the planning stage. Uneconomical and environmentally unsound design drafts can thus be recognized quickly, allowing development capacity to be concentrated on more promising approaches. The result of evaluation-based product development is a transmission shaft whose production exacts a lower price in monetary, energy-related and ecological terms.

Whatever the economic and ecological significance of such optimization of details, savings to be made on the domestic market as a result of the development in shaft manufacturing are considerable. A reduction of approx. 200 megajoules in energy requirement corresponds to a saving of about 5 litres of petrol in the life of each passenger car fitted with such a transmission shaft. The potential economic saving for the Federal Republic of Germany, where around 37 million vehicles are currently registered, is clearly substantial [17].

16.5 Conclusions and prospects

Ecology and economy, two contradictory corporate objectives? Whereas the answer to this question was, in the past, all too frequently affirmative, attitudes and the outlook for the future are changing. Increasing pollution and the resultant costs are encouraging and industry to adopt new strategies integrating environmental and economic considerations: ecologically and economically motivated action plans are almost identical in the medium to long term [19, 20].

Industry's demand for cost reductions and the demand for the economy as a whole that resources be preserved are confronting manufacturing companies in particular with the need to treat the environment with respect. The experiments required in order to achieve this can be conducted on only a few products/production processes which are particularly important for industry. Planning tools are, therefore, essential if material, design or engineering alternatives are to be evaluated objectively.

The evaluation system outlined permits a part, product or operation sequence to be evaluated comprehensively; i.e. the consumption of energy-related and ecological resources is considered alongside the cost in financial terms. This permits measures to be defined which will lead to a reduction in the requirement for resources (Fig. 16.22).

The analysis of product-specific operation sequences permits resource guzzling operations to be identified and contributes to the definition of

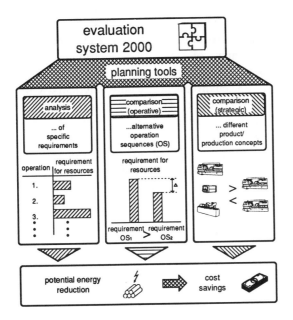

Fig. 16.22 Evaluation system 2000 - Potential uses

operations/materials which are less harmful to the environment (Fig. 16.22, left-hand side). On the other hand, alternative sequences of operations for the manufacture of a certain product can be compared on the basis of their requirement for resources (Fig. 16.22, middle). It is possible within the scope of strategic evaluation to draw conclusions as to fundamental requirements underlying various production and material concepts for one product 'family' (Fig. 16.22, right-hand side).

The concept behind 'Evaluation System 2000' is not to be interpreted at the final step in the development of an extended future-oriented evaluation strategy. The outline presents a challenge to science and industry to exploit its potential to the full.

Scientists are challenged to develop the strategies for the fields of research indicated in terms of both direction and detail. It is particularly important to ensure the efficiency of handling and application within a short space of time; this means that the application of the methods must be given adequate support by the development of practice-oriented software tools.

In contrast, the challenge now facing industry is the need to seize on existing strategies and to use them to shape their medium and long term oriented product and production planning. This will contribute to, or rather hasten the development of the tools required and will, at the same time, increase the efficiency of industry's contribution to environmental protection. Assessment of the results of the evaluation process will permit acute necessities for environment friendly alternatives to products and technologies to be recognized. These requirements form the basis of the information so urgently needed if environment friendly products and technologies are to be developed.

Science and industry are called upon to carry on and develop their already close cooperation; this offers the only hope of coping efficiently with the eco-financial challenges with which they are currently faced.

Appendix A: Members of the working groups

Members of the working group of chapter 1
Dipl.-Ing. H. Cronjäger, Mercedes-Benz AG
Prof. Dr.-Ing. W. Döpper, Hertel AG
Prof. Dr.-Ing. Dr. h.c. Dipl.-Wirt. Ing. W. Eversheim, WZL/FhG-IPT, Aachen
Dr.-Ing. G. Friedrich, FAGRO Preß- und Stanzwerk GmbH
Prof. Dr.-Ing. J. Herrmann, Carl Zeiss
Dipl.-Ing. H. Jansen, Scheidt & Bachmann GmbH
Dr.-Ing. W. Kalkert, Ford Werke AG
Dr.-Ing. E. Knorr, EX-CELL-O GmbH
Dipl.-Ing. Dipl.-Wirt. Ing. S. Krumm, WZL, Aachen
Dipl.-Ing. G. Miller, Scheidt & Bachmann GmbH
Dipl.-Ing. J. Schneewind, WZL, Aachen
Dr.-Ing. P. Stehle, Carl Freudenberg
Dr.-Ing. W. Wiedeking, Dr.-Ing.h.c. F. Porsche AG

Members of working group of capter 2
Dr.-Ing. K. Boddenberg, Mannesmann Demag Geschäftgruppe Verdichter, Duisburg
Dr.-Ing. J. Elzer, Robert Bosch GmbH, Stuttgart
Dr.-Ing. H. Golüke, FAG Kugelfischer KGaA, Schweinfurt
Dipl.-Ing. St. Hartung, FhG-IPT, Aachen
Dr.-Ing. D. Köppe, Ges. f. Qualitätssicherung, Aachen
Dr.-Ing. R. Kurr, Sony Europa, Stuttgart
Dr.-Ing. M. Lücker, WZL, Aachen
Dipl.-Ing. G. Orendi, FhG-IPT, Aachen
Prof. Dr.-Ing. Dr. h.c. T. Pfeifer, WZL/FhG-IPT, Aachen
Prof. Dr. K.J. Zink, Universität Kaiserslautern

Members of the working group of chapter 3
Dr.-Ing. G. Brüninghaus, Fa. Brüninghaus & Drissner, Hilden
Dipl.-Kfm. K.J. Ehrhart, Fa. Carl Edelmann, Heidenheim
Dipl.-Ing. G. Grzonka, Fa. Schlafhorst, Mönchengladbach
Dipl.-Ing. J. Heine, FhG-IPT, Aachen
Dipl.-Phys. H.-P. Jungen, IBM Werk Mainz

Dipl.-Ing. P. Klonaris, WZL, Aachen
Prof. Dr.-Ing. Dr.h.c. T. Pfeifer, WZL/FhG-IPT, Aachen
Dipl.-Ing. Th. Prefi, IPT-FhG, Aachen
Dr.-Ing. H. Rempp, KfK Karlsruhe, Projektträger Fertigungstechnik
Dr.-Ing. G. Schlechtriem, Fa. Scheidt & Bachmann, Mönchengladbach
Dipl.-Ing. N. Schmidt, IPT-FhG, Aachen
Dipl.-Phys. R. Tutsch, IPT-FhG, Aachen
Dr.-Ing. A. Uhlig, Fa. Barmag AG, Remscheid
Prof. Dr.-Ing. E. Westkämper, TU Braunschweig

Members of the working group of chapter 4
Dr.sc.math. ETH R. Boutellier, Leica Heerbrugg AG
Dr.-Ing. B. Dahl, ITT-Automotive GmbH
Dr.-Ing. A. Gohritz, Dörries Scharmann GmbH
Dr.-Ing. I. Kosmas, Bayerische Motorenwerke (BMW) AG
Dipl.-Ing. W.-D. Krause, AEG Electrocom GmbH
Dipl.-Ing. L. Laufenberg, WZL, RWTH Aachen
Dr.-Ing. Dipl. Wirt.-Ing. G. Marczinski, WZL, RWTH Aachen
Prof. Dr.-Ing. R. Noppen, Webasto AG
Dr.-Ing. R. Richter, Robert Bosch GmbH

Members of the working group of chapter 5
Dr. O. Abeln, Forschungszentrum Informatik (FZI), University of
Karlsruhe
Dipl.-Ing. M. Baumann, WZL, Aachen
Dipl.-Ing. B. Behr, WZL, Aachen
Dr. D. Berhalter, Transcat-Nord GmbH, Dortmund
Dr.-Ing. W. Budde, Exapt-Verein, Aachen
Prof. Dr.-Ing. H.-J. Franke, Institut für Konstruktionslehre, Maschinen-
und Feinwerkelemente, Braunschweig
Prof. Dr.-Ing. F.-L. Krause, Fraunhofer-Institut für Produktionsanlagen und
Konstruktionstechnik (IPK), Berlin
Prof. Dr.-Ing. Dr.-Ing h.c. H. Grabowski, Institut für Rechneranwendung in
Planung und Konstruktion, (RPK), Karlsruhe
Dipl.-Ing. S. Repetzki, WZL, Aachen
Dr.-Ing. S. Rude, Institut für Rechneranwendung in Planung und
Konstruktion (RPK), Karlsruhe

Dipl.-Ing. J. Schlingheider, Fraunhofer-Institut für Produktionsanlagen und
    Konstruktionstechnik (IPK), Berlin
Prof. Dr.-Ing. Dr.-Ing. E.h. M. Weck, WZL/FhG-IPT, Aachen
Dr.-Ing. G. Ye, IBM Deutschland GmbH, Stuttgart

Members of working group of chapter 6
Dr.-Ing. R.-J. Ahlers, Rauschenberger GmbH, Asperg
Dipl.-Ing. J. Chiang, WZL, Aachen
Dipl.-Ing. G. Ernst, Mercedes-Benz AG, Stuttgart
Dipl.-Inform. R. Grob, WZL, Aachen
Dr.rer.nat. B. Hohler, GSB GmbH, Stuttgart
Dipl.-Ing. A.T. Kwam, WZL, Aachen
Prof. Dr. M. Jarke, Lehrstuhl für Informatik V, Aachen
Dipl.-Ing. A. Neumann, WZL, Aachen
Prof. Dr.-Ing. Dr. h.c. T. Pfeifer, WZL/FhG-IPT, Aachen
Dr.-Ing. P. Plapper, WZL, Aachen
Dipl.-Inform. W. Ritschel, WZL, Aachen
Dr.-Ing. R.E. Scheibere, BMW AG, Munich
Dipl.-Ing R. Schmiedgen, Siemens AG, Nuremberg
Dipl.-Ing. R. Schmid, Siemens AG, Munich
Prof. Dr.-Ing. G. Warnecke, FBK, University of Kaiserslauten
Dipl.-Ing. T. M. Zenner, WZL, Aachen

Members of the working group of chapter 7
U. Baraldi, CRIF-Matériaux, Liège (B)
Dipl.-Ing. I. Celi, FhG-IPT, Aachen
Dipl.-Ing. H. Eugster, Hilti AG, Schaan (FL)
G. Kerzendorf, BMW Motorsport GmbH, München
Prof. Dr.-Ing. Dr. h.c. W. König, WZL/FhG-IPT, Aachen
G. Neu, Mercedes-Benz AG, Sindelfingen
Dipl.-Ing. St. Nöken, FhG-IPT, Aachen
Dr.-Ing. C. Schmitz-Justen, BMW AG, München
Dipl.-Ing. C. Ullmann, FhG-IPT, Aachen
Dr.-Ing. R. Umbach, FhG-IUW, Chemnitz
Dr.-Ing. P. Zeller, Thyssen Nothelfer GmbH, Wadern-Lockweiler

Members of the working group of chapter 8
Dr.-Ing. H. Degenhardt, Boge AG, Bonn
Dipl.-Ing. T.P. Dörken, WZL, Aachen
Prof. Dr.-Ing. Dr. h.c. Dipl.-Wirt. Ing. W. Eversheim, WZL/FhG-IPT, Aachen
H. Focks, Lingen
Dr.-Ing. P. Hedrich, MTU AG, Friedrichshafen
Dipl.-Ing. Th. Heuser, WZL, Aachen
Ing.-grad. H. Kenn, Heidelberger Druckmaschinen AG, Heidelberg
Dr.-Ing. F. Lehmann, WZL, Aachen
Dipl.-Ing.(FH) W. Popp, Barmag AG, Remscheid
Dr.-Ing. U. Ungeheuer, BMW AG, Munich
Dr.-Ing. G. Werntze, Maho AG, Pfronten
Prof. Dr.-Ing. Dr. E.h. H.P. Wiendahl, IFA, Hanover
Dr.-Ing. B. Wilhelm, Volkswagen AG, Wolfsburg

Members of the working group of chapter 9
Dr.-Ing. J. Fabry, Hertel AG, Fürth
Dipl.-Ing. R. Fritsch, WZL, RWTH Aachen
Dr.-Ing. e.h. W. Kirmse, Mercedes Benz AG, Stuttgart
Dipl.-Ing. T. Klumpen, WZL, RWTH Aachen
H. Kolaska, Krupp Widia GmbH, Essen
Prof. Dr.-Ing. Dr. h.c. W. König, WZL, RWTH Aachen
Dipl.-Ing. L. Krämer, Dittel, Landsberg
A. Locher, AGIE, Losone
Dipl.-Ing. D. Lung, WZL, RWTH Aachen
Dr.-Ing. H.R. Meyer, Ernst Winter & Sohn GmbH, Norderstedt
Prof. Dr. F. Reinke, AEG-Elotherm GmbH, Remscheid
Dr.-Ing. R. Zeller, Robert Bosch GmbH, Stuttgart

Members of the working group of chapter 10
Dr.rer.nat. P. Ballhause, Leybold AG, Hanau
Dr.rer.nat. G. Glomski, Puls-Plasma-Technik GmbH, Dortmund
Prof. Dr.-Ing. G. Herziger, FhG-ILT, Aachen
Prof. Dr.-Ing. Dr. h.c. W. König, WZL/FhG-IPT, Aachen
Dipl.-Ing. V. Sinhoff, FhG-IPT, Aachen
Dr.rer.nat. R. Wanke, Audi AG, Ingolstadt

Dr.-Ing. M. Woydt, Bundesanstalt für Materialforschung und -prüfung,
  Berlin
Dipl.-Phys. S. Zamel, FhG-IPT, Aachen

Members of the working group of chapter 11
Dr.-Ing. W. Beuck, WZL, Aachen
E. Bott, Volkswagen AG, Wolfsburg
Dr.-Ing. F. Ertl, Dr.-Ing. Höfler Meßgerätebau GmbH, Ettlingen
Dipl.-Ing. R. Flamm, FhG-IPT, Aachen
Dipl.-Ing. R. Freudenberg, WZL, Aachen
D. Gengenbach, KOMEG GmbH, Riegelsberg
Dipl.-Ing. B. Grün, Mauser-Werke GmbH, Oberndorf
Dipl.-Ing. K.-J. Lenz, Leitz Meßtechnik GmbH, Wetzlar
Prof. Dr.-Ing. Dr.h.c.(BR) T. Pfeifer, WZL/FhG-IPT, Aachen
Dipl.-Ing. C. Pietschmann, WZL, Aachen
Dr.-Ing. J. Thies, FAG Kugelfischer, Schweinfurt
Prof. Dr.-Ing. A. Weckenmann, Universität Erlangen -Nürnberg, Erlangen

Members of the working group of chapter 12
Dr.-Ing. W. Adam, IPK
cand.-ing. C. Beck, WZL
Dipl.-Ing. R. Bonse, WZL
Prof. Dr.-Ing. E. Doege, IFUM
Dipl.Ing. W. Folkerts, WZL
Dr.-Ing. P. Grund, Chiron Werke GmbH
Dipl.-Ing. W. Haferkorn, Waldrich Siegen GmbH
Dipl.-Ing. G. Hanrath, WZL
Dipl.-Ing. N. Hennes, WZL
cand.-ing. M. Hoyer, WZL
Dipl.-Ing. H. Ispaylar, WZL
Dipl.-Ing. A. Koch, WZL
Dipl.-Ing. W. A. Kuppe, Klüber Lubrication München KG
cand.-ing. R. Leonhardt, WZL
Dipl.-Ing. F. Michels, WZL
Dipl.-Ing. H. Nebeling, WZL
Dipl.-Ing. N. Seidensticker, Schiess AG
Dr.-Ing. H. Stave, Hüller Hille GmbH
Dipl.-Ing. T. Steinert, WZL

Prof. Dr.-Ing. K. Teipel, Schiess AG
Dipl.-Ing. W. v. Zeppelin, Traub AG
Prof. Dr.-Ing. E. h. Dr.-Ing. M. Weck, WZL/FhG-IPT
Dipl.-Ing. H. Wiesner, WZL
Dr.-Ing. H.-H. Winkler, Chiron Werke GmbH

Members of the working group of chapter 13
Dr.-Ing. D. Binder, Robert Bosch GmbH, Erbach
Dipl.-Ing. (FH) E. Bühler, Mercedes-Benz AG, Sindelfingen
Dipl.-Ing. U. Butz, Gebr. Heller Maschinenfabrik GmbH, Nürtingen
Dr.-Ing. B. Grünert, Pilz GmbH & Co., Ostfildern
Dipl.-Ing. K.-R. Hoffmann, Siemens AG, Erlangen
Dipl.-Ing. (FH) W. Klauss, Traub AG, Reichenbach/Fils
Dipl.-Ing. F. Klein, WZL, Aachen
Dipl.-Ing. A. Kohring, WZL, Aachen
Dr.-Ing. G. Krebser, ISW, University of Stuttgart
Prof. Dr.-Ing. G. Pritschow, ISW, University of Stuttgart
Dipl.-Ing. K. Ruthmann, Droop & Rein Werkzeugmaschinenfabrik,
    Bielefeld
Dipl.-Ing. F. Saueressig, IBH, Schwieberdingen
Dipl.-Phys. E. Schwefel, Dr. Johannes Heidenhain GmbH, Traunreut
Dr.-Ing. U. Spieth, Alfing Keßler Sondermaschinen GmbH, Aalen
Prof. Dr.-Ing. Dr.-Ing. E.h. M. Weck, WZL/FhG-IPT, Aachen

Members of the working group of chapter 14
J. Abler, Liebherr Verzahntechnik GmbH, Kempten
P. Beske, Volkswagen AG, Wolfsburg
Dr. H. Braas, Sony, Stuttgart
Dipl.-Ing. R. Dammertz, WZL, Aachen
Dipl.-Ing. Diedrich, ABB Roboter GmbH
Dipl.-Ing. K. Etscheid, Aachen
P. Früauf, NAM at DIN, Frankfurt
R. Hamm, Siemens AG, Nuremberg
Dr.-Ing. H. Heiss, BMW, Munich
Dr.-Ing. H.-J. Klein, Mannesmann Demag Fördertechnik, Wetter
Prof. Dr.-Ing. J. Milberg, IWB, Technical University of Munich
Dipl.-Ing. St. Peper, Aachen
W. Reis, Reis GmbH & Co. Maschinenfabriken, Obernburg

Dr.-Ing. F. Rühl, FhG-ILT, Aachen
W. Six, Howaldswerke Deutsche Werft AG, Kiel
Prof. Dr.-Ing. Dr.-Ing. e.h. M. Weck, WZL, Aachen

Members of the working group of chapter 15
Dr.-Ing. H.-J. Adlhoch, Herding GmbH, Amberg
Prof. Dr. W. Brandstätter, Ford Werke AG, Köln
Prof. Dr.-med. J. Bruch, Institut für Hygiene und Arbeitsmedizin, Essen
Dipl.-Ing. K. Gerschwiler, WZL, Aachen
Dipl.-Ing. P. Johannsen, Mercedes-Benz AG, Stuttgart
Ing.-grad. W. Kempf, K. Kässbohrer Fahrzeugwerke GmbH, Ulm
Prof. Dr.-Ing. Dr. h.c. W. König, WZL/FhG-IPT, Aachen
Dipl.-Ing. D. Lung, WZL, Aachen
Dr.-Ing. T. Mang, Fuchs Petrolub AG, Mannheim
Dipl.-Ing. G. Osterhaus, WZL, Aachen
Dipl.-Ing. S. Rummenhöller, FhG-IPT, Aachen
G. Szelag, Staatl. Gewerbeaufsichtsamt, Aachen
Dr.-Ing. K. Yegenoglu, Guehring Automation, Stetten a. k. M.

Members of the working group of chapter 16
Dipl.-Ing. Dipl.-Wirt. Ing. U. Böhlke, FhG-IPT, Aachen
Dipl.-Ing. K. Gressenich, Fa. Bosch-Siemens Hausgeräte GmbH,
      Traunreuth
Dipl.-Ing. Dipl.-Wirt. Ing. M. Hartmann, WZL, Aachen
Dipl.-Ing. Dipl.-Kfm. B. Katzy, WZL, Aachen
Dipl.-Ing. A. Klugmann, Fa. Thomson Video Europe GmbH, Villingen
Dr.-Ing. V. Lessenich-Henkys, IKV, Aachen
Dipl.-Ing. W. Noske, Fa. Bauknecht-Hausgeräte GmbH, Calw
Prof. Dr.-Ing. A. Weber, Fa. BASF AG, Ludwigshafen
H.-D. Welpotte, Fa. Miele & Cie. GmbH & Co., Gütersloh

Appendix B: References

References of chapter 1:

[1] Necker, T.: Veränderung der Märkte und ihre Auswirkungen auf den Produktionsbetrieb; Vortragsband zum fertigungstechnischen Kolloquium '91, Springer Verlag, Stuttgart (1991), S. 1-2

[2] Bullinger, H.-J., in: Vorgehensweisen und Praxisbeispiele zum Chancenmanagement in den90er Jahren; in: 10. IAO-Arbeitstagung, Stuttgart 19.-20. Februar 1991, Springer-Verlag (1991), S. 14-56

[3] Berger, R.: Local Hero; manager magazin 12 (1992), S.202-209

[4] Rommel, G.; Brück, F.; Diedrichs, R.; Kempis, R.; Kluge, J.: Einfach überlegen, Das Unternehmenskonzept das die Schlanken schlank und die Schnellen schnell macht; Schäffer-Poeschel Verlag, Stuttgart, (1993)

[5] Späth, W.: Neue Technologien und sich verändernde Qualifikations-anforderungen aus der Sicht eines Luft- und Raumfahrtunter-nehmens; in: Ingenieurqualifikation für das Jahr 2000, Leuchtturm-Verlag (1989), S. 11-18

[6] Leibinger, B.: Der deutsche Maschinenbau im internationalen Wettbewerb -wirtschaftliche und technische Positionen; FTK '91, Vortragsband, Springer-Verlag, Stuttgart (1991), S. 3-6

[7] Milberg, J.; Koepfer, Th.: Aufgaben- und Rechnerintegration - ein Gegensatz zur schlanken Produktion?; in: VDI-Bericht 990, VDI-Verlag (1992), S. 1-23

[8] Böhm, H.: Lean Management; Didacticum 15 (1993), S. 4-6

[9] Warnecke, H.-J.: Die Fraktale Fabrik, Zukunftsgerichtete Fertigungs-strukturen; CIM management 2 (1992), S. 27-32

[10] Eversheim, W.; Michaeli, W.: CIM im Spritzgießbetrieb, Wirtschaftlich Fertigen durch Rechnerintegration; Hanser Verlag (1993)

[11] Geiger, W.: Geschichte und Zukunft des Qualitätsbegriffs; Qualitäts-management 37 (1992) 1, S. 33-35

[12] Schulz, W.: Zertifizierte Qualität in den USA noch unterentwickelt; VDI-N, Nr.9 (1993), S. 20

[13] Rohe, E.-H.: Unternehmensziel Umweltschutz vor dem Hintergrund internationaler Umweltpolitik; ZfB-Ergänzungsheft 2 (1990), S. 23-40

[14] Barg, A.: Recyclinggerechte Produkt- und Produktionsplanung;
VDI-Z 133 (1991), Nr.11, S. 64-74

[15] Lingg, H.: Von der Bedeutung des Wettbewerbsfaktors Zeit;
Management Zeitschrift 61 (1992) Nr. 7/8, S. 73-77

[16] Brödner, P.; Schultetus, W.: Erfolgsfaktoren des japanischen
Werkzeugmaschinenbaus; RWK-Verlag, Eschborn 1992

[17] Simon, H.: Stein der Weisen; manager magazin 2 (1993), S. 134-140

[18] Theerkorn, U.: Problematik einer veränderten Produktion;
Werkstattechnik (wt), 81 (1991), S. 607-611

[19] Eversheim, W.; Böhmer, D.; Müller, St.; Tränckner, J.:
Reorganisation der technischen Auftragsabwicklung -
Rationalisierungspotentiale nutzen; Industrie-Anzeiger, 112. Jg.
(1990), Nr. 68, S. 10-18

20] Peters, G.: Ablauforganisation und Informationstechnologie im Büro
Konzeptionelle Überlegungen und empirisch-explorative Studie;
Müller Botermann Verlag, Köln (1988) (Reihe: Personalwesen,
Organisation, Unternehmensführung, Band 5), zugl. Dissertation,
Universität Köln, (1987)

[21] N.N.: Chancen und Risiken von CIM-Ergebnisbericht; Hrsg.:
Projektträger Technologiefolgenabschätzung, VDI-
Technologiezentrum Düsseldorf im Auftrag des Bundesministers
für Forschung und Technologie, Düsseldorf, September (1991)

[22] Striening, H.-D.: Prozeß Management: Versuch eines integrierten
Konzeptessituationsadäquater Gestaltung von
Verwaltungsprozessen in multinationalen Unternehmen;
Dissertation, Karlsruhe, (1988)

[23] Milberg, J.: Flexibilität braucht dezentrale Organisation;
VDI Nachrichten, 10. Januar (1992), S. 11

[24] Dunkler, H.: Auftragsabwicklung mit integrierter
Informationsverarbeitung unter Einsatz moderner
Produktionssystematik; REFA-Nachrichten, 3/(1985), S. 14-22

[25] Wildemann, H.: Auftragsabwicklung in einer computergestützten
Fertigung (CIM); Zeitschrift für Betriebswirtschaft (ZfB), 57. Jg.
(1987), Nr. 1, S. 6-31

[26] Schneider, M.: Die Quantifizierung organisatorischer Sachverhalte;
G. Marchal u. H.-J. Matzenbacher-Verlag, Berlin, (1981)

[27] Vallone, C.: Informationsmanagement wird zur Führungsbasis, 1.Teil,
io-Management-Zeitschrift, 59 (1990), Nr.1

[28] Eversheim, W.; König, W.; Weck, M.; Pfeifer, T.:
Produktionstechnik - Auf dem Weg zu integrierten Systemen,

Aachener Werkzeugmaschinenkolloquium, VDI-Verlag,
Düsseldorf, 1987

[29] Müller, S.: Entwicklung einer Methode zur prozeßorientierten
Reorganisation der technischen Auftragsabwicklung komplexer
Produkte; Dissertation, RWTH Aachen (1992)

[30] Kosiol, E.: Organisation der Unternehmung; 2. Auflage, Gabler-
Verlag Wiesbaden (1976)

[31] Scholz-Reiter, B.: Konzeption eines rechnergestützten Werkzeugs zur
Analyse und Modellierung integrierter Informations- und
Kommunikationssysteme in Produktionsunternehmen; Dissertation,
Berlin, (1990)

[32] Gaitanides, M.: Prozeßorganisation - Entwicklung, Ansätze und
Programme prozeßorientierter Organisationsgestaltung;
Verlag Franz Vahlen, München (1983)

[33] Milberg, J.: Flexibilität braucht dezentrale Organisation;
VDI-Nachrichten, 10. Januar (1992), S. 11

[34] N.N.: Kennzahlenkompaß, Informationen für Unternehmen und
Führungskräfte; Maschinenbau Verlag, VDMA Ausgabe (1992)

[35] Eversheim, W.; Müller, S.; Krumm, S.; Popp, W.; Montagegerechte
Produktsteuerung; VDI-Z 135 (1993), Nr. 1/2 Januar/Februar

[36] Mertens, P.; Holzner, J.: Wi - State of the Art, Eine
Gegenüberstellung von Integrationsansätzen der
Wirtschaftsinformatik, in: Wirtschaftsinformatik
34 (1992), S. 5-25

[37] N.N.: Anbieter-Recherche, CASE Tools für die CIM-Modellierung,
CIM-Management 4/92, S. 41-44

[38] Traenckner, J.H.: Entwicklung eines prozeß- und elementorientierten
Modells zur Analyse und Gestaltung der technischen
Auftragsawicklung von komplexen Produkten; Dissertation,
RWTH Aachen, (1990)

[39] Krogh, H.: Gefährliche Kreuzung; manager magazin 2 (1993), S.127-
132

[40] N.N.: Phantom in der Pipeline, Wie der Automobilzulieferer Carl
Freudenberg seine Teileproduktion beschleunigte;
manager magazin 11(1992)

[41] Kalkert, W.: Industrieller Entwicklungsprozeß von PKW-
Antrieben; Vorlesungmanuskript, RWTH Aachen, Oktober (1992)

[42] Seifert, H.: Zeit ist Geld; manager magazin 11 (1992)

[43] Modrich, G.; Kitzsteiner, F.: Integrierte Meßdatenrückführung im
FFS; Produktionsautomatisierung 1 (1992), S. 43-46

[44] Müller, G.: "Denkendes" Werkzeugsystem; MEGATECH 3 (1992)

References of chapter 2

[1]　N.N.: Statistisches Jahrbuch 1991 für das vereinte Deutschland. Hrsg.
　　　Statistisches Bundesamt. Erschienen im September 1991,
　　　Wiesbaden ISBN 3-8246-0078-1

[2]　Bemowski, K.: The International Quality Study. Quality Progress.
　　　November 1991, S. 33-37

[3]　N.N.: The single most important challenge for europe. European
　　　Quality Management Forum. Hrsg.: McKinsey & Company.
　　　Montreux 19. Oktober 1989

[4]　Pfeifer, T., u.a.: Untersuchung zur Qualitätssicherung, Stand und
　　　Bewertung, Empfehlung für Maßnahmen. Kernforschungszentrum
　　　Karlsruhe GmbH, PFT-Bericht 155, Karlsruhe 1990

[5]　N.N.: International Quality Study - Automotive Industry Report.
　　　Hrsg.: American Quality Foundation and Ernst&Young, Cleveland
　　　1992

[6]　Peacock, R.D.: Ein Qualitätspreis für Europa. Qualität und
　　　Zuverlässigkeit QZ 37 (1992) 9, S. 525-528

[7]　Büchner, U.: Bedeutung der Qualitätssicherung im internationalen
　　　Wettbewerb.In: Integration der Qualitätssicherung in CIM. Hrsg.:
　　　Verein Deutscher Ingenieure. VDI-Verlag, Düsseldorf 1991

[8]　Suzaki, K.: Modernes Management im Produktionsbetrieb. Carl
　　　Hanser Verlag, München Wien 1989

[9]　Womack, J.P., Jones, D.T., Roos, D.: Die zweite Revolution in der
　　　Automobilindustrie. Campus-Verlag, Frankfurt New York 1991

[10] Warnecke, H.J.: Innovative Produktionsstruktur. Vortrag zum FTK
　　　am 1. und 2. Oktober 1991 in Stuttgart. Springer Verlag, Berlin
　　　1991

[11] Köster, A.: Total Quality Management: Im internationalen
　　　Wettbewerb bestehen. Qualität und Zuverlässigkeit. QZ 37 (1992)
　　　7, S. 393-399

[12] Lücker, M.: Qualitätsorientierte Beschaffung. Dissertation RWTH
　　　Aachen. D82, Aachen 1992

[13] Danzer, H.H.: Quality-Denken stärkt die Schlagkraft des
　　　Unternehmens. Verlag Industrielle Organisation, Zürich 1990

[14] König, W.; Weck, M.; Eversheim, W.; Pfeifer, T.: Wettbewerbsfaktor
　　　Produktionstechnik. VDI-Verlag GmbH, Düsseldorf 1990

[15] Haist, F.; Fromm, H.: Qualität im Unternehmen. Carl Hanser Verlag,
[16] Pfeifer, T., u.a.: Qualitätsmanagement. Z. Zt. im Druck. Carl
  Hanser   Verlag, München Wien 1993
[17] Zink, K.J.: Qualität als Herausforderung. In: Zink, K.J.: Qualität als
  Managementaufgabe. Verlag moderne Industrie, Landsberg/Lech
  1989
[18] Runge, J.H.: Der steinige Weg zur Weltspitze. Qualität und
  Zuverlässigkeit QZ 37(1992) 11, S. 645-650
[19] Imai, M.: Kaizen, der Schlüssel zum Erfolg der Japaner im
  Wettbewerb. Wirtschaftsverlag Langen Müller/Herbig, München
  1992
[20] Shigeo Shingo: Poka-Yoke, Prinzip und Technik für eine Null-Fehler-
  Produktion. Hrsg.: gfmt-Gesellschaft für Management und
  Technologie AG, St. Gallen 1991
[21] Masing, W.: Qualitätspolitik des Unternehmens. In: Masing, W.:
  Handbuch der Qualitätssicherung. Carl Hanser Verlag; München,
  1988

References of chapter 3

[1] Pfeifer, T., Heine, J.; Orendi, G.: Stand der Qualitätssicherung und
  Handlungsbedarf in Deutschland - Vergleich mit
  Wettbewerbsnationen; Workshop Qualitätssicherung; Leipzig,
  1991
[2] Pfeifer, T.; Heine, J.; Köppe, D.; Lücker, M.; Orendi, G.:
  Länderspiegel Qualitätssicherung - Empfehlung für Maßnahmen;
  Carl Hanser Verlag, QZ, 4/91; Seite 201-206, München, 1991
[3] Büchner, U.: Bedeutung der Qualitätssicherung im internationalen
  Wettbewerb; in: Integration der Qualitätssicherung in CIM. HrsG.:
  Verein Deutscher Ingenieure; VDI Verlag; Düsseldorf, 1991
[4] Seitz, K.: Die Japanische Herausforderung. Blickpunkt 10/91, o.
  Verlag, o. Ort, 1991
[5] Pfeifer, T.; Heine, J.; Köppe, D.; Lücker, M.; Orendi, G.:
  Länderspiegel Qualitätssicherung - Qualitätssicherung in der
  Bundesrepublik Deutschland; Carl Hanser Verlag; QZ, 3/91; Seite
  135-140; München, 1991
[5] Qualitätssicherung - Qualitätssicherung in der Bundesrepublik
  Deutschland; Carl Hanser Verlag; QZ, 3/91; Seite 135-140;
  München, 1991

[6] Zink, K.J.: Qualität als Herausforderung; in: Zink, K.J.: Qualität als Managementaufgabe; Verlag moderne Industrie; Landsberg/Lech, 1989

[7] Pfeifer, T.; Heine, J.; Herter, K.; Ruegenberg, H.: Was der Produktionsingenieur von der Qualitätssicherung wissen muß; Die Bedeutung der Qualitätssicherung; VDI-Verlag; Düsseldorf (1991)

[8] Leibinger, B.: Der deutsche Maschinenbau im internationalen Wettbewerb wirtschaftliche und technische Position; Vortrag zum FTK (Fertigungstechnisches Kolloquium) am 1.-2. Oktober 1991 in Stuttgart; Springer Verlag; Berlin, Heidelberg, New York, 1991

[9] Pfeifer, T.; Heine, J.; Orendi, G.: Bedeutung der Qualitätssicherung; Prozeßüberwachung und Qualitätssicherung in der Lasermaterialbearbeitung; FhG-ILT; Aachen, 1991

References of chapter 4

[1] N.N.: Innovationsdynamik der Industrie sucht stabile Umsatzträger; VDI nachrichten 45/no. 18, 3rd May, 1991, p. 1

[2] Rommel, G., Brück, F., Diedrichs, R., Kempis, R.-D., Kluge, J.: Einfach überlegen. Das Unternehmenskonzept, das die Schlanken schlank und die Schnellen schnell macht; Schäffer-Poeschel Verlag, Stuttgart, 1993

[3] Wheelwright, S.C.j., Clark, K.B.: Revolutionizing Product Development - Quantum Leaps in Speed, Efficiency and Quality; Free Press (Maxwell Macmillan), New York, 1992

[4] Schurtzmann, B., et al.: Simultaneous Engineering (SE) in der Projektabwicklung/Produktentwicklung; Leitfaden der Arbeitsgemeinschaft Prozeßperipherie (Produktbereich Steuerungstechnik) des VDMA, published by VDMA, Frankfurt, 1991

[5] Eiff, W.: Prozesse optimieren - Nutzen erschließen, Simultaneous Engineering durch effizientes Informations-Management; IBM Nachrichten 41 (1991), issue 305, pp. 23 - 27

[6] Wildemann, H.: Entwicklungsstrategien für Zulieferunternehmen; ZfB, volume 62 (1992), issue 4, pp. 391 - 413

[7] Bronder, C. (publisher), Pritzl, R.: Wegweiser für strategische Allianzen: Meilen- und Stolpersteine bei Kooperationen; Allgemeine - Zeitung für Deutschland, Gabler Verlag, Wiesbaden, 1992

[8] Rotering, C.: F&E-Kooperationen zwischen Unternehmen, Poeschel Verlag, Stuttgart, 1990

[9] Stolz, H.: Automobilzulieferer gehen harten Zeiten entgegen; Industrie Anzeiger 22/92, p. 3

[10] Schneider, D., Zieringer, C.: Make-or-buy-Strategien für F&E: Transaktionskostenorientierte Überlegungen; Gabler Verlag, Wiesbaden, 1991

References of chapter 5

[1] Sebulke, J.: Durchdachte Produktlinien - Zukunftssicherheit in sich schnell wandelnden Märkten; Konstruktion 44 (1992), 12, pp. 398-406

[2] Arker, H.; Jüttner, G.: Ein integriertes Engineering System für den Maschinenbau; VDI-Berichte 993.1, Datenverarbeitung in der Konstruktion '92 - CAD im Maschinenbau; VDI-Verlag, Düsseldorf, 1992, pp. 1-14

[3] Spur, G.; Krause, F.-L.: CAD-Technik: Lehr- und Arbeitsbuch für die Rechnerunterstützung in Konstruktion und Arbeitsplanung; Carl Hanser Verlag, Munich, Vienna, 1984

[4] Grabowski, H.; Langlotz, G.; Rude, S.: 25 Jahre CAD in Deutschland: Standortbestimmung und notwendige Entwicklungen; VDI-Berichte 993, Datenverarbeitung in der Konstruktion '92, Plenarvorträge; VDI-Verlag, Düsseldorf, 1992, pp. 1-30

[5] Pfitzmann, J.; Jin, Z: Benutzungsorientierte Dialoggestaltung für die CAD-Anwendung; VDI-Berichte 993.1, Datenverarbeitung in der Konstruktion '92, CAD im Maschinenbau; VDI-Verlag, Düsseldorf, 1992, pp. 133-148

[6] Dressler, E.: Computer-Grafik-Markt 1992; Dressler-Verlag, Heidelberg, 1992

[7] Obermann, K.: CAD/CAM-Handbuch 1992; Verlag für Computergrafik GmbH, Munich

[8] Trippner, D.: ProSTEP - Die Einführung standardisierter Produktmodelle in die industrielle Anwendung am Beispiel der deutschen Automobilindustrie; VDI-Berichte 993.2, Datenverarbeitung in der Konstruktion '92, CAD im Fahrzeugbau und in Transportsystemen; VDI-Verlag, Düsseldorf, 1992, pp. 31-42

[9] Weck, M.: Produktentwicklung im Werkzeugmaschinenbau;

Produktionstechnisches Kolloquium; Berlin, 1992

[10] Krause, F.-L.: Leistungssteigerung der Produktionsvorbereitung; Productionstechnisches Kolloquium; Berlin, 1992

[11] Abeln, O.: CAD-Systeme der 90er Jahre - Vision und Realität; VDI-Berichte no. 861.1; VDI-Verlag, Düsseldorf, 1990, pp. 85-100

[12] Abeln, O.: Referenzmodell für CAD-Systeme, Bericht aus einem Arbeitskreis der GI; Informatik-Spektrum (1989) 12, pp. 43-46

[13] Marczinski, G.; Prengemann, U.; Holland, M; Mittmann, B.: Anwendungsorientierte Analyse des zukünftigen Schnittstellen-Standards STEP; ZwF 84 (1989) 8, pp. 456-461

[14] Franke, H.-J.; Peters, M.: Konstruktionsumgebung MOSAIK - eine grafisch-interaktive Benutzer- oberfläche zur Integration von Konstruktions-werkzeugen; VDI-Berichte no. 993.3; VDI-Verlag, Düsseldorf, 1992, pp. 175-191

[15] Franke, H.-J.; Mohmeyer, G.; Weigel, K.D.: Integrierter Produktentwurf in einer rechner-gestützten Konstruktionsumgebung - modularer Ansatz, Erfahrungen und objektorientiertes Anwendungsbeispiel; VDI-Berichte no. 861.2; VDI-Verlag, Düsseldorf, 1990, pp. 27-45

[16] Spur, G.; Sanft, C.; Schüle, A.: Ein Konstruktionssystem für Werkzeugmaschinen; ZwF 87 (1992) 8, pp. 434-438

[17] Eversheim, W.; Baumann, M.: Konfigurierbare Konstruktionssysteme. Ein neuer Ansatz zur anwendungsspezifischen Systemunterstützung; VDI-Z 134 (1992), no. 12, December, pp. 99-102

[18] Grabowski, H.; Anderl, R.; Schmitt, M.: Das Produktmodell von STEP; VDI-Z 131, 1989, issue 2, pp. 84-96

[19] Grabowski, H.; Anderl, R.; Schilli, B.; Scmitt, M.: STEP - Entwicklung einer Schnittstelle zum Produktdatenaustausch; VDI-Z 131 (1989), issue 9, pp. 68-76

[20] N.N.: VDI-Richtlinie 2221: Methodik zum Entwickeln und Konstruieren technischer Systeme und Produkte; VDI-Verlag, Düsseldorf, 1986

[21] Anderl, R.: Fertigungsplanung durch die Simulation von Arbeitsvorgängen auf Basis von 3-D Produktmodellen; Fortschritt-Berichte VDI; series 10, no. 40; VDI-Verlag, Düsseldorf, 1985

[22] Ottenbruch, P.: Entwicklung eines Systems zur Unterstützung der Konzept- und Entwurfsphase; Dissertation, RWTH Aachen, 1989

[23] Bauert, F.: Methodische Produktentwicklung für den rechnerunterstützten Entwurf; Schriftenreihe Konstruktionstechnik

(published by W. Beitz), no. 18; Technical University of Berlin, 1991

[24] Feldhausen, J.: Systemkonzept für die durchgängige und flexible Rechnerunterstützung des Konstruktionsprozesses; Schriftenreihe Konstruktionstechnik (published by W. Beitz), no. 16; Technical University of Berlin, 1989

[25] Grabowski, H.; Rude, St.: Methodisches Entwerfen auf Basis zufünftiger CAD-Systeme; VDI-Berichte no. 812, VDI-Verlag, Düsseldorf, 1990, pp. 203-226

[26] Weck, M., Repetzki, S.: Ein objektorientiertes Konstruktionswerkzeug; VDI-Z-SPECIAL CAD/CAM, APril, 1992, pp. 32-29

[27] Behr, B.: In kurzer Zeit; Spindel-Lager-Einheiten entwerfen mit Hilfe wissensbasierter Konstruktionssysteme; Maschinenmarkt; Würzburg 98 (1992), pp. 35-41

[28] Heckmann, A.: Zerlegungs- und Vernetzungsverfahren für die automatische Finite-Elemente-Modellierung; Dissertation, RWTH Aachen, 1992

[29] Kölsch, G.: Diskrete Optimierungsverfahren zur Lösung von konstruktiven Problemstellungen im Werkzeugmaschinenbau; Fortschritt-Berichte VDI; series 1: Konstruktionstechnik/Maschinenelemente; no. 213, 1992

[30] Eversheim, W.; Humburger, R.: CAD-Expertensystem-Kopplung - Systemverbund zur wissensbasierten Konstruktion von Baukastenvorrichtungen; VDI-Z-SPECIAL CAD/CAM, 1993

[31] IMPACT book; Trondheim, Norway; published by Tapir; 1992-ISBN 82-519-0973-2

[32] Eversheim, W.; Cremer, R.; Schneewind, J.: A Methodology for the Flexible Development of Integrated Process Planning Systems; Annals of CIRP, 1992

[33] Marczinski, G.: Verteilte Modellierung von NC-Planungsdaten - Entwicklung eines Datenmodells für die NC-Verfahrenskette auf Basis von STEP (Standard for the Exchange of Product Model Data); Dissertation, RWTH Aachen, 1993

References of chapter 6:
[1] Pfeifer, T. and Prefi, Th.: Die präventive Qualitätssicherung; Produktionsautomatisierung 1/92

[2]   Bergholz, H.-J.: Total Quality Management: Der Weg in die Zukunft; Qualität und Zuverlässigkeit, issue 7, 1991

[3]   Köster, A.: Total Quality Management: Im internationalen Wettbewerb bestehen; Qualität und Zuverlässigkeit, issue 7, 1992.

[4]   Zink, K. J.: Qualität als Managementaufgabe; 2nd revised edition, Verlag Moderne Industrie, Landsberg, 1992.

[5]   Röhrich, J.: Stand und Entwicklung objektorientierter graphischer Benutzeroberflächen; HMD journal, issue 160, 1991.

[6]   Pfeifer, T. and Scholz, C.: Verbrennungsmotoren im Schleppversuch prüfen; Industrie Anzeiger 76, 1990.

[7]   Niemann, H.: Klassifikation von Mustern; Springer Verlag, 1983.

[8]   Hinton, G.E.: Connectionist Learning Procedures; Artificial Intelligence, vol. 40, 1989.

[9]   Pao, Y.H.: Adaptive Pattern Recognition and Neural Networks; Addison Wesley, 1989.

[10] Ahlers, R.-J. et al.: Bildverarbeitung - Forschen, Entwickeln, Anwenden; Eigenverlag Technische Akademie, Esslingen, 1989 and 1991

[11] Conklin, J.: Hypertext: An Introduction and Survey; Computer IEEE, Sept. 1987.

[12] Arscyn, R.M., McCracken, D.L., Yoder.: KMS: A Distributed Hypermedia System for Managing Knowledge in Organizations; Communications of the ACM, vol. 31, no. 7, July 1988.

[13] Conklin, J. et al.: GIBIS: A Hypertext Tool for Exploratory Policy Discussion; ACM Transactions on Office Information Systems, vol. 6, no. 4, Oct. 1989, pp. 303-331.

[14] Ellis, C.A., Gibbs, S.J. and Rein, G.L.: Groupware: Some issues and experiences; Communications of the ACM, vol. 34, no. 1, Jan. 1991, pp. 18-29.

[15] Rich, E.: Artificial Intelligence; McGraw-Hill, Singapore, 1986.

[16] Harmon, P. and Sawyer, B.: Creating Expert Systems; John Wiley & Sons, 1990.

[17] Krems, J.: Expertensysteme im Einsatz; Oldenbourg, Munich, Vienna, 1989.

[18] Behrendt, R.: Angewandte Wissensverarbeitung; Oldenbourg, Munich, Vienna, 1990.

[19] Von Altrock, C.: Über den Daumen gepeilt; CT, issue 3, 1991.

[20] Zimmermann, H.-J.: Fuzzy Set Theory and Its Applications; 2nd edition, Cluwer Nijhoff Publish., 1990.

[21] Pfeifer, T. and Plapper, P.: Neue Perspektiven durch den Einsatz von

Fuzzy Logic; Produktionsautomatisierung, issue 3, 1993.

[22] Stroustrup, B.: Die C++ Programmiersprache; Addison-Wesley
Publishing Company, Bonn, 1992.

[23] Charniak, E. and McDermott, D.: Introduction to Artificial
Intelligence; Addison-Wesley Co. Ltd., 1985, pp. 9-11.

[24] Gimpel, B.: Qualitätsgerechte Optimierung von Fertigungsprozessen;
dissertation, RWTH Aachen, 1991.

[25] Owen, M.: Statistical Process Control, IFS Ltd. and Springer-Verlag,
Berlin - Heidelberg - New York - Tokyo, 1989.

[26] Itami, H.: Häufige Kontakte - Warum Nippon die Nase vorn hat.
Wirtschaftswoche, no. 39, 18th Sept., 1992, pp. 59-62.

[27] N.N.: ISO/TR 9007: Information processing systems - Concepts and
terminology for the coneptual schema and the information base,
1987.

[28] Huber, K.-P.: Diagnose von Automatikgetriebe-Schaltplatten mit
Expertensystemen; paper presented at the IAO forum
Expertensysteme in Produktion und Engineering, Stuttgart, March
'92.

[29] Mertens, P. and Legleiter, T.: Aufbau und Anwendung von
Diagnoseexpertensystemen; VDI-Z 130, no. 11, 1988.

[30] Pfeifer, T. and Zenner, Th.: Wissensbasierte Durchführung der
Fehlermöglichkeits- und Einfluß-Analyse FMEA; proceedings of
the research conference Qualitätssicherung, Frankfurt, 1992.

[31] Kersten, G.: FMEA - eine wirksame Methode zur präventiven
Qualitätssicherung; VDI-Z 132, no. 10, October 1990.

[32] N.N.: IQ-FMEA 2. Die zweite Generation von DV-Systemen für die
Fehlermöglichkeits- und Einfluß-Analyse. Informational brochure
from the CAP Debis company, 1992.

[33] Sullivan, L.P.: Der Erfolgreiche setzt Maßstäbe, QZ 36 (1991), pp.
681-686.

[34] Pfeifer, T.: Qualitätsprüfung im Wandel, Technisches Messen,
February 1990.

References of chapter 7

[1] Eversheim, W.: Simultaneous Engineering - eine organisatorische
Chance; VDI Berichte 758, VDI Verlag, Düsseldorf, 1989

[2] Evans, B.: Simultaneous Engineering; Mechanical Engineering, vol.
2, no. 2, 1988, pp. 38-39

[3]   Vasilash, G.S.: Simultaneous Engineering: Management's New
      Competitiveness; Tool Production, no. 7, 1987, pp. 36-41

[4]   Miyakawa, S.: Simultaneous Engineering and Producibility
      Evaluation Method; International Conference on the Application of
      Manufacturing Technologies; Conference Proceedings, Virginia,
      17th-19th April, 1991

[5]   Reimer, H.U.: Entwicklungszeit und -kosten reduzieren; Industrie-
      Anzeiger 91 (1991), pp. 26-27

[6]   Quast, H.: Schnell zum funktionsfähigen Modell; Laser-Praxis
      (1992), LS 116-118

[7]   N.N.: Modelle kurzfristig fertigen: Rapid Prototyping; Werkstatt und
      Betrieb 124 (1991) 7, p. 576

[8]   N.N.: Der schnelle Weg zum Muster; Automobil-Produktion 8
      (1992), pp. 102-104

[9]   N.N.: Mit Laser-Stereolithographie schneller zum serienreifen
      Dichtkonzept; VDI-Z 132 (1990) 8, pp. 17-23

[10] N.N.: Stereolithographie formt Modelle ohne Spezialwerkzeuge;
      Werkstatt und Betrieb 123 (1992), p. 796

[11] Kruth, J.P.: Material Incress Manufacturing by Rapid Prototyping
      Techniques; Annals of the CIRP, vol. 40 / 2 / 1991

[12] N.N.: Modelle nach Wunsch; Laser-Praxis (1992), LS 60

[13] Jacobs, P.: Rapid Prototyping & Manufacturing - Fundamentals of
      Stereolithography; Society of Manufacturing Engineers (1992)

[14] Stereolithographie beschleunigt Bauteilentwicklungen drastisch;
      Laser-Praxis (1992), LS 58-59

[15] N.N.: Unternehmensführung: Rapid Prototyping; Industrie-Anzeiger
      33 (1992), pp. 35-37

[16] N.N.: Third International Conference on Rapid Prototyping;
      Conference Proceedings, Dayton, 7th-10th June, 1992

[17] N.N.: 1st European Conference on Rapid Prototyping; Conference
      Proceedings, Nottingham, 6th-7th July, 1992

[18] Medler, D.; Jacobs, P.: DTM: What's new, what's next? Conference
      Proceedings, Chicago, 29th September, 1992

[19] Lumbye, K.: Desk Top Manufacturing; Danish Technological
      Institute, Aarhus (DK), 1992

[20] N.N.: Solid Freeform Fabrication Symposium 1992; Conference
      Proceedings, Austin, 3rd-5th August, 1992

[21] Colley, D.: Instant Prototypes; Mechanical Engineering, July 1988,
      pp. 68-70

[22] N.N.: 3D prototypes sintered from powders; Machine Design, 23rd January, 1992

[23] Muraski, St.: Make it in a minute; Machine Design, 8th February, 1990

[24] König, W.; Weck, M.; Herfurth, H.-J.; Ostendarp, H.; Zaboklicki, A.K.: Formgebung mittels Laserstrahlung; VDI-Z 135 (1993) 4

[25] Geiger, M.; Vollerstein, F.: Amon, St.: Flexible Blechumformung mit Laserstrahlung - Laserbiegen; Bleche Rohre Profile 38 (1991), no. 11, pp. 856-861

References of Chapter 8

[1] Deysson, C.; Handschuch, K.; Wolf-Doettinchem, L.: Kühne Experimente; Wirtschaftswoche no. 9, 26th Feb., 1993, pp. 14-18

[2] Pester, W.: Automobilhersteller unterliegen starkem Anpassungsdruck; VDI-Nachrichten, volume 47/no. 10, 12th March, 1993, p. 1

[3] Tränckner, J.: Entwicklung eines prozeß- und elementeorientierten Modells zur Analyse und Gestaltung der technischen Auftragsabwicklung von komplexen Produkten; dissertation, RWTH Aachen, 1990

[4] Roever, M.: Tödliche Gefahr, Überkomplexität I, Problem und Lösung; Manager Magazin 10 (1991), pp. 218-233

[5] Eversheim, W.; Müller, St.; Schares, L.: Funtionsbausteine der Montagevorbereitung, VDI-Z 135 (1993), no. 3, pp. 58-63

[6] Eversheim, W.; Dobberstein, M.; Fuhlbrügge, M.: Zeitgemäße Fertigungs- und Steuerungskonzepte, VDI-Z 136 (1993), no. 4

[7] Köblin, R.: Der Mensch im Unternehmen, Werkstatt und Betrieb 125 (1992) no. 12, pp. 885-887

[8] Stalk, J.; Hout, T.: Competing against Time, How Time-based Competition is Reshaping Global Markets; The Free Press, New York, Collier Macmillian Publishers, London, 1990

[9] Striening, H.-D.: Prozeß-Management: Versuch eines integrierten Konzeptes situationsadäquater Gestaltung von Verwaltungsprozessen in einem multinationalen Unternehmen; dissertation, University of Kaiserslautern, 1988

[10] Büdenbender, W.: Entwicklung von Anforderungen und Gestaltungsvorschlägen zur Konzeption einer ganzheitlichen Produktionsplanung und -steuerung für Unternehmen des

Maschinenbaus mit serieller inomogener
Auftragsabwicklungsstruktur, dissertation, RWTH Aachen, 1989

[11] Schnorbus, A.: Die Fabrik der Zukunft muß wie ein Organismus
arbeiten, Blick durch die Wirtschaft, volume 35, no. 189; 30th
Sept., 1992, p. D18

[12] Eversheim, W.; Müller, St.; Heuser, Th.: "Schlanke"
Informationsflüsse schaffen; VDI-Z 134 (1992), no. 11, pp. 66-69

[13] Müller, St.: Entwicklung einer Methode zur prozeßorienterten
Reorganisation der technischen Auftragsabwicklung komplexer
Produkte; dissertation, RWTH Aachen, 1992

[14] Owen, J.V.: Benchmarking World Class Manufacturing;
Manufacturing Engineering, March 1992, pp. 29-34

[15] Jacob, R.: How to Steal the Best Ideas Around; Fortune, October
19th (1992), pp. 86-89

[16] Wagner, D.; Schumann, R.: Die Produktinsel, Leitfaden zur
Einführung einer effizienten Produktion in Zulieferbetrieben,
Verlag TÜV Rheinland, Cologne, 1991

[17] Womack, J.P.; Jones, T.P.; Roos, D.: Die zweite Revolution in der
Autoindustrie, Campus Verlag, Frankfurt, New York, 1991

[18] Horn, V.; Trage, P.: Segmentierung steigert die Leistung; ZwF 87
(1992) 6, pp. 306-312

[19] Dörken, T.P.; Melchert, M.; Skudelny, Ch.: Flexible Fertigung,
Fachgebiete in Jahresübersichten; VDI-Z 134 (1992), no. 7-8,
pp. 58-76

[20] Göttker, A.: Teilefamilienbildung; Verlag TÜV Rheinland, Cologne,
1990

[21] N.N.: Tausend Teile von FFS im Fertigungsmatrix; Produktion 9
(1993), p. 12

References of chapter 9

[1] Mäscher, G.: Moderne Qualitätssicherungssysteme in der
Großserienfertigung - Welche Anforderungen ergeben sich daraus
für die Werkstofftechnik, HTM 2/93, to be published shortly

[2] Spur, G.: Entwicklungstendenzen in der spandenden Fertigung:
Maschine und Schneidstoffe im Wettlauf, Schweizer
Maschinenmarkt 39 (1991) 4, pp. 16-18, 20-21

[3] Tönshoff, H.K.; Brandt, D.; Spintig, W.: Hartbearbeitung in der
Praxis, wt Wissenschaft und Technik, no. 6 (1992), pp. 40-44

[4] Schmidt, J.; Kallabis, M.: Nacharbeit - Gehärtete Bauteile räumen mit Hilfe kristalliner Hartstoffe, Maschinenmarkt Würzburg 97 (1991) no. 36, pp. 50 ff

[5] Tönshoff, H.K.; Spintig, W.: Bohren gehärteter Stahlwerkstoffe: Fertigungsfolge vereinfachen, Ind. Anz. 113 (1991) 44, pp. 36-37

[6] Ackerschott, G.: Grundlagen der Zerspanung einsatzgehärteter Stähle mit geometrisch bestimmter Schneide, dissertation, RWTH Aachen (1989)

[7] König, W.; Link, R.: Optimierter Werkzeug- und Schneidstoffeinsatz beim Hartbohren, IDR 2/92, pp. 109-112

[8] König, W.; Klinger, M.: Räumen mit Hartmetall - Leistungssteigerung durch angepaßte Prozeßauslegung, 4. Karlsruhe Kolloquium 1992, published by wbk, University of Karlsruhe

[9] Fabry, J.: Cermet - Die wirtschaftliche Alternative zur Sicherung der Produktqualität bei Steigerung der Prozeßstabilität, Maschinenbau 3/93, pp. 34-39

[10] Johannsen, P.: Personal communication, March 1993

[11] Schlump, W.; Grewe, H.: Entwicklungsstand bei Herstellung und Anwendung von nanokristallinen Werkstoffen, Maschinenmarkt, Würzburg, 97 (1991) 13, pp. 84-89

[12] Kolaska, H.: Werkzeuge vor Leistungssprung, m+w, 26 (1991), pp. 50-52

[13] Kolaska, H.; Dreyer, K.: Verschleißfeste Schneidstoffe aus Hartmetall, Reibung und Verschleiß (Oct., 1992), pp. 65-85, published by H. Grewe, DGM Informationsgesellschaft mbH, Oberursel

[14] Kolaska, H.; Dreyer, K.: Hartmetalle, Cermets und Keramiken als verschleißbeständige Werkstoffe, Metall 45 (1991) 3, pp. 224-235

[15] N.N.: Mehr Zähigkeit wird verlangt, Produktion no. 1/2 (1993), pp. 6-7

[16] Christoffel, K.: Die weitere Entwicklung der Schneidstoffe, Werkstatt und Betrieb, 126 (1993) 1, pp. 15-17

[17] Kolaska, H.: Die wirtschaftliche Bedeutung der Hartmetalle und ihre Einsatzgebiete, Metall 46 (1992) 3, pp. 256-261

[18] Mason, F.: Cutting tools for the 90s, American Machinist (Feb., 1991), pp. 43-48

[19] Tönshoff, H.K.; Kaestner, W.: Temperaturbelastung von Hartmetallwerkzeugen bei Schnitt-unterbrechungen, VDI-Z 133 (1991) 9, pp. 85-91

[20] Kübel, E.: New developments in chemically vapour-deposited

coatings from an industrial point of view, Surface and Coatings Technology, 49 (1991), pp. 268-274

[21] Icks, G.: Naßfräsen mit beschichtetem Hartmetall, VDI-Berichte 762 (1989), pp. 221-232, VDI-Verlag GmbH, Düsseldorf, 1989

[22] Taberski, R.; van den Berg, H.: Neue Technologien entwickelt: Plasma-CVD-Verfahren für Hartmetalle, Ind. Anz. 72 (1991), pp. 36-42

[23] König, U.; van den Berg, H.; Tabersky, R.; Sottke, V.: Niedrigtemperaturbeschichtungen für Hartmetalle, proceedings, 12th Int. Plansee Seminar '89, vol. 3, pp. 13-25

[24] König, U.; Taberski, R.; van den Berg, H.: Research, development and performance of cemented carbide tools coated by plasma activated CVD, paper presented at the ICMCTF (22nd - 26th April, 1991), San Diego

[25] Knotek, O.; Löffler, F.; Krämer, G.: Substrate- and interface-related influences on the performance of arc-physical-vapour-deposition-coated cemented carbides in interrupted-cut machining, Surface and Coatings Technology, 54/55 (1992), pp. 476-481

[26] König, W.; Fritsch, R.: Performance and wear of carbide coated by physical vapour deposition in interrupted cutting, Surface and Coatings Technology, 54/55 (1992), pp. 453-458

[27] König, W.; Fritsch, R.: PVD- und CVD-beschichtete Hartmetalle im Leistungsvergleich, proceedings, 13th Int. Plansee Seminar '93, paper C2, to be published shortly

[28] Kolaska, H.: Cermets modern und leistungsfähig, Chem. Produktion 3/1992, pp. 37-39

[29] Kato, M.; Yoshimura, H.; Fujiwara, Y.: Mechanical Properties And Cutting Performance of TiC-TiN-Cermet Coated With TiN By PVD, Proc. 12th Int. Plansee Seminar '89, vol. 3, pp. 93-107

[30] N.N.: Weltneuheit beschichtetes Cermet UP 35 N, News 06/90 D, MMC Hartmetall GmbH

[31] Kopplin, D.; Großmann, G.: Beschichtete Cermets: Gewinn für das Drehen, VDI-Z 133 (1991) 2, pp. 16-20

[32] N.N.: New coated cermet insert is available, American Machinist (Feb., 1991), p. 15

[33] N.N.: Produktivitätssteigerung durch beschichtetes Cermet, Werkstatt und Betrieb 124 (1991) 1, p. 25

[34] N.N.: Die neue Bewegung bei Cermets, Werkzeuge (Dec., 1992), pp. 30-33

[35] N.N.: Arbeiten wie die Japaner, Werkzeuge (Dec., 1992), pp. 37-38

[36] N.N.: Schlanker und leistungsstärker, m+w 22-23 (1992), pp. 32-41

[37] Lux, B.; Haubner, R.; Pan, X.X.; Oakes, J.: Chemical vapour deposition diamond coatings on cemented carbide tools, Materials Science Monographs 73 (1991), pp. 600-607, Elsevier Science Publisher B.V., 1991

[38] Soederberg, S.; Westergren, I.; Reineck, I.; Ekholm, E.P.: Properties and performance of diamond coated ceramic cutting tools, Materials Science Monographs 73 (1991), pp. 69-76, Elsevier Science Publisher B.V., 1991

[39] Okuzumi, F.: Gaseous phase synthesis poly-crystalline diamond tool, Japan New Diamond Forum (1990), pp. 80-82

[40] Bachmann, P.K.; Messier, R.: Emerging technology of diamond thin films, Chemical & Engineering News, special report (15th May, 1989), pp. 24-39

[41] Leyendecker, T.; Lemmer, O.; Jürgens, A.; Ebberink, J.: Einsatz von kristallinen Diamant-schichten auf Werkzeugen und Verschleißteilen, Reibung und Verschleiß (Oct., 19992), pp. 215-225, published by H. Grewe, DGM Informations-gesellschaft mbH, Oberursel

[42] Westkämper, E.; Freytag, J.: PKD-Schneidstoffe zum Sägen melaminharzbeschichteter Spanplatten, IDR 25 (1991) 1, pp. 46-49

[43] Westkämper, E.; Licher, E.; Prekwinkel, F.: Hochgeschwindigkeitszerspanung von Holz und Holzwerkstoffen (2), HK 26 (1991) 3, pp. 20-25

[44] Ebberink, J.; Gühring, J.: Diamantbeschichtung für Zerspanungswerkzeuge, VDI Bericht 917 (1992), VDI-Verlag, Düsseldorf

[45] Malle, K.: Hochgeschwindigkeitsschleifen - Alternative zum Drehen und Fräsen?, VDI-Z 130 (1988), no. 7, pp. 50-56

[46] Meyer, H.-R.; Klocke, F.: High Performance Grinding with CBN, Society of Manufacturing Engineers 1991, Conference Superabrasives '91 (11th-13th June, 1991, Chicago, Illinois)

[47] König, W.; Ferlemann, F.: CBN-Schleifscheiben für 500m/s Schnittgeschwindigkeit, IDR 4/90, pp. 242-251

[48] König, W.; Ferlemann, F.: Eine neue Dimension für das Hochgeschwindigkeitsschleifen, IDR 2/90, pp. 66-72

[49] Meyer, H.-R.; Koch, N.: Richtiges Abrichten von Schleifscheiben mit Al O -, Diamant- und CBN-Körnungen als Voraussetzung für wirtschaftliche und hochproduktive Schleifprozesse, 7. Internationales Braunschweiger Feinbearbeitungskolloquium

(FBK) (1993)

[50] König, W.; Kammermeier, D.: Charakterisierung von TiN-, Ti(C,N)- und (Ti,Al)N-Hartstoffschichten anhand von Zerspan- und Simulationsuntersuchungen, Tribologie, Tagungsband zur 5. Präsentation Tribologie 1991, Coblence, pp. 17-34, Projektträgerschaft Material- und Rohstofforschung Forschungszentrum Jülich GmbH

[51] Kammermeier, D.: Charakterisierung von binären und ternären Hartstoffschichten anhand von Simulations- und Zerspanversuchen, D82 (dissertation, RWTH Aachen), Fortschritt-Berichte VDI, series 2: Fertigungstechnik, no. 271, VDI Verlag GmbH, Düsseldorf, 1992

[52] Link, R.: Gratbildung und Strategien zur Gratreduzierung, Dissertation, RWTH Aachen (1992)

[53] Communication from Robert Bosch GmbH, Stuttgart (1993)

[54] Meyer, H.-R.; Boos, J.; Rost, W.; Koch, N.: Nockenwellen-Schleifen mit CBN - Schleifscheiben, Jahrbuch "Schleifen, Honen, Läppen und Polieren", 57th edition (1993)

[55] König, W.: Fertigungsverfahren, vol. 3, Abtragen, 2nd edition, 1990, VDI-Verlag GmbH, Düsseldorf, 1990

[56] Dünnebacke, G.: High Performance EDM Using a Water Based Dielectric, proceedings of ISEM 10, Magdeburg, 1992

[57] König, W.; Siebers, F.J.: Funkenerosive Schmiede-gesenkherstellung mit wäßrigen Arbeitsmedien, Industrielle Gemeinschaftsforschung, in IDS no. 29, Hagen, 1992

[58] Siebers, F.J.: Thermoenergetische Verhältnisse im Arbeitsplatz bestimmen die Leistungsfähigkeit, dima 45 (1991) 9

[59] Rüling, R.F.: Technologische und physikalische Einflüsse auf das Funktionsverhalten von funkenerosiv geschnittenen Schneidwerkzeugen, dissertation, 1980, Technical University of Munich

[60] Hunzinger, I.: Schneiderodierte Oberflächen, Dissertation, 1986, iwb, Technical University of Munich, Springer-Verlag, Berlin, New York, Tokyo

[61] König, W.; Siegel, R.: Funkenerosives Feinstschneiden - Mit minimaler Energie zu guten Oberflächen, dima 9/92, pp. 26-30

[62] N.N.: Erodiertechnischer Multi-Knüller, Special Tooling 1/92, pp.16-21

References of Chapter 10

[1] Fleck, W.: Moderne Personenwagen-Schaltgetriebe -Mehr
     Leistungsdurchsatz bei gleichem Bauraum; ATZ
     Automobiltechnische Zeitschrift 92 (1990), pp. 744-753

[2] Todt, H.: Die Bedeutung der Mikro- und Feinwerktechnik in der
     heutigen Zeit; Feinwerk- und Meßtechnik F&M100, 1992, pp. 270-
     271

[3] Ehrfeld, W.: Fortschritt heißt Mikrotechnik; Feinwerk- und
     Meßtechnik F&M100, 1992, pp. 282-286

[4] O'Connor, L.: MEMS: Microelectromechanical Systems; Mechanical
     Engineering, Feb. 1992, pp. 40-47

[5] N.N.: Mikrosystemtechnik-Förderungsschwerpunkt im Rahmen des
     Zukunftskonzeptes Informationstechnik; Bundesministerium für
     Forschung und Technologie, 1990

[6] N.N.: Mikromechanik: Rasantes Wachstum; Markt&Technik no. 41,
     October 1992

[7] N.N.: Die Zukunft im Kleinen; Markt&Technik no. 50, December
     1992

[8] Linders, J.: Mikrosysteme-abgestimmter Einsatz von Mikro- und
     Systemtechniken; Feinwerk- und Meßtechnik 100, 1992, pp. 177-
     181

[9] Feiertag, R.: Mikrostrukturtechnik in der Feinwerktechnik; Feinwerk-
     und Meßtechnik 96, 1988, pp. 71-76

[10] Rosen, J.: Machining in the Micro Domain; Mechanical Engineering,
     March 1989, pp. 40-46

[11] Arzt, E.: Eigenschaften von Werkstoffen in kleinen Dimensionen;
     concise documentation on the papers presented at the 64th meeting
     of the scientific board of AIF, November 1992, p. 8

[12] N.N.: The LIGA Technique; company journal of MicroParts
     Gesellschaft für Mikrostrukturtechnik mbH, 1991

[13] Tönshoff, H.K.: Mikrotechnik zwischen Forschung und Anwendung;
     concise documentation on the papers presented at the 64th meeting
     of the scientific board of AIF, November 1992, p. 4

[14] N.N.: Dünnschichttechnologien, VDI-Technologiezentrum
     Physikalische Technologien 1989, p. 6

[15] Westheide, H.: Einfluß von Oberflächen-beschichtungen auf den
     Werkzeugverschleiß bei der Massivumformung; Berichte aus dem
     Institut für Umformtechnik, University of Stuttgart, Springer-
     Verlag, 1986

[16] Weist, Chr.: Verschleißminderung an Werkzeugen der

Kaltmassivumformung durch Ionenstrahltechniken; Berichte aus dem Institut für Umformtechnik, University of Stuttgart, Springer-Verlag, 1992

[17] Frey, H.; Kienel, G.: Dünnschichttechnologie, VDI-Verlag, Düsseldorf, 1987

[18] Haefer, R.A.: Oberflächen- und Dünnschichttechnologie, Teil I, Springer-Verlag, 1987

[19] König, W.; Fritsch, R.; Kammermeier, D.: New Approaches to Characterizing the Performance of Coated Cutting Tools; Annals of the CIRP, vol. 41/1/1992, pp. 49-54

[20] König, W.; Fritsch, R.; Kammermeier, D.: Physically vapour deposited coatings of tools: performance and wear phenomena; Surface and Coatings Technology, 49 (1991), pp. 316-324

[21] N.N.: Beschichtungstechnologie für die Tribologie TriTec; company journal of Leybold AG, 1992

[22] Ballhause, P.: Verschleißschutzschichten durch Sputter- und Ionenstrahltechnik; Werkzeuge, November 1991, pp. 70-75

[23] Straede, C.A.: Practical Applications of Ion Implantation for Tribological Modification of Surfaces; Wear 130, 1989, pp. 113-122

[24] Patir, N.; Cheng, H.S.: An Average Flow Model for Determining Effects of Three-Dimensional Roughness on Partial Hydrodynamic Lubrication; Transactions of the ASME, vol. 100, 1978, pp. 12-17

[25] Patir, N.; Cheng, H.S.: Application of Average Flow Model to Lubrication between Rough Sliding Surfaces; Transactions of the ASME, vol. 101, 1979, pp. 220-230

[26] Tripp, J.H.: Surface Roughness Effects in Hydrodynamic Lubrication: The Flow Factor Method; Transactions of the ASME, vol. 105, 1983, pp. 458-465

[27] Lietz, H.: Elektronenstrahlanlage zum Texturieren von Dressier- und Arbeitswalzen für den Einsatz in Kaltwalzwerken; company journal of Linotype-Hell AG, 1991

[28] Boppel, W.: Schnelles Elektronenstrahlgravier-verfahren zur Gravur von Metallzylindern; Optik 77, no. 22, 1987, pp. 83-92

[29] Wanke, R.: Personal communication, 1992

[30] Metz, N.: Entwicklung der Abgasemissionen des Personenwagen-Verkehrs in der BRD von 1970 bis 2010; ATZ Automobiltechnische Zeitschrift 92, 1990, pp. 176-183

[31] Schönwald, B.; Kuhn, J.: Technologie-Management am Beispiel

neuer Werkstoffe; Werkstoff und Innovation 5/6, 1990, pp. 37-46

[32] Glüsing, H.; Aengeneyndt, K.D.: Ingenieurkeramik im
Kraftfahrzeugmotor; Tribologie und Schmierungstechnik 3/1990,
pp. 132-138

[33] Woydt, M.; Habig, K.H.: Technisch-physikalische Grundlagen zum
tribologischen Verhalten keramischer Werkstoffe (summary of
relevant literature); Forschungsbericht 133 der Bundesanstalt für
Materialprüfung (BAM), Berlin, 1987

[34] Ashley, S.: Technology Focus: Materials and Assembly; Mechanical
Engineering, February 1992, pp. 16-20

[35] N.N.: Laserschweißen ohne Netz und doppelten Boden; wt-Report,
April 1992, pp. 30-35

[36] Sheasby, P.G.; Wheeler, M.J.: Überblick über die Technologie von
Aluminiumkonstruktionen im Fahrzeugbau; ATZ
Automobiltechnische Zeitschrift 92, 1990, pp. 258-265

[37] N.N.: Aluminium-Verkehr-Umwelt; informational brochure from
Aluminium-Zentrale e.V., Düsseldorf, 1992

[38] N.N.: HYCOT - The Automotive Tube; company journal of Hydro
Aluminium Automotive, Tonder (Denmark), 1993

[39] Glomski, G.: Magnetumformen von Aluminium; Aluminium-Journal
Fügetechniken, Aluminium-Zentrale Düsseldorf, volume 6 (1990),
no. 10, pp. 12-13

[40] Dengler, K.; Glomski, G.: Die Hochleistungsimpuls -Technik - Eine
zukunftsträchtige Hoch- geschwindigkeits-Umformtechnologie;
Bleche, Rohre, Profile 38 (1991) 4, pp. 285-286
Plasma-Technik GmbH, Dortmund, 1990

[42] Endriz, G.: High-Power Laser Diodes; IEEE-QE, vol. 28, no. 4
(1992), pp. 952-965

[43] Beach, R.: Modular microchannel cooled heatsinks for high average
power laser diode arrays, IEEE-QE, vol. 28, no. 4 (1992), pp. 966-
976

[44] Mundinger, D.: Laser diode cooling for high average power
applications; SPIE, vol. 1043 (1989), pp. 351-358

[45] Krause, V.; Treusch, H.G.; Beyer, E.; Loosen, P.: High-Power Laser
Diodes as a Beam Source for Materials Processing; conference
proceedings, ECLAT, 1992

[46] Auch, W.; Oswald, M.; Regener, R.: Fiber Optic Gyro Productization
at Alcatel SEL, company journal of Standard Elektrik Lorenz AG

[47] N.N.: Dreiachsiges Strapdown Kurs- und Lagemeß-Referenzsystem;
company journal of Helasystem, 1992

[48] Zeidler, G.: Telekommunikation im Wandel; Feinwerk- und Meßtechnik 100 (1992), pp. 275-278

[49] Auracher, F.; Plihal, M.: Photonik für breitbandige Übertragungs- und Vermittlungstechnik; Siemens-Zeitschrift Special FuE, spring, 1992

[50] Albrecht, H.: Integrierte optoelektronische Komponenten; Siemens-Zeitschrift Special FuE, spring, 1992

[51] Roß, L.; Hollenbach, W.; Wolf, B.; Fabricius, N.; Fuest, R.: Integriert optischer Abstandssensor auf Glas; company journal of IOT Entwicklungs-gesellschaft für integrierte Optik-Technologie mbH, Waghäusel-Kirrlach

[52] Fuest, R.: Integriert optisches Michelson-Interferometer mit Quadraturdemodulation in Glas zur Messung von Verschiebewegen; TM-Technisches Messen 58 (1991), issue 4, pp. 152-157

References of chapter 11

[1] Köppe, D.; Heid, W.: Möglichkeiten und Grenzen von SPC. Qualität und Zuverlässigkeit 34 (1989) 12, pp. 682-687

[2] Brinkmann, R.: Fähigkeit von Meßeinrichtungen. VDI-Bericht 1006, 1992, pp. 97-108

[3] Neumann, H.J.: Der Einfluß der Meßunsicherheit auf die Toleranzausnutzung in der Fertigung. Qualität und Zuverlässigkeit 30 (1985) 5, pp. 145-149

[4] Wortberg, J.; Häußler, J.: Moderne Konzepte der kontinuierlichen Prozeßüberwachung. Qualität und Zuverlässigkeit 37 (1992) 2, pp. 98-104

[5] Levi, P.; Lazslo, V.: Sensoren für Roboter. Technische Rundschau 20 (1987), pp. 108-122

[6] Heiler, K.-U.: Realisierung von Qualitätsregelkreisen durch Einsatz von Datennetzen. Dissertation, RWTH Aachen, 1989

[7] Elzer, J.: Automatische optoelektronische Erfassung des Bohrerverschleißes in Flexiblen Fertigungssystemen. Dissertation, RWTH Aachen, 1990

[8] Bott, E.; Kirstein, H.: Alternative Meßmöglichkeiten in der Karosserie-Fertigung. Messen Prüfen Automatisieren (1988) 7/8, pp. 374-377

[9] Krumholz, H.J.; Beuck, W.: Das Koordinatenmeßgerät im Qualitätsregelkreis. In: Pfeifer, T. (publisher):

Koordinatenmeßtechnik für die Qualitätssicherung; Grundlagen - Technologien - Anwendungen - Erfahrungen. VDI-Verlag GmbH, 1992, pp. 241-261

[10] Pfeifer, T.; Beuck, W.: Die Koordinatenmeßtechnik im Qualitätsregelkreis. In: Neumann, H.J. (publisher): Koordinatenmeßtechnik: Neue Aspekte und Anwendungen. Kontakt und Studium, volume 426, Expert Verlag, 1993, pp. 1-17

[11] Sigle, W.: Offline-Programmierung von Koordinatenmeßgeräten - Ein Erfahrungsbericht. VDI-Bericht 1006, 1992, pp. 1-21

[12] Weckenmann, A.: Koordinatenmeßtechnik im Wandel der Anforderungen. VDI-Bericht 751 (1989), pp. 1-17

[13] Beuck, W.: Entwicklung und Realisierung eines Kontur-Prüfkörpers für die aufgabenspezifische Überwachung von Koordinatenmeßgeräten. Dissertation, RWTH Aachen, 1992

References of chapter 12

[1] N.N.: Metalworking Engineering and Marketing, November 1992, pp. 28-36

[2] N.N.: Reihe 5, "Löhne, Gehälter und Arbeitskosten im Ausland", Fachserie 16, "Löhne und Gehälter", Statistisches Bundesamt, July 1992

[3] N.N.: Die HEYNUMATen, company journal of Heyligenstaedt GmbH & Co. KG, Gießen, 1991

[4] J. Thielcke: Gebündelte Arbeitsgänge, Industrieanzeiger 114 (1992) 45, pp. 49-51

[5] N.N.: Monorail MMS, company journal of W. Schneeberger AG, Roggwil/CH, 1993

[6] W. König, D. Lung, M. Klinger, A. Berktold: Spanende Bearbeitung gehärteter Werkstücke mit Schneidkeramik und CBN, paper presented at the seminar "Moderne Schneidstoffe: Cermets, Schneidkeramik", Technical University of Mannheim, 1991

[7] J.G. Endriz et al.: High Power Diode Laser Arrays, IEEE Journal of Quantum Electronics 28 (1992) 4, pp. 952 ff.

[8] M. Weck, C. Plewnia, H.-P. May: Schwingungsmessung auf rotierenden Bauteilen - Beispiele für angewandte Meßverfahren, VDI-Bericht no. 904, VDI-Verlag, Düsseldorf, 1991

[9] A. Koch: Experimentelle Untersuchung von Schrägkugellagern für Hochgeschwindigkeits-Spindellagersysteme, Lehrgang

"Konstruktion von Spindellagersysteme für die Hochgeschwindigkeitsmaterialbearbeitung", TAE Esslingen, March 1993

[10] W. Haas: Berührungsfreie Wellendichtungen für flüssigkeitsbespritzte Dichtstellen, dissertation, University of Stuttgart, 1986

[11] M. Weck: Werkzeugmaschinen, volume IV, "Meß-technische Untersuchung und Beurteilung", ISBN 3-18-400485-6, VDI-Verlag, Düsseldorf, 1992

[12] M. Weck, R. Bonse: Abnahmebedingungen an Werkzeugmaschinen, VDW-Forschungsbericht no. 0157, Frankfurt, 1992

[13] M. Weck, R. Bonse: Studie zum Thema Abnahmebedingungen an Werkzeugmaschinen - Bestandsaufnahme und Problemanalyse, Verein Deutscher Werkzeugmaschinenfabriken e.V., VDW-Forschungsbericht no. 0157, March 1992

[14] M. Weck, H. Ispaylar: Untersuchung von Wälzführungen zur Verbesserung des statischen und dynamischen Verhaltens von Werkzeugmaschinen, VDW-Forschungsbericht no. 0153, 1992

[15] A. Wieners: Vergleichende Untersuchung der erreichbaren Genauigkeit von Präzisionsführungen, internal, unpublished report from FhG/IPT, Aachen, 1992

[16] M. Brüstle: Beitrag zur adaptiven Verbesserung der Arbeitsgenauigkeit von CNC-Werkzeugmaschinen durch mikroprozessorgestützte Istwertmodifikation, dissertation, Hochschule der Bundeswehr, Hamburg, 1985

[17] M. Weck, I. Schubert, R. Bonse: Schneller und genauer; Industrie-Anzeiger 114 (1992) 47

[18] W. Haferkorn: Heavy Duty Portal Machining Centers, paper presented at the conference "The International Machine Tool Engineers' Conference", Osaka/Japan, 1990

[19] J.M. Dehmer: Prozeßführung beim funkenerosiven Senken durch adaptive Spaltweitenregelung und Steuerung der Erosionsimpulse, dissertation, RWTH Aachen, VDI-Verlag GmbH, Düsseldorf, 1992

[20] M. Weck, G. Kölsch: Finite-Element-Optimierung bei Berücksichtigung diskreter Parameterbeschränkungen, Konstruktion (1992) 44, pp. 237-241

[21] M. Weck, A. v. Arciszewski, T. Steinert: Without Limits, Fabrik 2000 7 (1991) 2

[22] H. Lorösch: Betriebssichere Schmierung von

Werkzeugmaschinenlagerungen, FAG Publikation no. 02 113 DA, Schweinfurt, 1985

[23] G. Pritschow, W. Philipp: Linear-Asynchronmotoren für hochdynamische Bewegungen: Oft die bessere Lösung, Industrie-Anzeiger 113 (1991) 21

[24] D. Lembke: Auf dem Weg zur einheitlichen Schnittstelle zwischen Maschinen und Werkzeugen, paper presented at the INFAG seminar "Tool Management", Schaffhausen, 1989

[25] D. Becker: Kühlschmiermittel - technologische und arbeitsmedizinische Aspekte, dissertation, University of Nuremberg, 1989

[26] F. Kalberlah: Kühlschmierstoffe auf Rapsöl- oder Mineralölbasis?, wt 82 (1992) 4

[27] W. A. Kuppe: Die Schmierung hochdrehender Spindellager unter besonderer Berücksichtigung der Fettschmierung mit Hochgeschwindigkeitsfetten, Lehrgang "Konstruktion von Spindellager-systemen für die Hochgeschwindigkeitsmaterial-bearbeitung, TAE Esslingen, March 1993

References of chapter 13
[1] Weck, M.[2] Weck, M.; Kohring, A.; Aßmann, S.: Bereichs-übergreifende Spezifikation von Werkzeugmaschinen; in: VDI-Z, issue 5/1992, pp. 111-115

[3] N.N.: Die Steuerung wird zur k.o.-Frage, in: Produktion, issue 42/1992, p. 3

[4] Pritschow, G.: Offene Steuerung - ein Gebot der Stunde? in: wt, issue 11/1992, p. 1

[5] Möller, W.: Offen und ehrlich; in: Industrie-Anzeiger, issue 30/1992, pp. 11-13

[6] Pritschow, G.: Merkmale eines offenen Steuerungs-konzeptes; in: Vortragsunterlagen zur Tagung "Offene Steuerung", 30.9/1.10.1992, Böblingen

[7] Pritschow, G.: Neuere Steuerungsentwicklungen (Teil 1); in: msr 34 (1991) 10, pp. 362-366

[8] Weck, M.; Friedrich, A.: MMS: Der Weg zum Companion Standard; collection of lectures for the MMS informational event on 3rd June, 1991 at the Standards Committee for Mechanical Engineering ('Normenausschuß Maschinenbau'), Frankfurt/Main

[9] Weck, M.; Friedrich, A.: NC-Verfahrenskette - Teil 4: Die Companion Standards zu MMS; VDI-Z 133 (1991), issue 10, pp. 64-67

[10] Mertens, R.: Akustische Sensorik und Bedienelemente mit taktiler Rückkopplung; in: Erfahrungsgeleitete Arbeit mit CNC-Werkzeugmaschinen und deren technische Unterstützung. IfA - Gh-Kassel, 1992, pp. 85-91

[11] N.N.: Anforderungen an einen Feldbus; in: Markt & Technik, no. 31/1992, pp. 54-55

[12] N.N.: Feldbussysteme für den Maschinenbau; in: Markt & Technik, no. 11/1991, pp. 68-70

[13] N.N.: SERCOS interface - Dokumentation; Förder-gemeinschaft SERCOS interface e.V., July 1990 issue

[14] Glantschnigg, F.: Zukunft und Wirtschaftlichkeit für Konstruktion und Herstellung komplexer Formen und Werkzeuge durch CAD-CNC-Kopplung; in: Werkstatt und Betrieb 123 (1990) 7, pp. 557-564

[15] Beckhoff, H.: PC - die Maschinensteuerung der nächsten Generation; in: Elektronik 17 / 1991, pp. 106-116

[16] N.N.: Die amerikanische Alternative? In: Fertigung, July 1992, pp. 12-15

[17] N.N.: Der amerikanische Vorstoß; in: Produktion, 23rd April, 1992, no. 17, p. 1

[18] Politsch, H.W.: Offensive mit offener Steuerung; in: Fertigung, July 1992, pp. 18-20

[19] Weston, R.H.; Harrison, R.; Booth, A.H.; Moore, P.R.: A new concept in machine control; in: Computer-Integrated Manufacturing Systems, vol. 2, no. 2, May 1989, pp. 115-122

[20] N.N.: Offenheit in Hard- und Software; in: Markt & Technik, no. 44/1992, p. 79

[21] Koch, D.: CNCs schneller projektieren; in: NC-Fertigung 4/91, pp. 131-136

[22] Buhl, H.: Offene Steuerungen gibt es nicht erst seit heute; in: wt November 1992, pp. 29-30

[23] N.N.: Bosch fordert den Schulterschluß heraus; in: Produktion, 15.10.1992, no. 42, p. 1

[24] Kreidler, V.: Flexible Automatisierung der NC-Bearbeitung; in: HARD AND SOFT, October 1988, pp. 58-65

[25] Pritschow, G.; Kugler, W.: Kommunikationskonzept für Steuerungen an Werkzeugmaschinen und Industrierobotern; in: wt, November 1992, pp. 69-70

References of chapter 14

[1] Mit der Pistole auf die Wanne, Kaldewei emailliert Bade- und Duschwannen mit Robotern; ROBOTER, November 1991, pp. 18-19

[2] Braas, J.: Japanische Montagestrategie für deutsche Produkte; company publication of Sony Europa GmbH

[3] Gezieltes Entgraten, Robotergestütztes Finishing in der Teilefertigung; ROBOTER, February 1991, pp. 52-53

[4] Schmidt, J.; Bott, K.: Hartes sanft entgratet, Flexible Fertigungszelle zum Entgraten von Keramik; ROBOTER, August 1991, pp. 28-30

[5] Lawo, M.: Handarbeit passé, Ein flexibles Robotersystem zum Gußputzen; Industrie-Anzeiger 61 / 1991, pp. 22-24

[6] Haspich, W.: Günstig in die Kleinserie, Robotik: Flexible Entgrat- und Schleifzelle; Industrie-Anzeiger 21 / 1991, pp. 46-48

[7] Grube, G.: Schliff vom Forscherteam, Uni Dortmund lehrt Robotern das Bandschleifen; ROBOTER, November 1991, pp. 48-52

[8] Merz, P.: Anspruchsvolle Aufgabe, Roboter beschichten PKW-Scheiben; Industrie-Anzeiger 75 / 1991, pp. 38-40

[9] Flexibilität pur, Roboterautomatisierte Getriebe-montage; ROBOTER, October 1992, pp. 20-22

[10] Warnecke, H.-J.; Würtz, G.: Einpressen leicht gemacht, Passives Toleranzausgleichsmodul für das flexibel automatisierte Einpressen; ROBOTER, May 1991, pp. 28-30

[11] Pritschow, G.; Rentschler, U.: Im µm-Bereich montieren, Sensoreinsatz für Roboter in der Montage; Industrie-Anzeiger 93 / 1991, pp. 25-27

[12] Schweizer, M.; Weisener, T.; Herkommer, T.F.: Robot assembly of pliable hoses; The Industrial Robot, December 1990, pp. 201-205

[13] Fichtmüller, N.; Hoßmann, J.; Kugelmann, F.: Richtschnur für Handling von Dichtschnur, Automatische Montage formlabiler Bauteile ist eine Frage der Systematik; ROBOTER, February 1992, pp. 34-37

[14] Wößner, J.F.: Automatische Fügung, Neues Montage-verfahren für biegeschlaffe Teile; ROBOTER, August 1992, pp. 32-33

[15] Gillner, A.; Nitsch, H.; Wolff, U.: 3D-Laser im Vergleichstest, Laser und Roboter. Grundlagen, Auswahl und Einsatzerfahrungen; ROBOTER, September 1991, pp. 55-59

[16] Formschluß mit Roboterzange, Druckfügen: Blechteile kostengünstig verbinden; Industrie-Anzeiger 5 / 1992, pp. 46-50

[17] Manz, D.; Gaul, M.: Alles im Lot, Roboterzelle für die Produktion

von Blechgehäusen; ROBOTER, February 1991, pp. 54-58

[18] Emmerich, H.: Meterweise Draht verlötet, Löten und Verlegen von Leitungen mit einem Werkzeug; FLEXIBLE AUTOMATION 5 / 1992, pp. 84-88

[19] Weiser, K.: Präzise Sensorhandführung, Induktionshärten von Formkanten mit einem Portalroboter; Industrie-Anzeiger 47 / 1991, pp. 12-18

[20] Ising, G.: Roboter steuern Fertigungsanlage zum Biegen; Werkstatt und Betrieb 125 (1992) 4, pp. 279-281

[21] Schweigert, U.; Herkommer, T.F.: Preise für Europa, Dienstleistungsroboter und Sensor-anwendungen im Blickpunkt der 22. ISIR; ROBOTER-Markt 1992, pp. 10-14

[22] Innovationen in Hülle und Fülle, Hannover Messe 92: MHI-Highlights; ROBOTER, May 1992, pp. 14-18

[23] Mobil gemacht, Schwerlastroboter mit eigenem Hydraulik-Fahrwerk; ROBOTER, October 1990, pp. 68-69

[24] Kreis, W.; Mehlan, A.; Rademacher, L.; Schoppol, M.: Montage und Handhabungstechnik, Industrieroboter, Fachgebiete in Jahresübersichten; VDI-Z 133 (1991), no. 4, pp. 55-59

[25] Schweers, E.: Abfälle vermeiden, Roboter in der Lackindustrie; Industrie-Anzeiger 69 / 1991, pp. 22-24

[26] Drews, P.; Arnold, St.: Garanten für Tempo und Spurtreue, Technische Hochschule Aachen entwickelt neue Robotersteuerung; ROBOTER, November 1991, pp. 22-24

[27] Beuthner, A.: Schalmeienklänge aus der Chipkiste, Transputer in der Automatisierungstechnik; ROBOTER, MAy 1991, pp. 37-40

[28] Prüfer, M.: Schnell, schneller, Transputer, Moderne Robotersteuerungen mit Parallelprozessen; ROBOTER, October 1990, pp. 26-28

[29] SIROTEC ACR, Steuerung und Programmiersysteme für Roboter und Handhabungssysteme; Siemens company journal, NC 51, 1991

[30] Entwicklung eines Systems zur interaktiven rechnergestützten Generierung von Bahnführungsdaten für Schweißroboter im Schiffbau; BMFT Abschlußbericht, Forschungs- und Entwicklungsvorhaben K4, September 1992

[31] Weck, M.; Weeks, J.: Montage modularer Spannvorrichtungen durch Industrieroboter: Programmierung in der Werkstatt, Schweizer Maschinenmarkt, 16th September, 1992, pp. 16-21

[32] Weck, M.; Stettmer, J.: Robot Design, Aachen Space Course, 1991

[33] Pritschow, G.; Spur, G.; Weck, M.: Maschinennahe

Steuerungstechnik in der Fertigung, Carl Hanser Verlag, Munich, Vienna, 1992, pp. 229-258

[34] Weck, M.: Werkzeugmaschinen, vol. 3: Automatisierung und Steuerungstechnik, 3rd edition, VDI-Verlag, Düsseldorf, 1989

[35] Schüller, H.: Steigerung der Einsatzmöglichkeiten des Messens auf Bearbeitungsmaschinen durch Verringerung der Meßunsicherheit; dissertation, WZL, RWTH Aachen, 1988

[36] Cannon, R.H.; Schmitz, E.: Initial Experiments on the End-Point Control of a Flexible One-Link Robot; The International Journal of Robotics Research, vo. 3, no. 3, 1984

[37] Henrichfreise, H.; Moritz, W.: Regelung eines elastischen Knickarmroboters; VDI-Berichte, no. 598, VDI-Verlag, Düsseldorf, 1986

[38] Futami, S.; Kyura, N.; Hara, S.: Vibration Absorption Control of Industrial Robots by Acceleration Feedback; IEE Transactions on industrial electronics, vol. IE-30, no. 3, August 1983

[39] Müller, P.C.; Ackermann, J.: Nichtlineare Regelung von elastischen Robotern; VDI-Berichte, no. 598, VDI-Verlag, Düsseldorf, 1986

[40] Stave, H.: Möglichkeiten und Grenzen der mechanische Optimierung von Portalrobotern; dissertation, WZL, RWTH Aachen

[41] Wittenstein, M.; Butsch, M.: Planetengetriebe für Roboter; Antriebstechnik 31 (1992) no. 6, pp. 36-39

[42] Kein Ende der Fahnenstange - Wachsender Markt für Automatisierungskomponenten, Robotertechnik 1992, pp. 8-10

[43] Rakic, M.: Multifingered Robot Hand with Selfadaptability, Robotics and Computer Integrated Manufacturing 2/3 (1989), pp. 269-276

[44] Weck, M.: Automatisierte Fertigung bei Losgröße 1, VDI-Nachrichten 8 (1991), p. 39

[45] Tsai, R.Y.: A Versatile Camera Calibration Technique for High-ccuracy Off-the-Shelf TV Cameras and Lenses, IEEE Journal of obotics and Automation, vo. RA-3, no. 4, pp. 323-344, August 1987

[46] Haberäcker, P.: Digitale Bildverarbeitung, Hanser, Munich, 1987

[47] Hättich, W.: Automatische Modellerstellung für wissensbasierte Werkstückerkennungssysteme, VDI-Verlag, Düsseldorf, 1989

[48] Erne, H.; Benz, A.: Leittechnik für Materialfluß-systeme in Montage und Fertigung; Bosch company publication; no. 3 842 506 814 IA 04/91

[49] Breuer, H.J.: Handhabung und Industrieroboter suchen breiten Einsatz; Werkstatt und Betrieb 125 (1992) 7; pp. 545-549

References of: chapter 15

[1] Schmidt-Bleek, F.: Ein universelles ökologisches Maß ? Gedanken zum ökologischen Strukturwandel; Wuppertal Papers Nr.1/1992

[2] Holl, F.-L; Rubelt, J.: Betriebsökologie; Bund Verlag Köln, 1993

[3] Rufer, D.; Dörler, H.: Ökologische Unternehmensentwicklung; IO Management   Zeitschrift 61 (1992) 1, S.70-73

[4] Autorenkollektiv: Wettbewerbsfaktor Produktionstechnik; Aachener Werkzeugmaschinen Kolloquium, 1990

[5] Johannsen, P.: Unveröffentlichte Informationen der Mercedes-Benz AG, 1993

[6] Becker, S.: Hautschutz beim Umgang mit Kühlschmierstoffen; in: Die gesundheitlichen Aspekte beim Einsatz der neuen biostabilen Kühlschmierstoffe; Sonderdruck der Deutsche Castrol Industrieoel GmbH, Hamburg

[7] König, W.: Fertigungsverfahren, Band 1 Drehen, Fräsen, Bohren; 4. Aufl. 1990, VDI-Verlag Düsseldorf

[8] DIN 51 385: Kühlschmierstoffe Begriffe; Hrsg. Deutscher Normenausschuß, Juni 1991, Beuth Verlag Berlin

[9] VDI 3397, Blatt 2: Entsorgung von Kühlschmierstoffen; Entwurf August 1991, Hrsg. VDI-Gesellschaft Produktionstechnik (ADB), Beuth Verlag Berlin

[10] Richtlinie für den Umgang mit Kühlschmierstoffen (ZH 1/248); Entwurf November 1992, Hrsg. Hauptverband der gewerblichen Berufsgenossenschaften, Fachausschuß Eisen und Metall II, St. Augustin

[11] BIA-Report 3/91, 2. Auflage Januar 1993: Kühlschmierstoffe Umgang, Messung, Beurteilung, Schutzmaßnahmen; Hrsg. Hauptverband der gewerblichen Berufsgenossenschaften, St. Augustin

[12] N. N.: Kühlschmierstoffe; Hrsg. Maschinenbau- und Kleineisenindustrie-Berufsgenossenschaft, Düsseldorf

[13] Martin, K.; Yegenoglu, K.: HSG-Technologie. Handbuch zur praktischen Anwendung; Hrsg. Guehring Automation GmbH, Stetten a.k.M.-Frohnstetten, 1992

[14] Boor, U.: Kühlschmierstoffe zum Räumen; Teil 2, VDI-Z 134 (1992) 2, S. 71-81

[15] N. N.: N-Nitrosamine in wassermischbaren bzw. wassergemischten Kühlschmierstoffen; Technische Regel für Gefahrenstoffe, Entwurf 1991, Hrsg. Bundesminister für Arbeit und Sozialordnung und

Bundesminister für Uwelt, Naturschutz und
Reaktorsicherheit

[16] Behrend, R.: Flüssige Gehilfen. Kühlschmierstoffe und ihre Zusätze
verbessern das Ergebnis beim Metallzerspanen; Maschinenmarkt
93(1987)39, S. 42-48

[17] Appelbaum, G.: Neue Rezepte. Additive in modernen
Kühlschmierstoffen; moderne fertigung April 1988, S. 88-96

[18] Lingmann, H.: Hygiene und Arbeitsschutz beim Einsatz von
Kühlschmierstoff; Technik Aktuell (1990) 2

[19] N.N.: Giftcocktail Kühlschmierstoffe; Gefahrstoffinformation der IG
Metall Bezirksleitung Baden-Württenberg, Stuttgart, 1990

[20] N.N.: Wäßrige Werkzeuge habens schwer ..; NC-Fertigung 5/92,
S.136-138

[21] Groß, H.H.; Simon, H.: Atemluftbelastung in der Fertigung; Werkstatt
und Betrieb 125 (1992) 9, S.721-725

[22] König, W.; Gerschwiler, K.; Fritsch, R.: Leistung und Verschleiß
neuerer beschichteter Hartmetalle; in: Pulvermetallurgie in
Wissenschaft und Praxis, Band 8, 1992, Hrsg.: H. Kolaska,
VDI-Verlag Düsseldorf

[23] König, W.; Klinger, M.: Räumen mit Hartmetall. Leistungssteigerung
durch angepaßte Prozeßauslegung; in: Räumen - Neue
Technologien. Tagungsband zum Karlsruher Kolloquium, Februar
1992, Hrsg.: wbk, Universität Karlsruhe

[24] König, W.; Klinger, M.: Beschichtetes Hartmetall - Leistungsreserven
beim Räumen nutzen; VDI-Z 4 (1993)

[25] König, W.; Peiffer, K.; Knöppel, D.: Kühlschmierstofffreie
Zahnradfertigung. Die Produktion umweltverträglicher gestalten;
Ind.-Anz. 94 (1991), S.23-25

[26] N.N.: Methoden zur Rohstoff- und Energieeinsparung für
ausgewählte Fertigungsprozesse; DFG-Sonderforschungsbereich
144, Arbeits- und Ergebnisbericht 1991, Hrsg. SFB-Geschäftsstelle
RWTH-Aachen

[27] Goldstein, M.: Optimierung der Fertigungsfolge Fließpressen -
Spanen; Dissertation RWTH Aachen, 1991

[28] Lowin, R.: Schleiftemperaturen und ihre Auswirkungen im
Werkstück; Dissertation RWTH Aachen, 1980

[29] Ackerschott, G.: Grundlagen der Zerspanung einsatzgehärteter Stähle;
Dissertation RWTH Aachen, 1989

[30] König, W.; Berktold, A.; Severt, W.: Spanende Bearbeitung ohne
Kühlschmierstoffe - Ein Beitrag zur Verringerung der

Umweltbelastung; PA - Produktionsautomatisierung

[31] Ihrig, H.: Umweltverträgliche Schmierstoffe in den 90er Jahren;
Tribologie und Schmierungstechnik 39 (1992) 3, S.121-125

[32] Kiechle, A.: Zukunft des Kühlschmierstoffeinsatzes; Vortrag anl. des
Schmierstoff-Forums, 1992

[33] Böschke, K.: Nichtwassermischbare Kühlschmierstoffe - heute und
morgen; Vortrag anl. des Seminars "Kühlschmierstoffe heute -
morgen", Nürnberg, 1992

[34] Müller, J.: Grundlagen, Anforderungsprofile von wassermischbaren
Kühlschmierstoffen heute und morgen; Vortrag anl. des
Schmierstoff-Forums, 1992

[35] Angerer, W.: Die gesundheitlichen Aspekte beim Einsatz der neuen
biostabilen Kühlschmierstoffe; Sonderdruck der Deutsche Castrol
Industrieoel GmbH, Hamburg

[36] N. N.: Unveröffentlichte Untersuchungen am WZL der RWTH
Aachen, 1992

[37] N.N.: Intern gekühlte Ab- und Einstechverfahren; NC-Fertigung 6
(1992), S.90-95

[38] Wertheim, R.; Rotberg, J.; Ber, A.: Influence of High-Pressure
Flushing through the Rake Face of the Cutting Tool; Annals of the
CIRP 41 (1992), S.101-106

[39] Mang, T.: Unveröffentlichte Untersuchungen der Firma Fuchs
Petrolub AG, 1992

[40] Baumgärtner, Th.: Null-Lösung. Mercedes-Benz will
Kühlschmierstoffe reduzieren, Ind.-Anz. 69 (1991), S. 10-11

References of: chapter 16

[1]   N.N.: Umweltschutz, Österreich und Deutschland an der Spitze;
Wirtschaftswoche, (17.4.1992) Nr. 17, S. 15

[2]   N.N.: Mineralische Rohstoffe; Bundesministerium für Wirtschaft,
Bonn, 1987

[3]   Rogal, H.: Strategien zur entsorgungsgerechten Gestaltung von
Produkten; Abfallwirtschaftsjournal, 3. Jg. (1991) Nr. 11, S. 704

[4]   Wicke, L., De Maizière, L., De Maizière, T.: Ökosoziale
Marktwirtschaft für Ost und West, München, 1990

[5]   N.N.: Umweltfreundliche Beschaffung - Handbuch zur
Berücksichtigung des Umweltschutzes in der öffentlichen
Verwaltung und im Einkauf; Umweltbundesamt, Wiesbaden,

Berlin, 1986

[6]   Benjamin, R., Blunt, J.: Informationstechnik im Jahr 2000 - ein
      Wegweiser für Manager; Harvard Business Manager, (1993) Nr. 1,
      S. 73ff

[7]   N.N.: Umweltökonomische Gesamtrechnung: Ein Beitrag zur
      amtlichen Statistik; Statistisches Bundesamt, Wiesbaden, 1990

[8]   Cäsar, C.: Kostenorientierte Gestaltungsmethodik für variantenreiche
      Serienprodukte - Variant Mode and Effects Analysis (VMEA);
      Fortschritt-Berichte VDI, Reihe 2, Nr. 218, Düsseldorf, 1991

[9]   Hartmann, M., Lehmann, F.: Fachgebiete in Jahresübersicht:
      Demontage; VDI-Z, 135. Jg. (1993) Nr. 1-3

[10] Eversheim, W., Hartmann, M.: Verursachungsgerechte Bewertung
      der Montage; VDI-Z, 135 Jg. (1993) Nr. 5, S. 135ff

[11] Eversheim, W., Schmetz R.: Energetische Produktlinienanalyse; VDI-
      Z, 134. Jg. (1992) Nr. 6, S. 46-52

[12] Razim, C.: Automobil und Werkstoff - Spannungsfeld von
      Technologie, Ökonomie und Ökologie; Beitrag zum gleichnamigen
      Colloquium, Salzburg, 1992

[13] N.N.: Statistisches Jahrbuch 1990 für die Bundesrepublik
      Deutschland; Statistisches Bundesamt, Wiesbaden, 1991

[14] Binding, J.: Grundlagen zur systematischen Reduzierung des Energie-
      und Materialeinsatzes; Dissertation, RWTH Aachen, 1988

[15] Eversheim, W., Binding, J., Schmetz, R.: In der Produktion Energie-
      und Materialkosten einsparen; VDI-Z, 132. Jg. (1990) Nr. 2, S.
      41ff

[16] Eversheim, W., Böhlke, U., Schmetz, R.: Erstellung von
      Substitutionskriterien für Verfahren und Werkstoffe - Methoden
      zur Energie- und Rohstoffeinsparung für ausgewählte
      Fertigungsprozesse; Arbeits- und Ergebnisbericht des
      Sonderforschungsbereiches 144, RWTH Aachen, 1991, S. 7ff

[17] Eversheim, W., Böhlke, U., Adams, M.: Die Auswahl des 'richtigen'
      Werkstoffes - Neue ökonomie- und ökologieorientierte
      Bewertungsmethoden; Beitrag zum Werkstoff-Forum-Seminar
      "Energieeinsparung bei der Herstellung, dem Einsatz und der
      Entsorgung von Werkstoffen", Aachen, 1992

[18] Weber, A.: Kunststoffe im Automobilbau; Beitrag zum Colloquium
      "Automobil und Werkstoff - Im Spannungsfeld von Technologie,
      Ökonomie und Ökologie", Salzburg, 1990

[19] Dyllick, T.: Ökologisch bewusstes Management; Die Orientierung,
      Schweizerische Volksbank, Bern 1990

[20] Voss, G.: Wettbewerbsvorteile von Morgen; Umwelt, (1988) Nr. 5, S. 240f

Index

"lean" concepts, 447

(fundamental fit), 104

`fractal' factory, 39

3D Welding, 194
3D-Printing (3DP), 181

7 M's, 52

Acceptance testing, 334
Accessibility to narrow spaces, 410
Actuator, 435
Adapting production lines, 394
Added value, 11
Air vibration, 307
Allergic contact eczema, 469
Aluminium, 280
Analysis of operation, 21
Analytical phase, 195
Application of adhesives, 397
application-oriented programming
    user interface, 425
Aqueous dielectrics, 260
Aqueous working medium, 258
Assembly, 399
Assembly costs, 509
Assembly strategies, 399
Automatic component
    optimization, 135
Automatic control, 263
Avoidance of faults, 147

Aackground noise, 302
Bacteria, 466
Ballistic Particle Manufacturing
    (BPM), 181
Belt grinding, 396
Benchmarking, 216

Biodegradability, 352
Burner-changing system, 391
Burring, 251
Bus system, 436
Business process, 23

Cable breakage, 410
Cables, 436
Cabling, 410
CAD, 111; 133; 167
CAD system, 438
CAD systems, 426
CAD-assisted programming, 432
CAD/CAM, 112
Calibrated ball plates, 311
Calibration, 311
Camera calibration, 440
Camshaft, 300
Carbides, 477
CBN, 247; 256
CBN grinding technology, 256
CCD arrays, 303
Ceramic, 280; 517
Ceramic balls, 332
Cermets, 242
Cermets, 477
Certification, 68
Checking, 489
Chlorine, 352
CIM, 5
Classical fields of applications, 387
Clean manufacturing, 453
C_m, 297
CNC coordinate measuring
    devices, 305
CO_2 laser, 187
CO_2 lasers, 403
Coated carbide, 240
Coated or uncoated, 477
Coating, 394
Collision prevention, 443
Collision-prevention system, 436
Collisions, 442
Communications interfaces, 421
Compensate tolerances, 392
Compensating mechanical
    weaknesses, 416

Competitive factor, 9
Competitive strategies, 3
Complete machining, 325
Compliant tool-holders, 396
Composition, 460
Concentrates, 466
Concepts, 15
Configure control systems, 422
Continuous improvement, 49
Continuously controlled
    processes, 305
Contractor-subcontractor
    relationship, 88
Control circuit, 299
Control systems, 417
Controlled compensation
    process, 416
Controlling processes, 295
Cooling, 458
Cooling lubricant, 349
Coordination, 13
Copy-milling, 180
Core process, 11
Corporate activities, 4
Corporate strategy, 42
Correction matrix, 310
Correction tables, 416
Cost, 508
Cost-consciousness, 312
Costs and benefits, 75
C_{pk}, 297
CSCW (Computer-Supported
    Cooperative Works), 166
Cultural fit, 104
Culture of consensus, 66
Customer and supplier, 39
Cutting ceramics, 477
Cutting speeds, 299
Cutting tool, 301
CVD coating, 240
Cycle times, 417
Cycloidal, 414

Data inconsistency, 8
Data processing, 145
Data storage, 13
DAX, 156

Debugging, 431
Deburring, 396
Defined cutting edge, 459
Defining objectives by
    consensus, 63
Deming, 65
Departmental boundaries, 13
Deploy a robot system, 407
Design measures, 401
Design models, 177
Desktop Manufacturing, 180
Development of process
    technology, 396
Diamond layers, 274
Diamond-coated tools, 245
Digital mockup, 115
DIN ISO 9004, 148
Diode laser, 283
Direct linear drives, 346
Directives, 471
Dismantling, 510
Disposal, 465
Disposal costs, 236
Dosing system, 398
Dry broaching, 474
Dry cutting, 479
Dry machining, 237; 474
Dry plain milling, 475
Dry tapping, 476
Dynamic performance, 413

Easy-care cooling lubricants, 484
Ecological Aspects, 464
Ecology, 5; 499
Economic aspects, 472
Economic efficiency, 306; 406
Economy, 499
Edge-tracking sensors, 392
Edges, 438
EDM-sinking, 258; 264
Electron beam, 269
Electronic components, 436
Electropolishing, 260
Elements of quality management,
    48
Emissions, 456
Emulsions, 462

Enamelling, 394
Energy consumption, 512
Engineering data base, 115
Ensors for tracking the seam, 390
Environment, 496
Environment-friendly, 483
Environmental compatibility, 349
Environmental cost, 500
Environmental impact, 463
Environmental problems, 454
Environmental protection, 495
EP additives, 462
Ergonomics, 177
Erosion processes, 257
Evaluation portfolio, 506
Evaluation system 2000, 521
Expert and knowledge-based
   systems, 153
Expert knowledge, 161
Expert system, 156
Expert systems, 120
Extrusion of polyurethane sealing
   lips, 398

Failure mode and effects analysis
   (FMEA), 41; 160
Fault detection, 292
Feature oriented programming, 307
FEM, 133
Fibre gyroscope, 285
Field bus level, 421
Filter systems, 353
Finite element optimization, 136
Fitting, 392
Fitting of cooling water tube, 402
Fitting of small parts, 393
Fixture Expert, 137
Flame cutting, 392
Flexibility, 405
Flexible parts, 401
Flowchart, 429
FMEA team, 162
Force-measuring sensors, 301
Forces, 298
Frame components, 412
Free-form surfaces, 311
Frequency sensitivity, 303

FTAM, 420
Function, 14
Functional prototypes, 178
Fundamental fit, 104
Fungi, 466
Fused Deposition Modeling, 189
Fuzzy logic, 154

Gantries, 436
Gauge blocks, 311
Gauges, 301
Gearbox assembly, 400
Gears, 413
Geometric modelling, 117
Geometrical deviations, 292
Geometrical prototypes, 177
Graphic display, 307
Grey-scale images, 438
Grinding, 180
Grinding grain, 459
Grinding machine, 300
Grinding sludge, 492
Groupware and computer-
   supported cooperative work
   (CSCW), 153
Guideway, 328; 336

Hand-held measuring instruments,
   305
Handling, 392
Hard machining, 252; 330
Hard turning, 254
Harmonic drive, 414
Health risk, 468
Health-impairing characteristics,
   466
Height profile, 443
High-level languages, 426
High-speed grinding, 246
High-Temperature Laser Sintering,
   193
Historical development, 3
Historical fields, 388
Hollow joints, 412
Hollow-shaft adaptor, 348
Human resources, 25
Hybrid programming, 426

Hypertext, 166
Hypertext and hypermedia, 151

Image processing, 437
Improved logistic flows, 39
Increasing efficiency, 219
Industrial Robot Language, 428
Infeed grinding, 254
Infinite rotatability, 410
Information, 19
Information technology, 6
Innovation, 5
Insulating mats, 401
Integrated product model, 111
Integration, 133; 219; 326
Integration, 147
Integrational planning phase, 195
Interdisciplinary team, 165
Interpolation cycles, 419
Introduction of a quality
    management system, 59
Introductory phase, 195
Ion beam assisted deposition
    (IBAD), 275
Ion beam mixing, 275
Ion implantation, 269
IRL, 428
ISO reference model, 420

Junior managerial level, 64

Key factor, 25
Knocking sensor, 303
Knowledge base, 162
Knowledge-acquisition, 159
Knowledge-based design system,
    132
Knowledge-based spindle-bearing
    design system, 132
Knowledge-based systems, 162

Labour market, 3; 5
Lacquering, 394
Laminated Object Manufacturing,
    185
Laser, 269; 330; 403

Laser beam, 442
Laser cutting,, 403
Laser generation, 193
Laser measuring system, 442
Laser welding, 403
Laser-assisted machining (LAM),
    282
Laser-beam bending, 194
Layer Manufacturing, 180
Layer technique, 181
Lead time, 11
Lean Production, 12
Legislation, 471
LIGA technology, 287
Light metal, 517
Lithographic process, 286
Long-term improvement in process
    quality, 158
Low concentrations, 464
Lubricant, 466
Lubrication, 458

Machine capability, 300; 334
Machine pallet, 310
Machining facilities, 293
Machining speed, 343
Machining temperatures, 301
Macroscopic surface damage, 303
Magnetic forming, 280
Maintenance, 489
Malcolm Baldrige Assessment, 68
Man-machine interface, 444
MAP, 369; 420
Markets, 5
Material consumption, 512
Material fatigue, 307
Material flows, 515
Material-flow systems, 446
Mathematical model, 300
Measurement result, 300
Measuring equipment, 294
Mechanical characteristics, 409
Mechanical structural components,
    412
Medium-temperature CVD coating,
    240
Meshing of the gearwheels, 400

Metal Spray Tooling, 191
Methods, 15; 39
Microorganisms, 466
Milling with robots, 396
Miniaturizing, 436
MMS, 369; 420; 421; 431
Models, 510
Modular systems, 410
Motor spindle, 326
Movement path, 443
Multi-point measuring devices, 305
Multifunction oil, 485

National standards, 313
Natural hydrocarbons, 460
NC programming, 367
NC-milling, 180
Nd:YAG laser, 403
Near net shape geometry, 302
Negative mask, 185
Neural networks, 151
Nitrosamines, 464
Non-contact data transmission, 411
Number of robots deployed, 387
Numerical controllers, 356

Object-oriented technique, 154
Objectives of quality management, 46
Obstacles, 442
Off-line monitoring, 309
Oil mist, 303
On-line measurement, 309
On-line monitoring, 309
Open control system, 422
Open controller, 362
Open controllers, 358
Open Programming Environment for Robot Applications, 429
Operating system, 423
Optimization, 520
Optoelectronic, 269
Organization, 6
OSACA, 377
Osing system, 398
Overload signals, 300
Overspraying, 394

Pallets, 447
Paper foils, 185
Passive assembly aids, 400
Path welding, 389
PDCA, 65
Peak-to-valley height, 300
Percentage bearing, 300
Periphery, 446
Periphery, 408
Personnel training, 306
Phase-oriented quality management, 42
Philosophies, 15
Photopolymer, 182
photopolymerization, 182
Physical vapour deposition processes (PVD), 272
Pick-and-place, 392
Piezoelectric, 301
Pin into a borehole, 399
Planetary gear, 414
Planning quality, 169
Planning tools, 504
Plant layout, 408
Plasma cutting, 391
plasma CVD coating, 240
plastic, 517
Plastic Vacuum Casting, 190
Point-to-point controllers, 389
Poka-Yoke, 50
Pollution, 516
Portfolio technique, 511
Positional deviation at the tool centre point, 416
Post-process metrology, 307
Power supply, 436
Power trains, 412
Pre-production components, 179
Precision, 334
Precision machining, 261
Primary energy, 513
Problem-solving process, 71
Procedures for generating RC programmes, 424
Process, 60; 14
Process capability, 294; 305

Process design, 249
Process forces, 301
Process Modelling System, 139
Process moments, 301
Process monitoring, 231
Process optimization, 300
Process orientation, 11
Process reliability, 230
Process selection, 249
Process variables, 299
Process-integrated control, 292
Process-integrated sensors, 301;
    304
Process-monitoring sensors, 298
Processing capacities, 419
Product development, 22; 177; 495
Product differentiation, 82
Product liability, 148
Product life cycle, 496
Product life cycles, 4
Production, 495
Production sequences, 253
Production engineering, 9
Production process, 268
Production quality, 298
Production sequences, 233
Production technology, 268
Production-integrated metrology,
    291
Process orientation, 11
Programming, 423;408; 429; 434
Programming errors, 426
Programming languages, 417
Programming of a weld seam, 425
Programming tools, 429
Programming welding robots, 432
PVD coating, 240

QFD, 167
QM system, 59; 66
Quality, 5
Quality capability, 293; 307
Quality characteristics, 299; 312
Quality circle, 150
Quality control circuits, 313
Quality management, 60
Quality management, 6; 38; 145

Quality management as a
    continuous process, 43

Reactive Laser Sintering, 193
Real time requirements, 420
Recycling, 490
Reduction in costs, 423
Reference model, 119
Reference temperature, 307
Reliability, 307; 331
Relocation, 219
Remote centre compliance (RCC),
    400
Removal of burrs, 396
Requirement for resources, 521
Responsibility, 496
Restructuring, 218
Robot welding plant, 391
Robot-assisted assembly, 439
Robot-loading of presses, 392
Rolling bearings, 344

SCARA robots, 393
Sealing, 332
Segmentation, 219
Selective Laser Sintering, 187
Self-fulfilment, 3
Sensor, 331; 435
Sensor guidance, 444
Sensor signals, 304
SERCOS interface, 372
Signal filtering, 303
Significance of Quality
    Management, 36
Similarity identification, 433
Simulation, 394
Simulation system, 432; 434
Simulation systems, 427
Simultaneous engineering, 104;
    165; 175
Skin diseases, 469
Slices, 181
Small batch production, 18
Snap-on connections, 399
Software, 145
Software training, 168
Solid Freeform Manufacturing, 180

Solid Ground Curing, 184
Solutions, 462
Spark erosion, 257
Spark-erosion, 263
Specimen workpiece, 310
Spindle-bearing systems, 132; 332
Spot welding, 388
Spot-welding control systems, 389
Staff audit, 74
Standard interfaces, 417
Standardization, 437
Standardized interfaces, 423
State sensors, 301
Static and dynamic deformation, 413
Statistical process control, 293
Statistical test planning, 300
Steady rest, 297
Steel, 517
Stereolithography, 182
Strategic elements, 51
Strategic fit, 104
Strategic success factors, 37
Structured programming, 429
Supports, 182
Surface layer, 268
Surface microstructure, 268
Surface roughness, 268
Surveillance processes, 295
Synthetic esters, 460
Synthetic hydrocarbons, 460
System of objectives, 501
System of structured consensus, 67

Tacking, 390
Tail stock, 297
Tape-laying, 403
Target consensus, 64
Target prices, 105
Task- oriented programming, 434
Taylor's division of work, 63
Teaching process, 438
Teaching-in, 426
Teamwork, 148; 161; 165
Technical innovations, 4
Technical prototypes, 178
Technological planning phase, 195

Technology, 28
Temperatures, 298
Testing, 426
Testing equipment, 310
Testing plan, 300
Thermoelastic deformation, 339
Thin-layer technologies, 271
Time to market, 175
Tolerance limits, 308
Tool-changer, 328
Tools, 15
Tools of quality management, 45
Topology optimization, 136
Total Fluid Management, 491
Total Quality Management (TQM), 41; 145
Traceability, 313
Transfer of knowledge, 147
Triangulation, 182; 303
Trim cutting, 261
Turning, 180
Turning station, 297

Ultra-fine-grain carbide, 239
Ultraviolet lamp, 185
Ultraviolet laser, 182
Undetected rejects, 294
Uniform design concept, 323
User interface, 150; 370; 423
User-friendliness, 417
Utility value analysis, 511
Utilization phase, 195

Value added process, 16
Value creation, 220
Various pressure forces, 397
Vegetable esters, 460
Vibrations, 301
Vibratory conveyors, 447
Virtual prototyping, 115
Voice data entry facility, 445

Wall thickness, 344
Waste disposal, 495; 513
Water-immiscible cooling lubricants, 460